ROCK MECHANICS IN SALT MINING

ROCK MECHANICS IN SALT MINING

By
M.L. JEREMIC
Laurentian University, Sudbury, Canada

A.A.BALKEMA/ROTTERDAM/BROOKFIELD/1994

CIP-DATA KONINKLIJKE BIBLIOTHEEK, DEN HAAG

Jeremic, M.L.

Rock mechanics in salt mining / by M.L. Jeremic. -
Rotterdam [etc.] : Balkema. - Ill.
With index, ref.
ISBN 90-5410-113-X bound
ISBN 90-5410-103-2 pbk.
Subject headings: civil engineering / mining.

Published by
A.A. Balkema, P.O. Box 1675, 3000 BR Rotterdam, Netherlands
A.A. Balkema Publishers, Old Post Road, Brookfield, VT 05036, USA

ISBN 90 5410 113 X hardbound edition
ISBN 90 5410 103 2 student paper edition

Contents

Preface

Archaeological evidence has indicated that the first settled cultures arose where salt was available. The oldest settlements found in China are along the Yellow River, whose great bend embraces a salt swamp where ancient salt pans were still in operation until present time. The early villages of the Jordan valley, and especially the site of Jericho, dated about 8000 BC, and considered the oldest known agricultural settlement, are near the salty Dead Sea and the salt mountain of Mt. Sodom. The Mesopotamian cities transported their salt up the Tigris and Euphrates Rivers from sea-salt panning centres at the river mouths. In this regard it is quite understandable that first civilization has been developed around the Mediterranean Sea.

In Europe neolithic salt-boiling settlements arose around salt springs in the Tyrol, in the Moselle and Franche-Compte regions of France, at Saale and Lüneburg in Germany, in the Tuzla basin of former Yugoslavia, Droitwich in England and others. Through thousands of years, entire forests were consumed in supplying wood fires to evaporate the water from salt brines. Among ancient salt mines and quarries are those established by the Phoenicians in Spain and the apparently Wieliczka mine near Cracow in Poland. In this mine underground salt mining started as early as the thirteenth century.

Canada's salt reserves alone could supply the world's needs for over thousand years, yet there are few major regions in the world that do not have proportionately at least as much salt as Canada does. The astronomically large amounts of salts are available from sea, which in warm and dry regions could be source of salt production, by solar evaporation of sea water. The sea contains several tens of millions of cubic kilometers of salt of all mineral compositions (halite salt, trona, potash, anhydrite and others) but is a very secondary source for their production. The main sources of salt production, however, are the subsurface salt deposits, which are exploited by dry mining (underground mining) or by solution mining, as discussed in this volume.

With regard to salt production either by dry mining or solution mining the rock mechanics have been an important part of their technologies for many decades, because it plays an essential role in mine stability and strata control. In this volume consideration has been given to the mining of several facial types of deposits.

– *Halite salt*, also known as rock salt (Europe) or common salt (N. America), has been used from the beginning of human civilization to the present. Through history, the main sources of halite were playa lakes or marine lagoons and brines outcropping on the surface. Rock salt has played and is still playing an important role in the life of all mankind. Halite salt, as well as trona, made an impact on the chemical industry for soda production.

– *Trona*, also called soda ash, has been known since 4000-3000 BC, and it was used by the Egyptians to make glass containers and to preserve mummies by an embalming process. In 1880, the total world production of soda ash was about 75 000 tonnes (mostly produced from brines) but a hundred years later, world output is about 31.2 million tonnes (mostly produced from underground mining). Unfavourable changes in the damand of the soda ash altered expectations of an appreciable increase in production of trona ore.

– *Potash* is the general term applied to various potassium salts both naturally occurring and commercially derived. The potassium content of potash is measured in units K_2O equivalent. Potash mining in Canada accounts for up to 30% of world production and is second to the former USSR industry which accounts for over 30% of world capacity. The primary use of potash salts is for fertilizer and it is expected to keep its favourable trend in production.

– *Gypsum* occurs all over the world, including Canada, which is a large producer. There are thirteen open pits and three underground mines producing about 9 million tonnes per year of crude gypsum. Halifax County (Nova Scotia) has a single open pit producing 3 million tonnes gypsum per year. It should be noted that of all the evaporate deposits only gypsum could be mined by open pit, but halite, trona and potassium salts, due to their high solubility, are mined only by underground mining.

Rock mechanics as a scientific tool has been first introduced in salt mining due to unusual properties, and behaviour of the salts. In the early days it was recognized that creep deformation and failure of salt mines could be explained by the principles of mechanics and could also be controlled by the same principles.

There have been several valuable books published on this topic, each one with a specific approach and concept. This is also the case with the present volume, *Rock mechanics in salt mining*, where the author described his experiences and personal views on this subject. The author wrote this volume primarily as a contribution to the spread of knowledge and understanding of rock salt mechanics and mining. He is a mining engineer with a background in rock mechanics and has prepared the book for use by professionals in the salt mining industry and students of geological and mining engineering. This volume uses the same philosophy and engineering point of view as those in the author's earlier books *Strata mechanics in coal mining* and *Ground mechanics in hard rock mining*. What makes this volume particularly unique is the fact that it contains a great deal of first-hand data acquired by the author himself.

In order to appeal to a wider professional audience, the first five chapters of this volume consider general geology, folding and faulting structures, composition of salt and form of salt bodies with the simplifications which could be used for engineering solutions in rock salt mechanics and mining. The next three chapters deal with the exploration and opening of salt deposits with the aspect of design of safe and stable mine structures, and risk of water inflow into the mine. The following three chapters analyze deformation and failure of the salt due to elasto-plastic, creep and outbursts loading conditions. Five chapters discuss strata mechanics and control for different mining systems of flat, inclined and massive salt bodies, as well as solution mining and excavation for storage. The last chapter presents the stability analyses to the mine structures in regard to salt mining subsidence.

The objective of the author was to illustrate the advantages of the application of the rock salt mechanics concept to underground excavations in the salt bodies and, in doing so, emphasize that present mining technology, organized into an intellectual structure based upon scientific principles, is preferable to practical rules. In the new technological era of underground salt mining it is essential to utilize all salt mechanics effects in the system of mining to achieve a high production, and hence profitability, under difficult marketing conditions. Under these circumstances, there is no room for practical approximations. Rather, the application of the particular concept of rock salt mechanics, which are integrated in mining engineering design, is essential to achieve safe and productive mine operations.

The author acknowledges the assistance given by the mining students of Laurentian University, Sudbury, Canada, and Tuzla University, Tuzla, former Yugoslavia. In particular he greatly appreciates their contributions with regard to mathematical modelling and graphics.

It is a great pleasure to record the financial and moral support given by Dr D.E. Goldsack, Dean of Faculty of Science and Engineering at Laurentian University, without whose help the publishing of this book would not have been possible.

I am also greatful to Dr E. Mandzic, Dean of the Faculty of Mining and Geology at Tuzla University, for providing information on some underground phenomena important for rock mechanics of salt mining.

For reading the manuscript, I wish to express my deepest appreciation to my colleagues: Dr R.A. Cameron, Dr R.S. James and D.A. Trotter, Professors of Laurentian University: Mr M.L. Ames, Chief Technical Engineer of the Ministry of Labour; Mr C.B. Graham, Managing Director of Mining Research Directorate; Mr D. Morrison, Superintendent of Rock Mechanics, INCO, Sudbury and Mr P. Oliver, Manager of Solar Neutrino Observatory, Sudbury.

January 1, 1991

Dr. M.L. Jeremic
Professor of Rock Mechanics and Mining Engineering

CHAPTER 1

General geology

The geological characteristics of evaporite deposits are due to the particular primary depositional conditions and post-depositional alteration. The lithostratigraphy and facial development of evaporite sedimentation should be studied because they have some influence on geotechnical and mining considerations.

1.1 DEPOSITION OF EVAPORITES

Kant, the German philosopher, is responsible for the theory that explains rock salt deposition. He stated that the various minerals are the product of the evaporating marine water in shallow seas. Later, from the middle of the nineteenth century to the present, several hypotheses for the origin of evaporite deposits were proposed. Here, only three of these theories will be briefly discussed.

1.1.1 *Bar-basin theory*

This concept has been postulated by Ochsenius on the basis of studying the Gulf of Karabuhghas in the Caspian Sea. The principles of formation of evaporite deposits by the barrier theory is illustrated in Figure 1.1.1. The process of evaporite sedimentation is represented by five phases discussed below.[1,2]

1. An arm of the sea is isolated by a barrier or sill over which periodic influxes of fresh sea water occur; the rate of water evaporation in the region is typical of a hot climate. Under these circumstances water in 'bar-basins' becomes concentrated in salt solutions, as the process of evaporation progresses. At the point when a water reaches a density of 1.070, flora and fauna die out in the basin. Vegetation in surrounding land masses is unaffected.

2. With further increase in mineral concentration in the brines, the conditions for sedimentation calcium carbonate and iron oxide occur. When a brain of the basin achieves a density of 1.129, precipitation of gypsum begins. At present, the water of the Gulf of Karabuhghas has just this density. Vegetation in surrounding land mass is not affected.

3. Continuous supply of fresh sea water into this basin causes further salt concentration and increases in brine density. When the solution density achieves 1.218, precipitation of halite and gypsum begins; this process causes the basin to begin to fill with major amounts of chemical sediments and fill up of the basin is under way. In this phase of sedimentation vegetation in the surrounding land mass dies out and is transported into the basin. This results in iodin becoming a component of the halite-rich salt deposits.

1

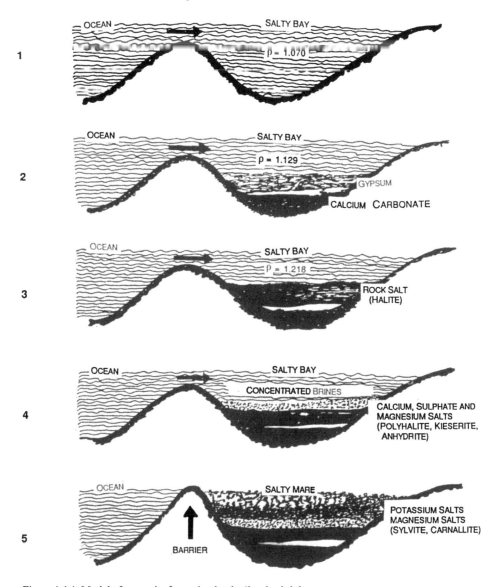

Figure 1.1.1. Model of evaporite formation by the 'bar-basin' theory

4. After deposition of halite salt layers, the precipitation of anhydride and gypsum starts and the basin becomes quite shallow due to the sedimentation process. Also, at this time the residual solution may be removed over the barrier due to reversal flow. This might explain why in some salt-bearing strata, potassium and magnesium salts are also non-existent.

5. Due to uplift or other geological factors, the basin may become totally separated from the ocean, and inflow or outflow of water ceases. In this case the residual solution may be present and from it are deposited the very soluble salts of potassium and magnesium, which represent the final horizons of evaporite deposits. Erosion of rock material from surround-

ing land, and its transportation into the basin form layers of clay, silt and sand sediments, which protect evaporite from dissolving due to local atmospheric events.

The theory of 'bar-basin' does not explain the existence of large and thick salt deposits.

1.1.2 *Ring theory*

A modification of the bar-basin concept by Branson Walther and others postulates the existence of series of semi-isolated sub-basins. The water temperature in them is thought to be between 15° and 35°C due to an arid and hot climate which facilitates intensive evaporation.[1]

This theory proposes that new supplies of sea water first enter the outer basin and overflow into the inner basin(s). Chemical differentiation of the mineral solution occurs on the basis of their relative solubilities. Under these circumstances gypsum would be deposited in the outer ring (basin) and halite in the inner ring (basin). The influxes of new sea water spilling into the inner basin also bring bitter salts into this basin. Due to the different degrees of solubility of individual salt, mineral deposition results in contouring of individual facies, where the most soluble salts are deposited at the end of the sedimentation process (Figure 1.1.2).

Postulation of sedimentation on the basis of solubility, suggests that carbonate layers are deposited first, Zechstein's limestone in the salt formation of Central Europe. Further evaporation results in the deposition of gypsum layers and finally salt deposits. The interfacial development of salts deposits is exhibited firstly by rock salt and gypsum strata, followed by rock salt and anhydrite, rock salt and polyhalite and finally the highly soluble salts of potassium and magnesium.

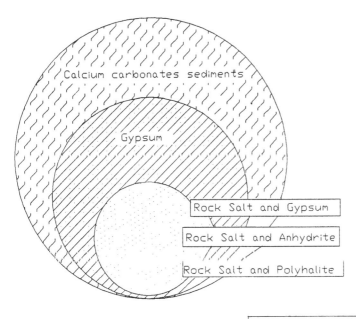

Figure 1.1.2. Concept of evaporite deposition by the 'ring' theory

1.1.3 *Fractional sedimentation theory*

This is a recent modern theory which is most often used in the analysis of evaporite deposits. It is postulated in a shallow epeiric sea, with a gently sloping deposition surface, not in closed marine environment. The epeiric sea is separated from the open ocean by a shallow shelf.[1, 2, 3]

The sedimentation of the individual facies of the evaporites has a particular order, governed by mineral concentration in the sea water. For example, during marine transgression, if sea water has a low mineral concentration, precipitation of $CaCO_3$-Mg beds may result. During marine regression, however, if sea water has a high mineral concentration, precipitation of NaCl-KCl beds can occur. From sea water of moderate mineral concentration $CaSO_4$ may precipitate.

Due to water currents it is possible that initially the density of stratified water at the massive of a continent can be altered. For example, the most dense mineral concentration may be displaced towards the ocean bottom and less dense mineral concentration shifted towards a land mass (Figure 1.1.3).

A final transgressive phase is marked by transportation of clay and fine silica material from land into an epeiric sea. From this material mudstone, claystone, siltstone, sandstone and other beds can form cover of an evaporite complex.

The theory of fractional sedimentation is the most comprehensive postulation for the origin of large and thick evaporite deposits. Epirogenic movements, which relate to marine transgression and marine regression, are important factors contributing to the formation of cyclic evaporite deposits. Each cycle of deposition is marked by a particular evaporite sequence, which is normally different from all others in terms of strata thickness and mineralogy.

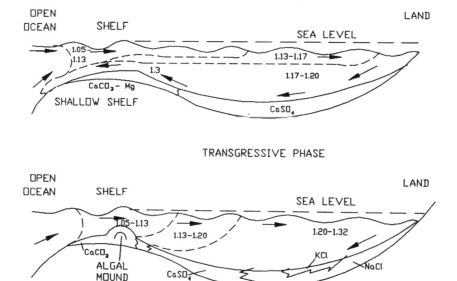

Figure 1.1.3. Model of fractional sedimentation of evaporite deposits

1.1.4 *Cyclic sedimentation*

The idea that cyclical sedimentation is responsible for the formation of evaporite sequence is the most common hypothesis for the origin of salts deposits. For example, Bamnet Narong basin, North-eastern Thailand, contains three cycles of evaporite deposits within the Maha Sarakham salt-bearing strata.[3, 4, 5] Geological and geophysical data for these strata are given below:

    – *The lower clastic cycle* (the marine transgression). It has been formed as a substratum to the evaporite deposits and consists mainly of ferruginous clastic sediments and calcareouo oandotono.

    – *The lower evaporite cycle* (the marine regression). After a brief progression causing penesaline conditions to occur anhydrite precipitated; by further concentration of the saline water halite and finally the sylvite (KCl) were deposited marking the termination of the restricted environment. At this point marine influx of non-saline water resulted in brine dilution forming penesaline conditions.

    – *The lower middle clastic cycle* (the marine transgression). The deposition of the lower evaporite cycle ceased and was replaced by sedimentation mudstone and claystone.

    – *The middle evaporite cycle* (the marine regression). Under restricted marine conditions the brine concentration increased to saline conditions and the lower middle halite sub-facies reprecipitated. Increase in the salinity of basin water was interrupted by marine influxes as evidenced by thin anhydrite layers. The rejuvenation of brine concentrations resulted in precipitation of the upper middle halite sub-facies.

    – *The upper middle clastic cycle* (the marine transgression). The deposition of the halite sub-facies ceased because the depositional basin was open for sedimentation of fine grain clastic sediments, namely, mudstone and claystone.

    – *The upper evaporite cycle* (the marine transgression). It marked the end of both the evaporite and clastic sedimentation.

    A simplified depositional profile of evaporitic-bearing strata is given in Figure 1.1.4.

    The sequence of the lower evaporite cycle is almost exclusively equivalent to lower middle, and some parts of the higher orders of theoretical marine evaporite depositions. The middle evaporite cycle and upper evaporite cycle are represented by the alternation of halite and anhydrite layers. These two cycles could be interpreted as lower and middle orders of a theoretical evaporite facies.[3, 4]

1.2 POST-DEPOSITIONAL CHANGES

Almost all evaporite deposits are to some degree exposed to post-depositional changes due to the high reactivity of water-soluble evaporite minerals and high susceptibility of evaporite beds to ductile deformation. Several possible post-depositional changes of evaporite deposits can occur; some are listed below.

1.2.1 *Mechanical alteration*

This type of post-depositional change of evaporite strata is so significant that in this book two separate chapters are devoted to it under the titles 'Folding structures' and 'Faulting structures'. At this point only a brief introduction is presented, using an example of post-depositional deformation of the sedimentary sequences of the Maha Sarakham formation (Thailand).

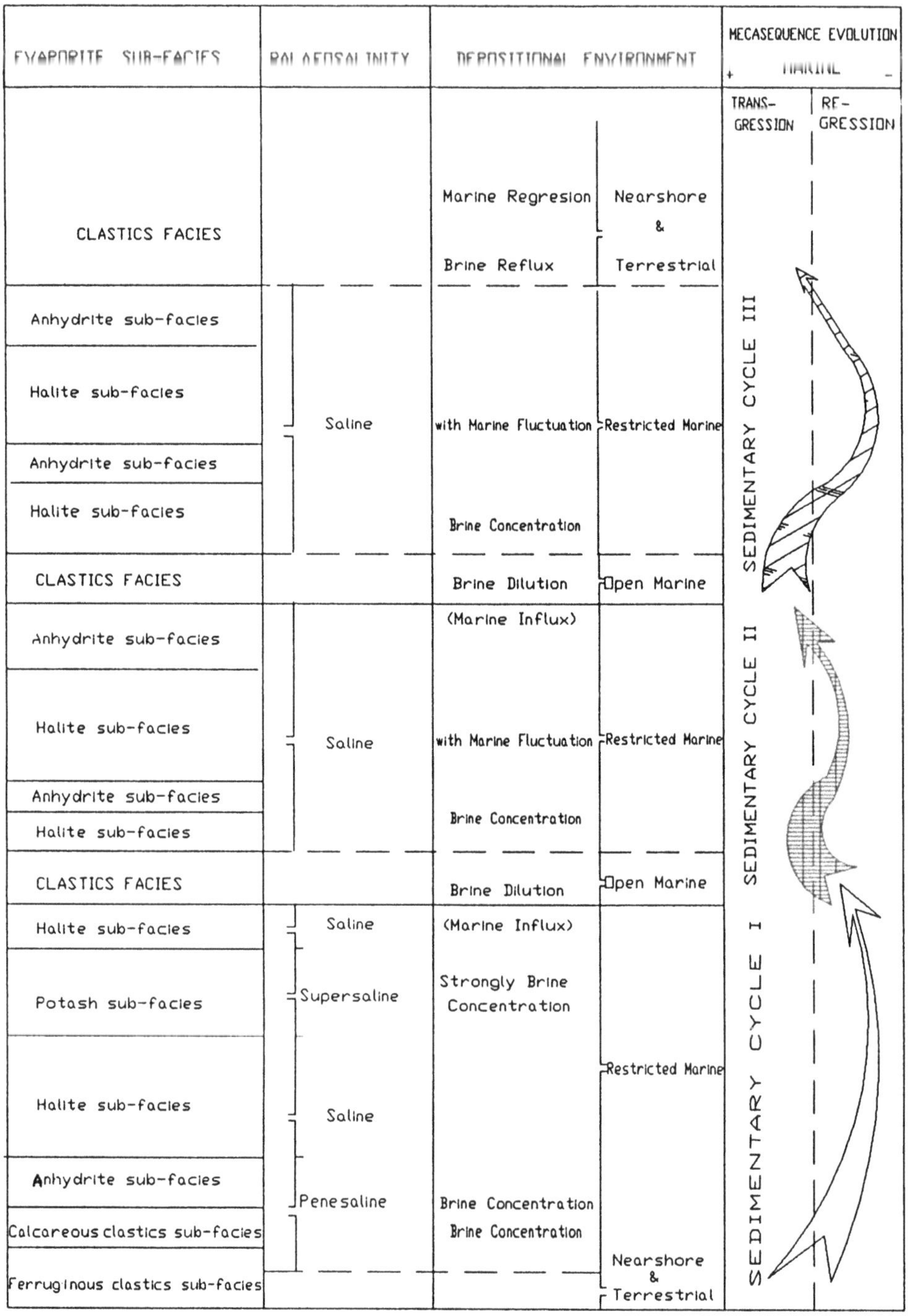

EVAPORITE SUB-FACIES	PALAEOSALINITY	DEPOSITIONAL ENVIRONMENT		MEGASEQUENCE EVOLUTION + MARINE –	
				TRANS-GRESSION	RE-GRESSION
CLASTICS FACIES		Marine Regresion Brine Reflux	Nearshore & Terrestrial		
Anhydrite sub-facies	Saline	with Marine Fluctuation	Restricted Marine		
Halite sub-facies					
Anhydrite sub-facies					
Halite sub-facies		Brine Concentration			
CLASTICS FACIES		Brine Dilution	Open Marine		
Anhydrite sub-facies		(Marine Influx)			
Halite sub-facies	Saline	with Marine Fluctuation	Restricted Marine		
Anhydrite sub-facies		Brine Concentration			
Halite sub-facies					
CLASTICS FACIES		Brine Dilution	Open Marine		
Halite sub-facies	Saline	(Marine Influx)			
Potash sub-facies	Supersaline	Strongly Brine Concentration			
Halite sub-facies	Saline		Restricted Marine		
Anhydrite sub-facies	Penesaline	Brine Concentration			
Calcareous clastics sub-facies		Brine Concentration			
Ferruginous clastics sub-facies			Nearshore & Terrestrial		

Figure 1.1.4. Depositional profile of the evaporite cycles of the Bamnet Narong basin (NE Thailand)

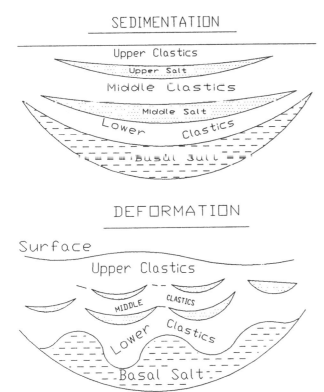

Figure 1.2.1. Model of mechanical alteration of the evaporite-bearing strata (modified after Hite)

Evidence from sub-surface data, detailed structures deducted from drill core samples, as well as other geological and geophysical data, strongly indicate that the evaporite facies in the Maha Sarakham formation had been subjected to mechanical deformation.[3, 4]

The mechanisms which caused this deformation of salt beds might be either the differential loading of the overlying sediments and the high plasticity of the evaporite strata or regional tectonic movement or a combination of both.[5] A simplified structure deformation model of the three evaporite cycles of the Maha Sarakham formation is illustrated in Figure 1.2.1.

1.2.2 *Palaeo-alteration by water solution*

Some of the massive evaporite deposits most likely formed as a result of post-depositional processes which caused water solution of salt beds and subsequent brine precipitation. For example, Walther suggested that during heavy rainfalls, newly formed, large, lenticular salt deposits were dissolved and transported in the direction of gravity into the large tectonic depression where reprecipitation occurred.[1] The process of formation of primary deposits, their solution and decantation in the same depression could be repeated many times. Of course each cycle will be separated by sediments of marine transgression (pelite and sand deposition).

It is interesting to note that in Slanic-Prahova (Romania) there is a locality called 'Salt-mountain', where the salt deposit is located in the Maramuresh's depression; it might

Figure 1.2.2. Outcrops of salt beds at Slanic-Prahova (Romania)

have a post-secondary origin. In this case, palaeo-solutions were precipitated in depressions and solidified, but due to the action of recent dissolution the salt has been dissolved and moved further from its original source. As a result of recent solution activity the salt outcrops exhibit a rigid topography within which there are numerous caverns (Figure 1.2.2).[6]

1.2.3 *Recent alteration by water solution*

The post-depositional changes of salt by water solution at the present time are mostly indicated by the cavities and missing parts of evaporite beds. In contrast to the activities of palaeo-solutions which in some cases cause the formation of secondary deposits, recent mineral solutions are removed and diluted by percolating ground water.

Firstly, changes of the chemical composition of percolating water throughout the evaporite basin are strongly influenced by mineralogical composition of salt layers.[7] This phenomenon can be observed by particular relationships between the zonation of evaporite facies and chemical composition of ground water (Figure 1.2.3). The geochemistry of percolating water can be utilized for hydrochemical prospecting of evaporite deposits. It should be mentioned that besides changes in chemical composition of water and possible chemical changes in the evaporite beds themselves (briefly discussed in the next subsection), it can also be shown that exhibited physical changes in the deposits occur in the formation of systems of caverns, due to removal of salt by solution processes.

Secondly, changes of surface topography occur due to subsidence and caving of the cavities within the salt deposits. This is observed in the field as 'depression zones' which are usually filled with water (small lakes), and also the occurrence of land slides which are located at the contacts between sediments of different solubility (Figure 1.2.4). The ground surface deposition due to the solution of salts by percolating water is named 'topographical geology' and it might be used for prospecting of salts deposits.[8]

It should be mentioned that at the present time deposition of evaporites from ground

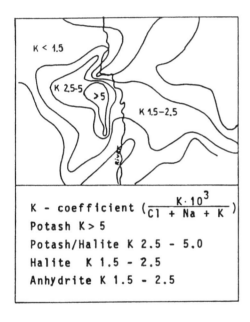

K - coefficient ($\frac{K \cdot 10^3}{Cl + Na + K}$)
Potash K > 5
Potash/Halite K 2.5 - 5.0
Halite K 1.5 - 2.5
Anhydrite K 1.5 - 2.5

Figure 1.2.3. The contour map of K coefficient distribution and zonation of evaporite facies (Sibosnica basin, former Yugoslavia)

Figure 1.2.4. Landslides due to salt solution by percolating water (Sibosnica basin, former Yugoslavia)

water in economical quantities is not known, but formation of nitrate deposits of economic importance does occur in Chile.

1.2.4 *Chemical alteration*

There are numerous factors which indicate that chemical alteration of an evaporite deposit has occurred since the present characteristics of the mineral deposit are different from those which primary processes would have formed.

The study of the evaporite facies of the Maha Sarakham formation, Thailand, indicated that textural characteristics of the evaporite facies correspond to chemical alteration in the salt within the anticline flanks of the deposits. This is confirmed by the bromine content of the minerals and mineral zonation within the evaporite facies.[3, 4] The chemical alteration of the evaporite facies of this formation is due to the percolation of ground water from the cap of anticline salt structures down the flanks of the deposits; this has resulted in the transformation of carnallite to sylvite through the process of incongruent alteration (Figure 1.2.5).[5] It should be noted that the problem of the $MgCl_2$ solution resulting from this incongruent alteration reaction has not yet been solved. The formation of anhydrite caps at the crest of anticlines is the result of a residual accumulation after the leaching of interbedded halite and anhydrite by ground water.

The overall model of post-depositional chemical alteration has been deduced from information obtained from a study of the Maha Sarakham formation. It represents a modification of Hite's[5] work and is presented in Figure 1.2.6. Special reference, however, is made with respect to the origin of secondary sylvite which is the most economical potash mineral.[1, 4, 5]

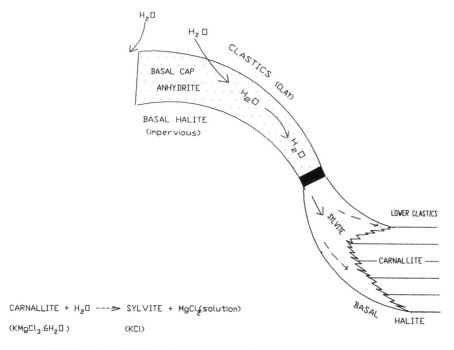

CARNALLITE + H_2O ---> SYLVITE + $MgCl_2$(solution)

($KMgCl_3.6H_2O$)   (KCl)

Figure 1.2.5. Secondary sylvite from incongruent chemical alteration of carnallite (modified after Hite)

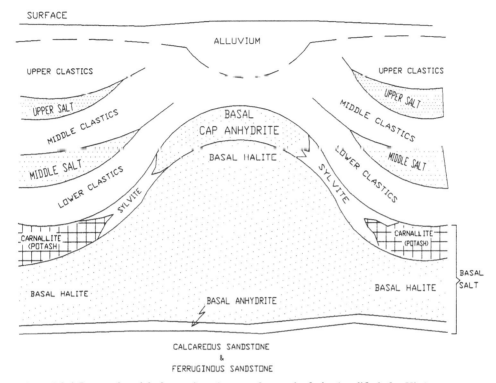

Figure 1.2.6. Proposed model of secondary changes of evaporite facies (modified after Hite)

In conclusion it is important to remember that the secondary sedimentary processes as well as original sedimentation events are important factors in determining the geometry, mechanical behaviour and underground mine stability of this type of ore body. These topics are discussed in this book.

1.3 LITHOSTRATIGRAPHY OF EVAPORITES

A knowledge of the sub-surface lithostratigraphy is based on the lithological analysis, mineralogical and textural studies and chemical as well as geophysical characteristics of the sedimentary sequences from the ground surface to below the level of evaporite deposition. In the following sections details of each of the major lithological units from the base upwards are presented.

1.3.1 *Lithostratigraphy of the depositional basin prior to evaporite deposition*

The lithology and stratigraphy of the depositional basin is of least importance because it presents no major problems when mining of salts deposits.

The sediments of the palaeo-basin may have formed in either a marine environment or terrestrial environment. The sediments of the depositional basin at Tuzla (former Yugoslavia), however, belong to vulcanogenic terrestrial facies formed in a lake. These sediments

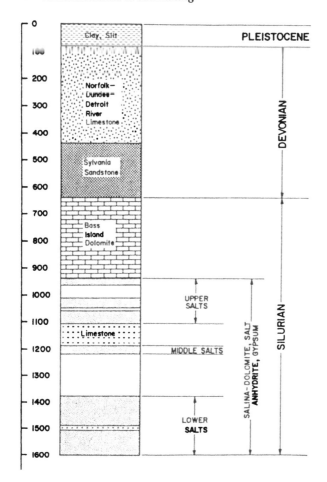

Figure 1.3.1. Lithostratigraphic of the evaporite-bearing strata and the overburden sediments (at Windsor, Michigan basin, Canada)

are composed of red mudstone, sandstone and conglomerate with interbeds of tuffite and they are called 'red series'; their thickness is in the range of 250 m. The evaporite-bearing strata overlie these rocks in a conformity manner; these rocks are referred to as the 'banded series' and are characterized by fine parallel bedding planes of the different sedimentary layers,[9] as illustrated in Figure 1.3.1.

1.3.2 *Lithostratigraphy of evaporite deposits*

The lithostratigraphy and thickness of the salt deposits coincide with the evaporite sequences present in the basin. The evaporite sequences may be complete or incomplete, as for example the Maha Sarakham formation which is discussed by Yumuang[3] and is illustrated in Figure 1.3.2.

The basal salt member has a nearly complete evaporite sequence. The lithostratigraphy of the basal salt member is represented by transitional ferruginous sandstone, calcareous sandstone, basal anhydrite, basal halite, potash, coloured halite.

The middle salt and upper salt members have incomplete evaporite sequences. Generally, the middle salt member is represented by four beds, namely, lower middle halite,

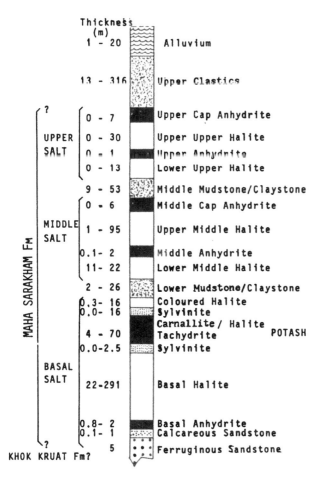

Figure 1.3.2. Lithostratigraphy of evaporite sequences at the Maha Sarakham formation (Thailand)

middle anhydrite, upper middle halite, and middle cap anhydrite. The upper salt member is also subdivided into four beds, notably, lower upper halite, upper anhydrite, upper halite and upper cap anhydrite.

The lithostratigraphy of evaporite sequences in the Maha Sarakham formation varies from the lowest to near highest order of theoretical evaporite deposition. The evaporite deposits exhibit distinctive lithological features at each stratigraphic level.

1.3.3 *Lithostratigraphy of interbedded sediments*

There is a qualitative difference between lithology of the evaporite sediments and inter-bedded sediments. As discussed in Subsection 1.1.4 (cycle sedimentation at Bamnet Narong basin), in contrast to marine regression which produced evaporite facies, during marine transgression fine grained clastic sediments were formed. These interbedded sediments are represented by mudstone and claystone: however, basal sediments are ferruginous sandstone and calcareous sandstone as a transitional series between the sediments of the depositional basin and the evaporite deposits.

The halite-bearing strata of the Trnovac salt deposits in the Tuzla basin, former

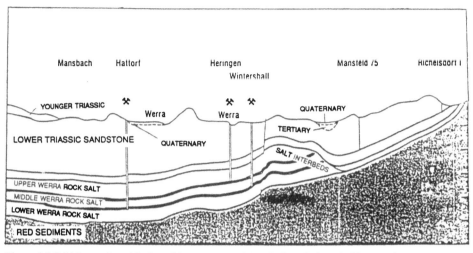

Figure 1.3.3. Lithostratigraphic depositional profile of the Flache salt deposit (Germany)

Yugoslavia, are represented by a 'banded series', which consists of dolomite and calcareous claystone. These sediments enclose and are interbedded with evaporite deposits represented by four halite sequences. It should be noted that some halite layers are more or less interlayered with thinner calcareous claystone, which suggests a frequent oscillation within the marine basin during halite precipitation. The banded series contains marine micro and macro fauna which defines their stratigraphic position in the lithostratigraphic profile. The lithostratigrahic profile is given for the Flache salt deposit, as illustrated in Figure 1.3.3.

It is interesting to note that interbedded strata of the Sifto salt mine in Goderich, Ontario, which belongs to the Michigan basin, have a thick lithostratigraphical sequence represented by bedded carbonate sediments, namely, shaly dolomite and dolomite. These sediments enclose and are interbedded with seven salt sequences.

1.3.4 *Lithostratigraphy of overburden strata*

A knowledge of the lithostratigraphy of the overburden strata and the position and nature of ground-water horizons is necessary for both solution and underground mining of salt deposits.

The sediments of the overburden strata are recognized as a rock formation separate from the evaporite sequence (Figure 1.3.4). For example, in the Tuzla and Sibosnica basins[8, 9] the overburden strata are thick (100 and 1000 m); they are described below.

– *Tortonian*: composed of massive and compacted claystone, altered claystone and fine grained sandstone; Middle Miocene in age.

– *Sarmatian*: composed of massive claystone interbedded with fine grained sandstone, sandstone and conglomerate, claystone, sandstone with intercalation of conglomerate and laminated claystone; Upper Miocene in age.

– *Pannonian*: composed of claystone of various thicknesses with or without intercalations of other beds; Lower Pliocene in age.

Figure 1.3.4. Outcrop of the overburden strata (Sibosnica basin, former Yugoslavia)

The lithostratigraphic profile of the overburden is usually complemented with the geotechnical and hydrogeological characteristics of the strata if the evaporite deposits are considered for mining exploitation.

1.4 SALINE FACIES

Four distinctive facies of evaporites can be required due to specific characteristics of the depositional environment. Each facies has a different appearance and mineralogy. The individual facies are distinctive not only mineralogically but also in their economic value.

1.4.1 Calcium sulphate facies

The chemical sedimentation of the calcium salts is represented by two facies, namely, anhydrite ($CaSO_4$) and gypsum ($CaSO_4$ 2 H_2O). Some calcium sulphate deposits due to digenetic processes may result in the formation sedimentary sulphur deposits (Piesechnie, Poland).[10]

Calcium sulphate is the product of second stage precipitation of sea water evaporation; gypsum separates first (below 34°C) and is followed by anhydrite (above 34°C). Some researchers believe that about one half of all the calcium sulphate present was deposited as gypsum before anhydrite formed. The marine beds of pure anhydrite imply either that the previously deposited gypsum was converted to anhydrite, or that deposition occurred above 34°C. It should be noted that if the sea-water solution is reduced to 9.5% of its volume by evaporation, most of the calcium sulphate induction will be deposited. Halite then begins to precipitate together with gypsum at temperatures below 7°C, and with anhydrite above that temperature.[11]

Anhydrite and gypsum occur in thick sedimentary sequences; for example, in Nova Scotia, they reached a thickness of 75 m and in South-eastern New Mexico over 1370 m. These two numerals are associated with salt deposits they underlie, cap, and are interbedded with the halite. The thickness and frequency of the interlayed gypsum/anhydrite beds are very variable.

The lateral development of the anhydrite facies is very extensive in the prairie evaporite strata in Saskatchewan (Figure 1.4.1). Bank-flanking anhydrite with the bedded halite and an extension sequence of anhydrite and dolomite (the shell lake member) thinly caps the banks in the Saskatoon area and slopes away from their summits into the adjacent salt.[12] It should be noted that the northern anhydrite facies pinches out in Alberta as well as in Manitoba, but the southern anhydrite facies extends into the USA. General depth of burial of the northern anhydrite is below 800 m and of the southern anhydrite is below 1500-2500 m. The bulk of the prairie evaporite strata is halite; beds and laminae of anhydrite are common in the lower part, and potash-rich beds are common in the upper part of the formation.

Finally, it should be stated that sometimes the gypsum and anhydrite facies are formed in the absence of salt deposits; as well the salt deposits may occur without the calcium sulphate facies. These occurrences are not explained by theories on the origin of evaporites.

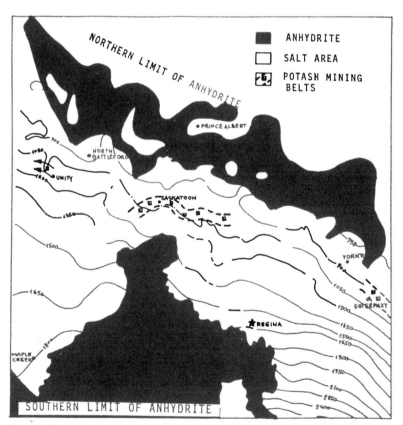

Figure 1.4.1. Contour of the anhydrite facies of the prairie evaporite (Saskatchewan)

The beds of the calcium sulphate facies are of great interest; as mineral deposits particularly of gypsum (in which extensive mining is currently carried out), or as strata interbedded with halite or sylvite in which case they could be important factors with respect to ground control in underground mining.

1.4.2 *Sodium salt facies*

The chemical deposition of the sodium salts most likely occurred in restricted basins such as lakes without water outlets. Water flowing into the lake was rich in sodium. The sodium originated during weathering of surrounding bedrock and leaching of volcanic ash. Sodium-rich salts which are concentrated within the twin solutions are directly from brines containing sodium chloride, sodium carbonate, sodium bicarbonate and sodium sulphate. Trona is the major sodium-rich mineral that precipitates from these brines. Associated minerals include shortite ($Na_2CO_3.2\ CaCO_3$), northupite ($Na_2CO_3.MgCO_3.NaCl$), pirssonite ($Na_2CO_3.CaCO_3.2\ H_2O$), and bradleyite ($Na_3PO_4.MgCO_3$). In some trona deposits halite (NaCl) is present in lesser or greater quantities. The precipitation of these minerals was most likely caused by an increase in brine concentration, a drop in temperature, or both. There is the fossil evidence that a sub-humid, savanna type flora existed in the surrounding terrain during sodium salt deposition; in contrast, a flora typical of a humid climate is common when sodium salt deposition is missing.[13]

There exists a hypothesis that deposition of the sodium salt facies was caused by fractionial sedimentation within a system of sub-basins; brines were concentrated by evaporation in each and were fed from other areas of the lake. This theory postulates relatively deep water in the lake.[13] A second hypothesis proposes that a formation of the sodium salt facies is due to the presence of playa lakes; it uses as an example the modern inactive playas such as Lake Natron, Tanzania. For example in Searles Lake (Figure 1.4.2), an area of about 50 km^2 in the midst of the larger playa is underlain by spongy salts. The lake formerly was 200 m deep, but now it is only marsh and at times it is dry. The bitter and alkali lakes might be used for exploitation of natural sodium compounds. For example at the Searless Lake, brines located 3-6 m below the surface are exploited by wells; beneath of 25-60 m of solid intermingled by saline minerals (borea sodium sulphate, potash, sodium salt) they are exploited by excavation.[14] The depositional rate of sodium minerals in this area is quite high: 0.4 m of solid per year.

In North America the main concentration of sodium salt facies is in Wyoming, where large deposits of trona were sedimented. During the Eocene period a large body of water called Lake Gosiute (the modern Green River basin) underwent many stages of filling and drying with associated deposition of sodium salts. This precipitation of the evaporite minerals could only have occurred in the presence of considerable amounts of carbon dioxide, which must have been provided in the lake by decaying plants and animals.[14] The sodium salt-bearing strata belong to the Green River formation, which consists of lacustrine, claystone, siltstone, malstone and a few interbeds of buff limestone, dolomite, and sandstone. Wilkins peak member contains the saline facies, which formed when the lake shrank to about 1/3 of its original length (Figure 1.4.3). The Wilkins member occurs as a large lenses surrounded by the fluvial sediments of the Wasatch formation. The main body of the salt deposit consists of trona mixed with halite. The beds of the trona deposits frequently contain the mineral shortite. The Wilkins peak member is an evaporite-bearing unit sandwiched between the Laney member on the top and the Tipton member – Luman member on the bottom. The Laney member starts with light grey or white dolomitic

Figure 1.4.2. Sodium carbonate-bearing brines at Searles Lake

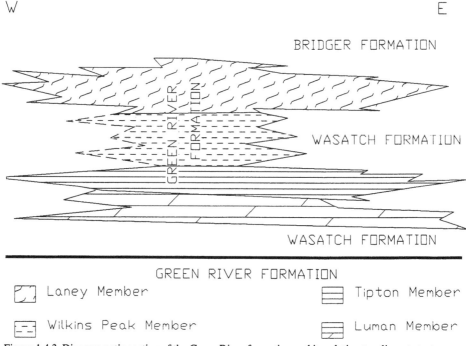

Figure 1.4.3. Diagrammatic section of the Green River formation and its relation to adjacent strata

marlstone and chippy shale to chalky buff laminated and varved limestone. The base of the Laney member is easy to recognize since oil shale occurs as a prominent marker. The Wilkins member itself is characterised by the abundance of dolomite and saline minerals. The member is thick, between 285 and 370 m and consists of more than 42 seams of trona, or mixed halite and trona. The trona beds which are economic are 1-12 m thick (25 beds).[15]

1.4.3 *Chloride facies*

The chloride facies is represented by halite - NaCl, which occurs as aggregated grains and sometimes as idiomorphic crystals. The associate minerals with halite are ardite (Na_2SO_4), glanberite ($Na_2Ca(SO_4)_2$) and bloedite ($Na_2Mg(SO_4), 4H_2$)). The chloride facies, commonly known as rock salt deposits, provides the source for about 3/4 of all salt used.

The halite deposits originated by sedimentary processes as discussed in Section 1.1. In many instances it is associated only with the calcium sulphate facies because of a lack of bittern salts (potash) as a result of incomplete evaporation of the sea water. Most likely, the evaporation process was interrupted, and the bittern solutions were drained off or diluted by an influx of sea water or fresh water.[11]

The chloride facies in the Tuzla basin, former Yugoslavia, originated in small, shallow basins with salt solutions located on the periphery of Pannonian Sea – Paratetis. The

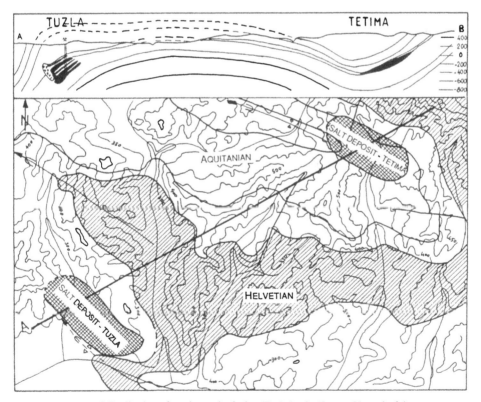

Figure 1.4.4. Map of distribution of natrium salts facies, Tuzla basin (former Yugoslavia)

evaporite-bearing strata were deposited 18 million years ago during a marine regression (salt facies), and marine transgression (marl, claystone, dolostone). The evaporite-bearing strata are called the 'banded series' (Figure 1.4.4). It is not clear, however, if deposition of evaporites occurred in a marine lagoon or salty lakes, but it is certain that at this time and in this region the climate varied from moderate tropical to aerial which facilitated the deposition of the chloride salts facies in the beds with a thickness over 30 m. Further sedimentation of marine sediments (Tortonian) formed a thick and protective cover over the halite deposits. The original salt deposits experienced chemical alteration (solution of salt minerals) and mechanical alteration (folding and multiplication of original beds)[16].

Investigation of the internal structure of rock salt deposits in the Tuzla basin indicates that the size of halite grains varies from millimeters to centimeters in maximum dimension. It can be assumed that the lower part of the natrium salt facies consists of white fine grained halite, but towards the roof of the deposit presence of gray coarse grained halite becomes common. The thickness of the individual layers varies from centimeters to decimeters in magnitude. The banded texture of the deposits is caused by interbeds of variously coloured halite or interlayers of claystone. The surfaces of bedding planes are not flat, because the halite grains intrude the claystone units, whose thickness is between millimeters to decimeters in magnitude.[9]

The thickness of common salt deposits varies from 10 m in the Michigan basin to 1000 m in the Siberian platform. The economic salt facies of evaporites are buried, because deposits which protrude to surface are exposed by erosion, or loosely dissolved. However, influx of ground water into buried salt deposits also affects their minability. Depending upon the depth and effect of groundwater on a deposit, the technology chosen to mine them is either underground mining or solution mining.

1.4.4 *Potassium facies*

After the sedimentation of the chloride facies, magnesium and potassium chloride and sulphates are deposited. Presence of the potassium facies indicates a complete cycle of evaporite sedimentation, and it results in the precipitation of evaporite solution almost to dryness.[10] This final cycle of evaporite deposition (represented by the potassium facies) is not common in the stratigraphic record. The main potassium minerals are carnallite ($KCl.MgCl.6\ H_2O$), kainite ($MgSO_4.KCl.3\ H_2O$), langbeinite ($K_2SO_4.2\ MgSO_4$), poly-halite ($K_2SO_4.2\ CaSO_4.2\ H_2O$), sylvite (KCl). They often occur in association with the mineral kierserite ($MgSO_4H_2O$).

At the Maha Sarakham formation (Thailand) it has been established that its basal salt member (as discussed in Subsection 1.4.3) consists of the almost complete evaporite sequence; calcareous facies – anhydrite facies – halite facies – potassium facies. The mineralogical composition of the potassium facies is close to the theoretically predicted mineral association sylvinite, carnallite, halite, tachyhydrite, sylvite.

The potassium salt facies in Romania occurs in Central Moldavia in intimate association with magnesium salts. The kalium and magnesium salts are an integral part of the rock salt deposits, which are located at the margin of the evaporite basin. The present spatial relationship between halite deposits and the potassium facies is irrelevant to the original sedimentary structure because of intensive tectonic deformation. The potassium facies are complex; it is not due to any post-depositional tectonic deformation and does not coincide with the model of a great majority of potash deposits. For example, after sedimentation of anhydrite-polyhalite beds carnallite strata were deposited on which were superimposed halite-sylvite beds. Finally, after kieserite deposition, the anhydrite-polyhalite beds completed the sedimentation cycle of the potassium facies (Figure 1.4.5). The deposition of carnallite immediately after the formation of polyhalite-anhydrite strata is not common and deviates from the chemical model of sedimentation for potassium beds. This must be due to changes in the chemical composition of the brines during the evolution of the depositional basin[17].

The potassium salt facies of the Canadian prairies was formed in Middle Devonian time as a sea advanced from the northwest and covered a large portion of Saskatchewan. As the sea water evaporated, salt crystallization on the sea bottom resulted in the deposition of layers about 200 m thick. The saturated potassium salt solutions, being more soluble than sodium salts, were deposited in regionally low areas and in layers of variable thickness (Figure 1.4.6). The potash facies occurs within the 60 m of the upper-most beds of the Elk

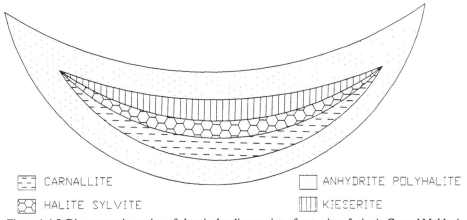

CARNALLITE   ANHYDRITE POLYHALITE

HALITE SYLVITE   KIESERITE

Figure 1.4.5. Diagrammatic section of chemical sedimentation of potassium facies in Central Moldavia (Romania)

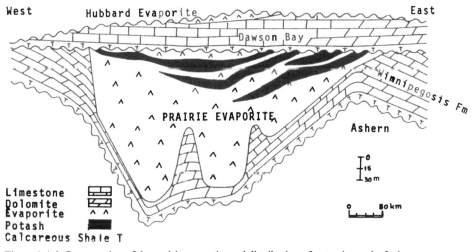

Figure 1.4.6. Cross section of the prairie evaporite and distribution of potassium salts facies

Point Group of Middle Devonian strata. The mineable potash beds, however, are 2-3 m thick, with a dipping angle of several degrees. The potash facies is mostly overlaid by a 20 m thick, competent salt roof, which has optimal mechanical properties and pervious characteristics. The principal minerals of the Canadian prairies potash facies are: sylvanite with isometric crystals of halite (13 mm to 50 mm), sylvite, and carnallite; clay forms the matrix. The potash facies extend for kilometers with only minor mineralogical changes due to localized depositional parameters such as sea water current, temperature, evaporation rate etc.

From the economic viewpoint sylvite is the most valuable salt of the potash facies, because it has the highest K_2O content, is a simple soluble chloride, and an easy mineable mineral. The beds of the pure potash salt are rare; in order to excavate them it is necessary to define the mining thickness, which means determining the economics of the mining process and establishing the existence of economical mining, satisfactory ground control.

REFERENCES

1. Jeremic, M.L. 1965. *Sedimentary deposits*. Sarajevo: Sarajevo Univ., 216 pp. (in Yugoslav).
2. Braitsch, O. 1971. *Salt deposits – their origin and composition*. Berlin/Heidelberg/New York: Springer Verlag, 297 pp.
3. Yumuang, S. 1983. On the origin of evaporite deposits in the Maha Sarakham formation in Bamnet Narong area, Changwat Chaiyaphum. Unp. M.Sc. thesis, 277 pp. Bangkok, Thailand: Chulalongkorn Univ.
4. Yumuang, S. & C. Khantaprab 1986. *On the evaporite deposits in Bamnet Narong area, Northeastern Thailand*. Bangkok: Chulalongkorn Univ., Geology Department.
5. Hite, R.J. 1982. *Progress report on the potash deposits of the Khorat plateau, Thailand*. USGS, Open File Report 82-1096, US Dept. of Int. Geol. Survey, 70 pp.
6. Jeremic, M.L. 1964. Possibility of the use of Romanian experience for exploration and development of salt deposits in former Yugoslavia. *Arch. Technol.* (Tuzla), 3-4: 15-28 (in Yugoslav)
7. Jeremic, M.L. 1964. Hydrochemical characteristics of evaporite basins in NE Bosnia. *Min., Geol. & Metall. Bulletin* (Belgrade), 5: 95-100 (in Yugoslav).
8. Jeremic, M.L. 1964. Topographical geology of Sibosnica near to Tuzla. *J. Geol.* (Warsaw), XII (12): 489-492 (in Polish).
9. Stojkovic, J. et al. 1986. Basic geological characteristics of Tetima rock salt deposits. Manuscript, Research Centre for Mining Investigation. Tuzla (in Yugoslav).
10. Jeremic, M.L. 1964. Salt deposits of evaporite bearing strata in Poland. *Min. & Metall. Bulletin* (Belgrade), 3: 60-66, 4: 79-82 (in Yugoslav).
11. Jensen, M.L. & A.M. Bateman 1981. *Economic mineral deposits* (3rd ed.). New York: Wiley, 593 pp.
12. Fuzesy, A. 1982. *Potash in Saskatchewan*. Saskatchewan Energy and Mines, Report 181, p. 25.
13. Harris, R.A. 1985. Overview of the geology and production of Wyoming trona. *Min. Engng. AIME* 37 (10): 1204-1208.
14. Kostick, D.S. 1985. U.S. Soda ash industry – the next decade. *Min. Engng. AIME* 37 (10): 1205-1212.
15. Post, L.N. 1985. FMC's Westvaco soda ash operation uses a variety of mining techniques. *Min. Engng. AIME* 37 (10): 1200-1203.
16. Stojkovic, J. et al. 1987. Salt deposits of Tuzla basin and prospect of future geological exploration. *J. Min. & Geol.* (Faculty of Mining and Geology, Tuzla Univ.), 15/16: 9-18 (in Yugoslav).
17. Jeremic, M.L. 1965. Geological structures of the salt deposits in Romania. *Min. & Metall. Bulletin* (Belgrade), 3: 57-65 (in Yugoslav).

CHAPTER 2

Folding structures

Folding deformation of evaporites is predominantly the product of orogenic movement, but in some cases may also be the product of differential overburden loading. It is possible to classify several groups of tectonic structures of salt deposits. The structural morphology represents an imprint of direction and magnitude of the acting geological stresses as well as the mechanics of resistance of evaporites to these stresses. Knowledge of structural geology of the evaporite-bearing strata is important because it determines the choice and design of mining methods and ground control.

2.1 FLEXURE FOLDING

Flexure folding is referred to the true folding. The folded salt deposits of flexure structure are usually products of lateral tectonic stress. The true folding is simple, because it is a product of the bending deformations with movement of the beds along planes of stratification. The slipping occurs along folded incompetent beds. Generally speaking, the applied load on the evaporite strata during folding had been of lower magnitude, but due to high plasticity of the salt beds, moderate deformation was produced. Primarily due to morphology of flexure deformations, four structural types are delineated.

2.1.1 Undulatory folds

The undulatory folded salt facies belongs usually to deeper deposits buried at depths of 1000 m or more. The most likely reason for the undulatory folding is due to differential loading of the overburden and high plasticity of rock salt and potash beds.

The morphological model of gentle undulatory folding is illustrated in Figure 2.1.1. It should be mentioned that in the great majority of cases of undulatory folded deposits a reasonable approximation of its geometry is made as a flat or gently dipping ore zone.

The undulatory folding is very well exhibited in Canadian potash mines. The undulatory folding of potash beds in the prairies is more gentle and regular, most likely due to limited action of external forces probably only because of overburden loading. The undulatory folding of potash beds in New Brunswick originated during differential overburden loading; later due to lateral tectonic stress they have been refolded (Fig. 2.1.2). The limb of the overturned fold, however, still exhibits an undulatory flexure deformation, but the angle of the dip, originally folded structure has been variably changed, as for example, starting at 10°, going to 15°, and finally dip at 25°.

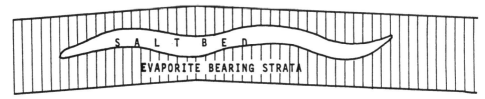

Figure 2.1.1. Model of slight undulatory folding

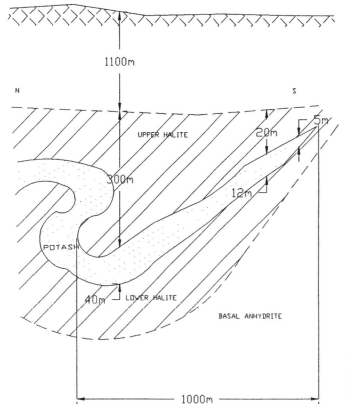

Figure 2.1.2. Model of undulatory fold, deformed by
refolding deformation
(New Brunswick)

It should be noted that undulatory folding is predominant for thinner potash deposits
which are sandwiched in thick rock salt strata, usually of syncline structure. This is most
likely due to the physical and mechanical differences between potash and rock salt, as for
example thickness of deposits, degree of plasticity, density of sedimentary planes.

The undulatory morphology of the salt beds is convenient for continuous mining technology, and adequate for strata control, where the variations of thickness and angle of inclination of beds could be accommodated by delineating of individual panels as separate extraction units.

2.1.2 *Syncline folds*

The morphology of syncline folds is represented by wide flexure structure, achieving maximum thickness at the maximum radius of carvatore. The thickness of the syncline gradually decreases toward the periphery until it finally pinches out. The syncline folds as well as undulatory folds of rock salt deposits retain more or less original sedimentary structures, and for this reason they might contain a lot of partings and be considered a 'dirty salt'.

The main characteristic of syncline folds is that they occur frequently in the basins which were located further from the main orogenic events, as for example the Canadian prairies in relation to the Rocky Mountains. The small degree of deformation of syncline folds suggests that they have been exposed either to differential gravity forces or limited tectonic forces or both.

The rock salt deposits controlled by syncline folds could be approximated as thick and more or less flat deposits. For example, in the Dez salt mine in Romania mining is carried out in beds of an average thickness of 100 m (maximum thickness is 150 m), as illustrated in Figure 2.1.3. The location of the portion of the rock salt deposit with decreased thickness is at the short syncline limbs. It is interesting to note while undulatory folding is characteristic of potash deposits, syncline folding is characteristic of rock salt deposits.

The limited influence of orogenic movement and uplift means that the majority of syncline salts deposits are located at greater depths, for example 1000 m and more. Also syncline folds are greatly influenced by depression morphology of the sedimentation basin. Anticline structures, in contrast, are due to stronger influence of orogenic movement, and they are not influenced by the palaeo relief of the basin. They are located at shallower depths, and in some instances outcrop at ground surface. The deep rock salt deposits of syncline structures might be unfavourable for underground mining and strata control because at this depth overburden load could be critical for the plastic salt pillars with regard to their bearing capacity.

Very often rock salt deposits are not considered for underground mining, but for solution mining due to their depth and appreciable waste partings. An example is the Tetima salt deposit at Tuzla, former Yugoslavia.

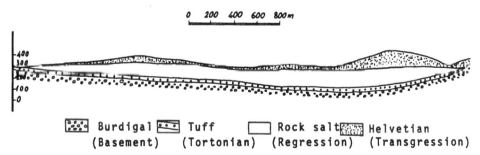

Figure 2.1.3. Syncline fold of Dez salt mine shaft (Romania)

2.1.3 *Anticline folds*

The anticline folds could be formed as a simple flexure structure or as a flexure complex of folded and refolded strata. In this subsection a simple anticline structure is considered, but flexure-complex folds will be discussed under a heading of complex anticline folds.

Anticline folds are often the product of higher lateral tectonic pressure, where the evaporite strata have been bent and moved upwards. The anticline folds are developed in the Slavik-Prahova area, Romania (Figure 2.1.4). Some anticlines of rocksalt deposits outcrop on the ground surface as shown in photographs in the first chapter of this book.

The anticline flexure structure is due to flow of rock salt beds along slips; this results in moderate thickening of the salt deposit whose thickness is at least double that found in syncline folds. The maximum thickness of an anticline fold is at the axis of folding. Due to thickening of the evaporite strata during anticline folding the width of rocksalt deposits is decreased (Figure 2.1.5).

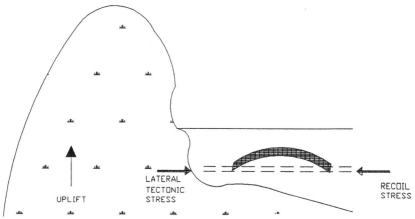

Figure 2.1.4. Anticline fold in the Slavik-Prahova area (Romania)

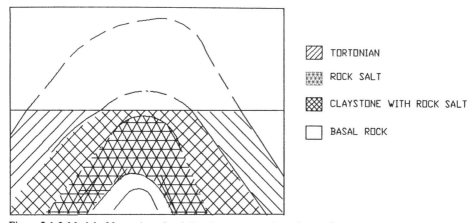

Figure 2.1.5. Model of formation of anticline flexure due to lateral tectonic stress

During orogenic movements and uplift, mass flow in the salt deposits occur; this causes the majority of anticline rock salt and potash beds to be concentrated to a certain degree; waste layers of claystone, gypsum and others are displaced. This phenomenon is not well exhibited in the syncline structures, where salt beds are usually interbedded with the waste rock layers. Under these circumstances only small portions of salt deposit or none are mineable.

Finally, the anticline represents a higher degree of true folding, which facilitates increase in thickness of salt deposits as well as their mechanical purification. The mining and ground control should still be adequate, compared to the complex of flexural folding in overturned folds which are discussed in Section 2.2.

2.1.4 *Syncline and anticline structure*

The flexure of the salt beds exhibited by concentric zonal deformation will result in the formation of anticlines and synclines, which represent the highest degree of folding compared to undulatory folding.

The integrated anticlines and synclines in one structure will produce a differential stress field within a salt deposit. The redistributed and concentrated internal stresses at the hinges of folds will oppose each other and as a result develop tensile stresses at the crest of anticlines, and compressive stresses at troughs of synclines (Figure 2.1.6). The folded anticlines and synclines of the competent beds tend to be of the same thickness, regardless of their position on the fold. The incompetent beds such rock salt and potash are typically thin at the limbs and thicken at the hinges. The geological cross-sections of some rock salt deposits in the Carpathian salt province of Europe indicate that there is a similar magnitude of thickening and thinning associated with both anticline and syncline structures, when they are integrated into one domical fold.[1, 2]

In Chapter 1 the concentric zonal structure of an integrated syncline and anticline fold is given for the model of evaporite deposits in the Bamnet Narong area, North-eastern Thailand. The model of syncline and anticline structure also suggests a distribution of rock salt and potash minerals, which have been affected by processes of mechanical and chemical alteration.[3, 4]

The concentric zonal deformation, particularly of plastic beds, influences to a great extent the physical and mechanical characteristics of their environment for mining and ground control. Besides stress and mineral assemblage redistribution and concentration reflecting the anticline-syncline geometry, displacement along bedding planes are with separate sedimentary surfaces. Finally, it should be mentioned that massiveness of the

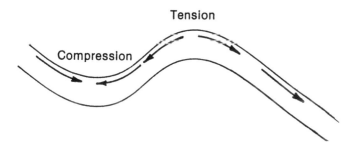

Figure 2.1.6. Stress distribution within syncline and anticline fold structure

formation is an important factor. For example for the same kind of rock salt, if one formation is 2 m thick and another is 200 m, the thick-bedded formation will be more competent and have significantly greater self-supporting ability.

2.2 COMPLEX FOLDING

This is a common tectonic structure of the many evaporite deposits, particularly of the Miocene of Europe. The flexural complex folding is represented by overturned inverted formations or refolding of units. Under these circumstances the evaporite-bearing strata assumed an anticline form with a complex internal structure. The main characteristic of complex folding is that flow of the salts mass is followed by intensive flexural deformation and mass concentration.

2.2.1 Overturned folds

By definition the overturned folds have the axial planes inclined and both limbs dip in the same direction, but at different angles. The overturned fold is the one that has been rotated up to 90° to attain its present attitude.[5]

Overturned folds have been studied in Europe, particularly in Romania, where a hypothesis to explain their origin for deposits in the Carpathian province has been proposed. As illustrated in Figure 2.2.1, the primary structure exhibited is the sedimentary feature of the undisturbed salt beds. The coast of the shallow marine basin is represented by the Carpathian flish of the Eocene age. Due to orogenic movement and uplift of the Carpathian Mountains, lateral tectonic stress has occurred in the direction of the sedimentary basin.

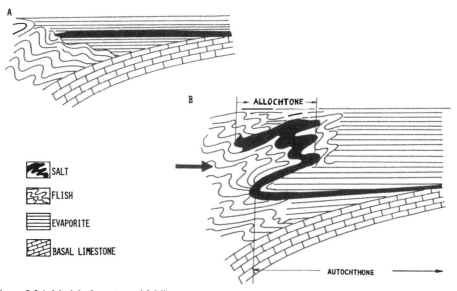

Figure 2.2.1. Model of overturned folding
a. original deposition; b. position after refolding

Figure 2.2.2. Overturned fold of rock salt beds, interbedded gypsum and anhydrite (Bochnia mine, Poland)

As a consequence of this stress action the rock salt strata with high plasticity were folded and pushed in the direction of the front of the lateral tectonic stress. Under these circumstances the Carpathian flish partially covered and intruded the salt deposits. The plastic flow of the salt deposits is manifested by intensive flexural deformation causing repetitions of many beds.

Further deformation and movement of the salt strata were stopped by clastically deformed gypsum and chemical limestone, which played the role of a retaining wall. Under these circumstances, two zones of rock salt structure can be distinguished. First, the autochthons were beds that are not deformed, but they are thin and stratified at great depth. Second, the allochthons, were beds that are intensively folded and thrust upwards towards ground surface.[5, 6]

In underground mines overturned folds have been observed where axes of folding are perpendicular to the direction of lateral tectonic stress (Figure 2.2.2). Together with the salt beds interbedded claystone and anhydrite had been folded simultaneously. The claystone and anhydrite act with clastic properties and so are fractured by tensile radial cracks.

In the case of overturned as well as other flexural folds, it is difficult to delineate the salt beds by the roof strata and floor strata as in the case of domical folding. The final structure of the folding is an important factor to be considered when determining the mining and rock mechanics factors of a deposit.

2.2.2 Recumbent folds

This type of fold has its axial plane horizontal or nearly so. Large-scale folds of this type are especially well exposed in the Alps. The strata in the inverted limb are usually thinner than the corresponding layers in the normal limb. The term 'arch-bend of overfolds' has been used for the flexural part of the fold between the normal and inverted limbs.[5]

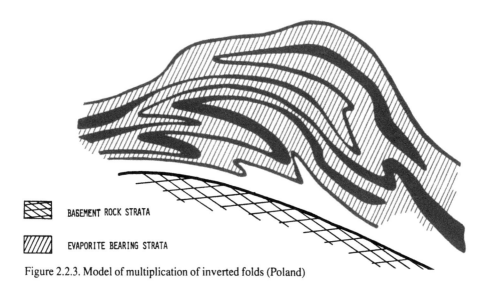

BASEMENT ROCK STRATA

EVAPORITE BEARING STRATA

Figure 2.2.3. Model of multiplication of inverted folds (Poland)

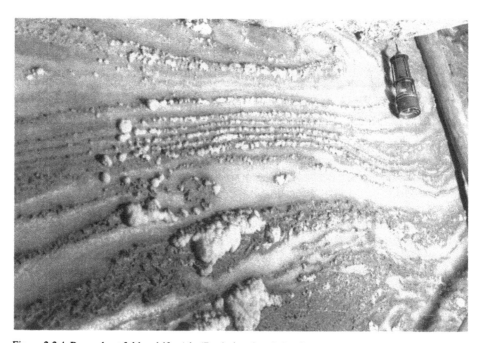

Figure 2.2.4. Recumbent fold at drift scale (Bochnia mine, Poland)

Increase in the number of overfolds of rock salt or potash beds will decrease the width of deposits, but increase their thickness. For example, at the Bochnia salt mine in Poland, due to recumbent folding the width of the deposit is up to 200 m, but the strike is 7000 m (Figure 2.2.3). In this mine the overturned folds could be observed at the scale of the underground mine openings; they exhibit the same features as those formed by regional recumbent tectonics (Figure 2.2.4).

The general picture of this type of deposits creates the impression that these are thick deposits formed by the repetition of the same series of sediments. From a mining and rock mechanics viewpoint the resulting deposit can be treated as a salt block of flat or gently dipping strata.

2.2.3 *Refolded folds*

As an example, refolding of concentric, zonally folded sediments could occur as follows: in the direction of the primary folding due to rejuvenation of the same orogenesis and in a direction which differs from the one of primary folding due to a younger orogenic movement in this particular part of the earth crust.

The refolding of the evaporite strata in the direction of the original folding is governed by shear folding (also known as slip folding) which results from minute displacements along closely spaced discontinuities.[6] In this refolding the mechanics of three types of deformations can be observed:

1. Displacement along bedding planes which had been separated by flexural slip during concentric zonal folding. Shearing and folding take place particularly in highly plastic beds (rock salt and potash), which pushed upwards of clastic beds.

2. Deformations of rock salt and potash beds by minor refolding such as drag folds could be related to the contemporaneous major phase of folding. The acute angles between the axial plane of drag folds and the more competent bed point in the direction of differential movement.

3. Displacement of clastic beds (sandstone, gypsum and other) along a radial fracture formed during bending deformations. This mechanism could be inferred from the geometry of micro and macro folds.

The beds of evaporite strata may exhibit shearing along sedimentation planes and dense refolding (plastic rocks), or shearing along the curvature of the fold (elastic rocks) forming small undulatory folds whose frequency depends on the density of radial cracks and the amplitude and magnitude of the slip (Figure 2.2.5). It should be noted that within underground mine structure the shear refolding is most readily demonstrable, and the extent to which major folds may be of this origin remains conjectural.[5] A hypothetical section of the rock salt deposit at Tuzla basin, however, suggests that shear refolding might occur at larger scales (Figure 2.2.6).[7]

The refolding of evaporite strata in the direction different from the primary folding, known as cross-folding, is recognized by the formation of large salt anticlines with limited strike extension.

The refolding of primary anticline and syncline structures of the evaporite strata at Bamnet Narong area (Thailand) has been analyzed with computer, by plotting data from exploration drill holes.[3, 4] The printed graphics indicate that there are gentle and steep anticline and syncline structures, as illustrated by the structure contour map of the top of the salt beds (Figure 2.2.7). The thickness of the evaporites increases south-eastwardly and the thickest part is generally associated with the anticline zones (Figure 2.2.8). The upper

Figure 2.2.5. Micro and macro asymmetrical refolding of the different evaporite beds (room – 190 level, Tusanj rock salt mine, former Yugoslavia).

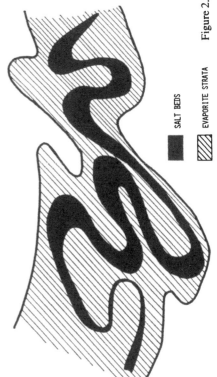

SALT BEDS

EVAPORITE STRATA

Figure 2.2.6. Model of refolded fold of rock salt beds (Tuzla basin, former Yugoslavia)

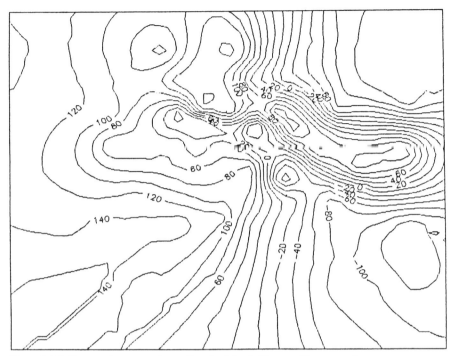

Figure 2.2.7. Structural contour of top of basal salt beds (Bamnet Narong area)

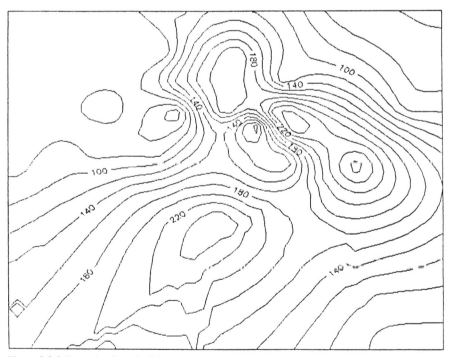

Figure 2.2.8. Isopach of basal salt beds (Bamnet Narong area)

surface of the basal salt beds is also illustrated, but as a three-dimensional diagram in various lines of sight by the rotation angle from the horizontal axes in clockwise direction and by tilt of angle (Figure 2.2.9). The structural contour maps and particularly the three-dimensional diagram illustrate two directions of folding of the evaporite strata.

1. Primary folding was of greater magnitude and it is represented by concentric zonal folds of anticline and syncline structure. The salts facies were thickened to a limited degree at the crest of an anticline.

2. Secondary folding was of the same or lower magnitude and it is represented by almost perpendicular refolding of the original structure. This type of folding is known as cross-folding, which results in maximum thickness of the strata at the anticline area of intersections of the two fold axes.

It is interesting to note that structural features of the refolded fold, to a certain degree,

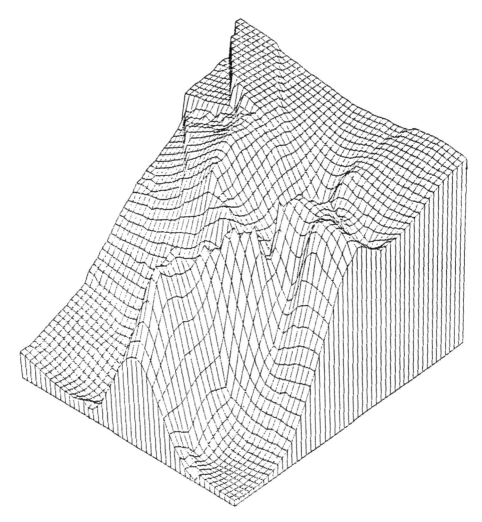

Figure 2.2.9. Upper structural surface of basal salt beds (rotation 45°/45°).

control the distribution of the mineral assemblage sylvite and carnallite. The sylvite/ carnallite zone is present in the gentle syncline flanks of rock salt anticline structure. The potash bed of rather pure sylvite is located only in the gentle flanks of syncline structure.[4] However, the present mineral distribution does not depend only on the mechanical alteration of the salt series, but also on its chemical alteration, as discussed in the first chapter of this book.

2.2.4 *Complex anticline folds*

In some cases the final product of the overturned, recumbent and intersected folding is an anticline when compacted rock salt and potash were thrust upwards.[8] Of course, there is a fundamental difference between anticlines formed by salt beds bending and anticlines formed by mass flow of salt.

In Romania the maximum thickness of the complex anticlines of the salt deposits is between 600 and 800 m; this compares with the result of concentric folding where the maximum thickness of the syncline is up to 150 m and the anticline is up to 300 m.[7] As a result of the increased thickness of the anticline, the length of the deposits is shortened (Figure 2.2.10). The enlarged thickness of complex anticline is caused mainly by pushing of the salt mass upwards into the anticline core. Due to their high plasticity the rock salt and potash easily assume the general anticline shape of the evaporite strata.

The anticline structures of evaporite strata result in thick salt deposits with a good quality of salt. Folding, refolding and other compression phenomena compacted the salt beds, unified and reinforced strength properties and behaviour, so that it is possible in this type of deposit to excavate rooms up to 50 m height. Also the quality of the salt is appreciably increased because during folding, due to differential mechanics, the waste rock partings (gypsum, claystone and others) were squeezed out and the rock salt body becomes almost mono-mineral. Under these circumstances some portion of the mineral deposits consists up to 99% NaCl.[9] Due to mechanical alteration and purification of the salt body, the halite

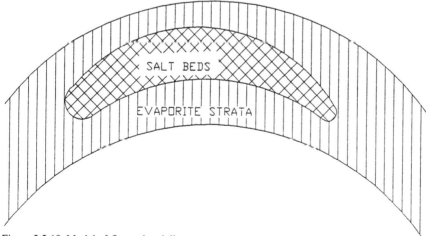

Figure 2.2.10. Model of flexural anticline

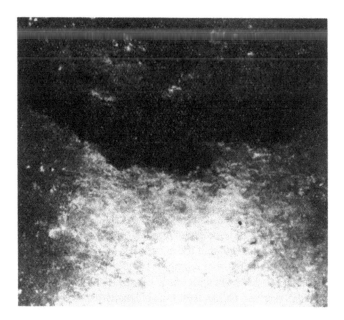

Figure 2.2.11. Contact of
clay/rock salt beds (dark)
and mechanically purified
rock salt mass (white)

crystals exhibit coarse grained texture (Figure 2.2.11). Also, besides the increase in the
quality of the salt, the recrystallized salt is white in colour, which is most desirable for
mining.[7]

The internal composition of the complex anticlines is very complicated because the beds
are intensively folded and arranged in various structural positions. In addition, the shear
refolding of internal beds is present at a small scale. The difference between a simple
anticline and a comples anticline is that the transverse profile in the first case follows
original geological sedimentary features; but in the second case, due to overturning of
evaporite strata, the same sedimentary profile could be in the footwall and hanging wall of
mineral deposits, e.g. the roof strata sediments are older than salt beds. The mining of a
complex anticline is complicated and requires careful planning and preliminary work
before a particular mining method is implemented.

2.3 DIAPIR FOLDING

This is a common structural type of the salt deposit for many evaporite basins as for
example Zechstein's deposits of Central Europe. The diapir folding is represented by
structural characteristics of transitional type of the complex to real piercing folds and salt
dome.

2.3.1 *Saddle fold*

A saddle fold is a type of fold which shows an additional flexure deformations in its crest at
right angles to that of the parent fold and of much larger radius. It should be considered as a
transitional type of piercing folding.

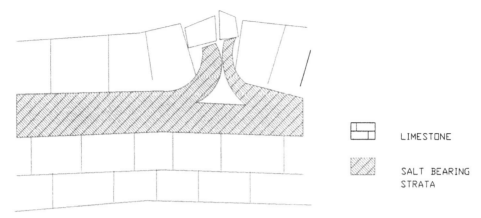

LIMESTONE

SALT BEARING
STRATA

Figure 2.3.1. A saddle folding (Wiesbaden, Germany)

The theory of formation of saddle folds is based on observations of salt deposits in the vicinity of Wiesbaden in Germany. The evaporite-bearing strata represented within beds of salts, claystone and mudstone exibit a soft and plastic medium, which is sandwiched between thick limestone strata of Triassic age (Figure 2.3.1).[1] Due to lateral pressure within the basin, the tertiary formation was tectonically deformed. The limestone in the roof strata acts as a clastic rock and was fractured into structural blocks, but the salt-bearing strata reacted plastically and were folded. The salt flowed in the direction of least resistance, e.g. into the fractured zone, which has been folded in the form of a saddle.

It should be mentioned that the saddle folds form as a result of lower lateral stresses, but with their increase of what formed more complicated structure as discussed in Subsection 2.3.2.

2.3.2 *Piercement fold*

The theory of piercement folding of salt bearing strata has been developed by H. Stille, on the basis of studying the Zechstein salt deposits in Germany. The concept of his theory is illustrated in Figure 2.3.2, which represents the gradual deformation of salt-bearing strata until a final piercement fold results in the form of a salt dome; the process is described as follows:[1]

1. salt beds, south of Hertz, Turyngshieg;
2. anticlinal fold, south of Hanover;
3. salt anticline excem, in the vicinity of Hildesheim;
4. salt anticline body, around Bentheim;
5. real salt dome, north of Hanover.

The degree of salt deformation depends on its plasticity and the lateral tectonic pressure. Salt minerals have a great ability for translation deformation when they are exposed to lateral pressure. Also with the increase of temperature, salt deforms with increasing ease, as can be seen by the mobility of salt minerals.

The evaluation of the piercement fold indicates specific tectonic deformation of salt deposits. The main feature of this deformation is the flow of the salt mass upwards in the direction of the fractured zone of the overlying sediments. The degree of intrusion of the

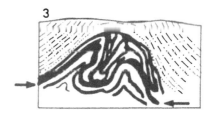

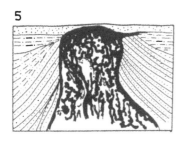

Figure 2.3.2. Tectogenetic structures of salt deposits in Germany (after H. Stille)

salt mass is variable; when it achieves its maximum then salt dome is formed, as discussed in Subsection 2.3.3. Under these circumstances the salt beds from depth are translated close to or onto the ground surface. Outcropping salt deposits formed in this way could be preserved in arid climates without rainy seasons.

2.3.3 *Domal fold*

Salt domes are folds in which beds dip radially away, or toward a point respectively. The domes are a very important structure in oil geology and in salt geology.

The uplift of a salt cupala from a depth greater than 2000 m is followed by fracturing and pushing upward of younger overlying strata. These strata, at present, rest on the sides of the dome. The tectonic contact between salt domes and younger sediments is exhibited by their upward position. The pull of overlying sediments upwards is uneven.[1, 2] The younger sediments could have great economic importance because some of them bear oil and gas, as illustrated in Figure 2.3.3. The structure of the salt domes is of particular interest because of the prolific oil accumulations associated with some of them. In many cases salt domes are more important as an indicator of oil and gas deposits than as a source of salt deposits.[10]

The surface expression of salt domes could be absent, as they may be indicated by depressions. The salt domes have circular or elliptical horizontal sections, whose width could be in range between 1000 and 10 000 m. The salt plug is usually several thousands of meters thick and the final depth of many salt domes has not yet been determined.[10]

The internal folding of salt domes is of a complex nature, particularly in the case of multi-salts mineral deposition. For example, a salt dome of a length of 25 km and a width of 2 km at Klodawa mine (Poland) exhibits a complexity of internal structure of rock salt/potash facies, which can be seen in Figure 2.3.4. This is the only deposit in Poland where potash mining is carried out. Due to the importance of inner folding for mining and strata stability, this topic has been discussed separately in Section 2.4.

Within the dome the salt mass is usually not differentiated from the layers of waste rocks as in the case of flexural folding, because all sediments during uplift had been compressed and elastically and plastically deformed. As a result of this, within the structure of the dome dislocation can be observed. Some domes contain rock salt and potash deposits of inferior

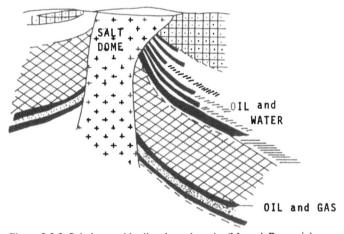

Figure 2.3.3. Salt dome with oil and gas deposits (Moreni, Romania)

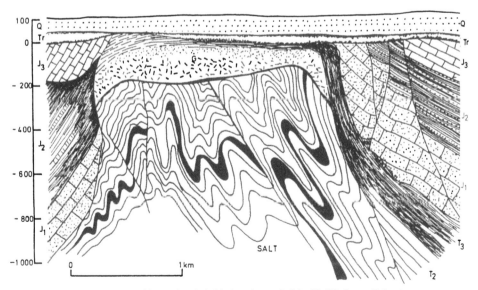

Figure 2.3.4. Dome composed by rock salt (white) and potash (black), Klodawa, Poland

quality, where the maximum content of NaCl is 50-60%. The domes of complex inner structure are mineable only under optimal mining-geological conditions as well as satisfactory strata behaviour.

From the discussion in this section it is obvious that fold structures have a great influence on rock mechanics and mining technology of rock salt and potash deposits, particularly in the case of complex anticline fold and salt domes.

2.4 INNER FOLDING

The internal folding of salt deposits considers a complex anticline folds and dome structure of salt deposits. The inner folding described at this place is based on the J. Poborski and K. Slizowski geological investigations and mining experience in the Zechstein domes in Central Poland.[11] The four topics of inner folding as described by the authors are discussed.

2.4.1 *Predisposition for halotectonic*

As it appears, the inner structure of domes is extremely complex and entangled. However, the authors were able to establish certain regularities in that respect, taking into account petrophysical bipartition and of course their tectogenetic characteristics.

Regional distinctness of the Zechstein series column in the zone of Polish anticlinorium and the adjacent synclinoria was emphasized by the different lithostratigraphic development of the Zechstein stages Z3 and Z4. Starting in about the middle of the Z3 sequence and extending up to the 'variegated sandstone', sedimentation took place within continental water regime, thick halitic lutites had formed, so called zubers: brown zubers (Z3) and red zubers (Z4). This is why a petrophysical bipartition of the whole Zechstein column is strongly marked. This is why the sedimentary column can be divided into two contrasting segments from the point of view of rock mechanics, as illustrated in Figure 2.4.1 and described below.

1. a lower segment (Gl) comprising stages Z1, Z2 and the lower part of stage Z3, where thick massive chlorides predominate; and

2. an upper segment (Gu) consisting of the upper part of stage Z3 as well as of the whole stage Z4 with its prevailing salty clays.

Such a lithological and petrophysical bipartition must necessarily be reflected in the tectonic processes.

Rocks of the lower segment of the column (Gl) were continuously mobile owing to the potential plasticity of the chloride rocks. The upper segment (Gu), on the other hand, gives evidence of the opposite behaviour, as the clayey rocks remained passive, inflexible and easily crushed and brecciated.

Reviewing the petrological succession in detail in this Zechstein column, attention was paid to some stratigraphic horizons where these different rock members were in contact. By this we mean the contrasting petrophysical properties of one member against another. Such horizons were predisposed to the development of dislocation planes during the structural development of the salt domes. These were clearly indicated at the column edge by eight-branched stars (Figure 2.4.1).[11]

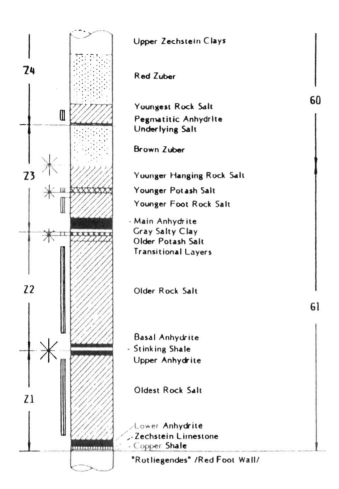

Upper Zechstein Clays

Z4

Red Zuber

Youngest Rock Salt   60
Pegmatitic Anhydrite
Underlying Salt

Brown Zuber

Z3   Younger Hanging Rock Salt

Younger Potash Salt
Younger Foot Rock Salt

- Main Anhydrite
Gray Salty Clay
Older Potash Salt
Transitional Layers

Z2   Older Rock Salt

                                                                61

Basal Anhydrite
- Stinking Shale
Upper Anhydrite

Oldest Rock Salt
Z1

Lower Anhydrite
- Zechstein Limestone
- Copper Shale

"Rotliegendes" /Red Foot Wall/

Figure 2.4.1. Zechstein
stratigraphic column in the
region of Central Poland,
petrophysical bipartition (Gl
and Gu) and predispositions
for structural subdivision
(after Poborski)

2.4.2 Features of inner folding

The salt domes are sometimes considered monolithic massives of rock salt, including some intercalations. In fact, they are built of the whole sequence of the salt series strata. The most striking features of the internal tectonics within the Zechstein salt domes may be mentioned as follows:

1. Intense and very steep folding of the layers, with a very high wave amplitude when compared with the wave length (Figure 2.4.2).

2. Quasi-plastic thickening of the salt fold bends (Figure 2.4.2).

3. Contrasting mechanical relation of the salt layers to the adjoining barren rock (anhydrite, dolomitic marl, shale); the best example being observed at the contact with a thick anhydrite stratum where quasi-plastic bulging of the layered salt against a sharp anhydrite bend is a typical phenomenon (Figure 2.4.3).

4. The anticline cores built of the older salts pierce upward across the overlying younger evaporite sequence; for this reason, several diapiric structures are to be surveyed (Figure 2.4.4).

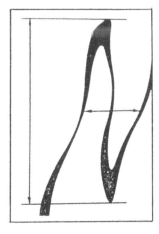

Figure 2.4.2. Folding of high wave amplitude with salt thickening

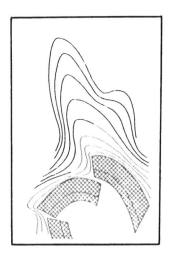

Figure 2.4.3. Differential deformation of salt (plastic) and anhydrite (clastic)

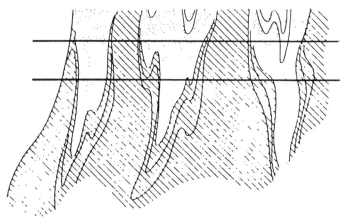

Figure 2.4.4. Inner diapiric structures of older salt

5. A common structural phenomenon observed within the anticline limbs. Certain stratigraphic chloride members are very much thinned from their normal thickness. At the same time the intercalating barren strata are torn up and squeezed as well as pinched completely out, the remaining joints marking their way upward.

2.4.3 *Inner structural pattern*

A general pattern of the internal structure may be briefly outlined, taking into consideration holotectonic phenomena.

First, and most important, is the mutual contrasting mechanical behaviour of the two Zechstein column seqments, denominated as the lower one (Gl) opposed to the upper one (Gu). During halokinetic folding the concentrated chloride strata of the lower segment (Gl) form the anticlines piercing upward through the overlying strata of the upper segment (Gu), as shown in Figure 2.4.4. So, secondary diapiric domes occur within the single main salt dome.

J. Poborski and K. Slizowski pointed out that the hard and stiff strata subdividing the lower column segment (Gl), i.e. 'upper anhydrite' (Z1), 'the stinking shale' and 'basal anhydrite' (Z2), which are underlain and overlain by very thick rock salt successions. We recognize that these strata were rigid and resisted the structural doming. In the majority of the Zechstein salt domes, which are circular or oval shaped, those strata were uparched although intersected and partly dislocated. From such a barrel-like vault the older rock salt massif was strongly deformed, thus forming two or three very high digitation anticlines (Figure 2.4.5).

Sub-surface geological methods as well as geophysical ones helped the authors to reconstruct the inner structure pattern of the salt domes taken as a whole in their height, as roughly outlined in Figure 2.4.5. Their, three structural stages have been revealed, each being graphically marked and explained.

Figure 2.4.5. General sketch pattern of the salt dome inner tectonics
1. Zechstein substratum ('Rotliegendes'); 2, 3, 4 etc. the first, second and third structural stages (after Poborski)

2.4.4 *Inner structural outlines*

The mineable salt beds of the Zechstein series are indicated at the margin of the stratigraphic column (Figure 2.4.1). The stratigraphic position of the salt beds is known with a tolerable accuracy, depending on the quality and the kind of the salt facies. The mineable salts are potash salt composed of potassium and magnesium salts as well as rock salt composed by halite.

Following the internal structure boundaries, the main rock salt deposits are developed during the second structural stage (Figure 2.4.5) and they are represented by the enormous concentration of the older rock salt (Z2) and the anticline cores of the first order (Figure 2.4.4). On the other hand, the older potash salt (Z2) as well as the younger one (Z3) occur in the anticline limbs, thus forming layered deposit, even there partly deformed. Moreover, the younger potash (Z3) concentrates within the second order fold bends (Figure 2.4.4).

In the third structural stage the youngest and purest rock salt (Z4) forms some mineable structure within the syncline bed.

REFERENCES

1. Jeremic, M.L. 1965. *Sedimentary deposits*. Sarajevo: Sarajevo Univ., 216 pp. (in Yugoslav).
2. Jeremic, M.L. 1965. Geological structures of the salt deposits in Romania. *Min. & Metall. Bulletin* (Belgrade), 3: 57-65 (in Yugoslav).
3. Yumuang, S. et al. 1986. *On the evaporite deposits in Bamnet Narong area, North-eastern Thailand*, p. 22. Bangkok: Chulalongkorn Univ., Geology Department.
4. Yumuang, S. 1986. Computer in potash and rock salt; post-depositional structural models at Bamnet Narong area, North-eastern Thailand, p. 33. In *Proc. of Conf. on Occurrence, Exploration and Development of Fertilizer Minerals*, held by UNDP-ESCAP, Bangkok, August 25 - September 2, 1986.
5. Hills, E.S. 1966. *Elements of structural geology*. London: Associate Book Publ., 482 pp.
6. Billings, P.M. 1972. *Structural geology*. London: Prentice-Hall, 606 pp.
7. Jeremic, M.L. 1964. The economic importance of rock salt deposits of North-eastern Bosnia. *Technol. Arch.* (Tuzla), 3: 71-78 (in Yugoslav).
8. Jeremic, M.L. 1964. Salt deposits of evaporite bearing strata in Poland. *Min. & Metall. Bulletin* (Belgrade), 3: 60-66 (in Yugoslav).
9. Jeremic, M.L. 1966. Mineability of the salt deposit Tusanj, Tuzla. *Geol. Bulletin* (Warsaw), 11: 494-499 (in Polish).
10. Jensen, M.L. & A.M. Batemen 1982. *Economic mineral deposits*, p. 593. New York: Wiley.
11. Poborski, J. & K. Slizowski 1983. Geological prediction of the mining feasibility within the salt domes of Upper Permian Zechstein, Central Poland. In *Proc. of 6th Int. Symp. on Salt*, The Salt Institute, Virginia, Vol. 1, pp. 303-310.

CHAPTER 3

Faulting structures

Faulting structures are not as dominant as folding structures but they are of certain importance for salt mining and strata stability, particularly in the phase of mine opening and development. As understandable, the faults are most frequently present in the clastic sediments (gypsum, marl, limestone and others) than in ductile sediments (rock salt, trona and others). The tectonic of clastic structures of evaporite strata are represented by four topics: normal faults, reverse faults, tectonic breccia and structural defects.

3.1 NORMAL FAULTS

Normal faults are formed by the effects of gravity forces in the earth's crust. They are not common in the salt geological structures as in the case of other lithology strata. Normal faults are defined as dislocations in which the hanging wall is displaced downwards relative to the footwall. Some normal faults may have an almost zero hade, whilst others may have over 50 degrees. Breccia or gauge is commonly developed and clean-cut fractures are rare.[1] The throw of normal faults might be significant to negligable small that cannot be traced throughout a vertical sequence. In regard to salt mining four normal fault structures are represented as follows.

3.1.1 Shear faults

They have been identified during potash exploration at Malagawatch, Cape Breton, Nova Scotia, by L. Dekker, who described them as further discussed.[2] The oblique shear faults have been discovered in the basement of the evaporite-bearing strata by geophysical surveying. Extensive gravity surveys on land, lake bottom, on ice, as well as vibriosis surveys were carried out simultaneously with exploration drilling. However, positions of oblique shear faults have been derived from interpretation of aeromagnetic data (Figure 3.1.1).

The brittle deformation of salt-bearing strata appears to coincide with oblique shear faults in the underlying basement. The salt-bearing strata are divided in tectonic blocks which are separated from each other by north-westerly striking, steeply dipping, gravity shear faults (Figure 3.1.2). The throw of shear faults has been also appreciable.

The block of hole M-5A is displaced 600 m downward in respect to the block with holes M-8 and M-9 (Figure 3.1.3). As L. Dekker pointed out, therefore only the upper potash zone was reached and the main potash horizon was not intersected in this hole as drilling

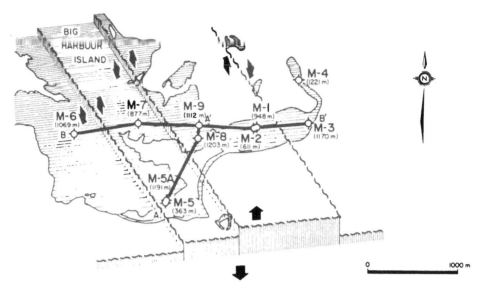

Figure 3.1.1. Inferred basement faults in relation to drill holes

stopped well above its inferred stratigraphic position. Holes M-3 and M-4 on the one hand and holes M-5/5A and M-7 on the other hand, are located on down-dropped blocks on either side of the block containing M-1, M-2, M-8 and M-9. A longitudinal section (Figure 3.1.2) shows that holes M-7 and M-3 also stopped well short of reaching the ore zone because of complex fold repetitions in the added hanging wall section, particularly on the block with M-3 and M-4. A considerable amount of stratigraphic section remained to be penetrated on these downthrown blocks before the objective would have been reached.[2]

3.1.2 *Interslicing faults*

At the edge of the dome, the salt or caprock may be positioned directly against the displaced sediments, where their separation might be occupied by a shale sheath or gauge. There is an opinion that the width of interslicing fault zone might vary from very thin (several centimeters) to perhaps very thick (several thousands of meters). The nature of the edge of some domes is not very well known, for example the Gulf region in the USA, but some opened by underground workings, for example in Germany, are better known. In most cases, the mine workings intersected a thin (1 cm to 21 m) mantle of anhydrite caprock which sharply separated the salt from the sediments. In one case, the salt and sediments were separated by 50 to 100 m of salt-mudstone diapiric sheet.[3]

P.C. Kelsall and W. Nelson described the edge of the salt at the Bayou Choctaw dome as illustrated in Figure 3.1.4. The edge is seen to be a complex melange with large blocks of salt detached from the main mass and blocks of sediment included within the salt. Although the complexity seen at Bayou Choctaw may be exceptional, it is well known from the Gulf region mines that the proportion of sedimentary inclusions increases toward the edge of dome. This indicates that some degree of interslicing of the salt and adjacent sediments might be common.

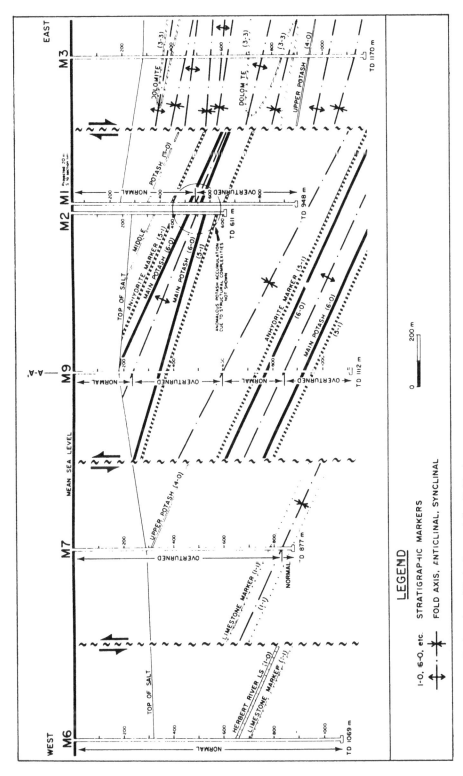

Figure 3.1.2. Longitudinal section of faulted strata (after Dekker)

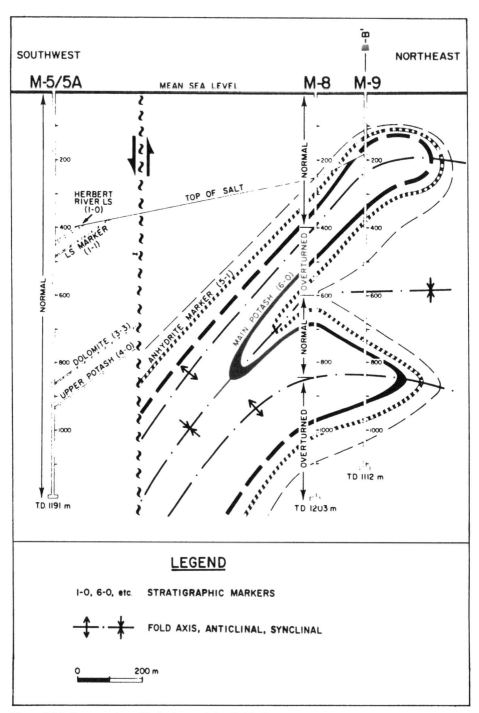

Figure 3.1.3. Cross-section of faulted strata (after Dekker)

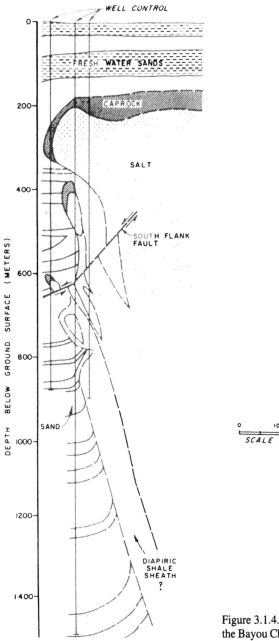

Figure 3.1.4. Inferred intersecting fault of the edge of the Bayou Choctaw dome

3.1.3 *Step faults*

The tectonic investigation of the Tuzla basin indicated that step faulting is a part of a large structural unit delineated on the north by Majevica Horst and on the south by an ophiolite zone. This large structural unit is represented by several sub-structural units such as the Dokanj sinclinale, Jala-Pozarnica anticlinale, Dolovi anticlinale and others. During de-

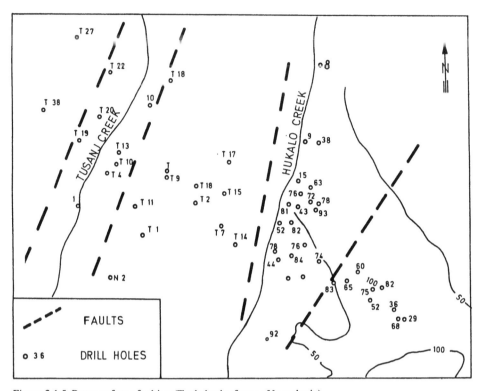

Figure 3.1.5. Pattern of step faulting (Tuzla basin, former Yugoslavia)

tailed geological mapping of the Tuzla salt basin several faults NNE-SSW strikes have been identified as a series of almost parallel step faults of similar displacement (Figure 3.1.5). The faults of gravitational origin belongs to a displacement of the eastern tectonic blocks. The degree of displacement is relatively small, and could be observed only in the clastic sediments (claystone, gypsum and others), because in the rock salt beds this phenomenon is exhibited as drag folds whose axes are perpendicular to the direction of displacement. Typically the normal faulting of evaporite-bearing strata should not have any appreciable influence on the underground mining and rock mechanics of the salt beds. These faults, however, have been a water feeder to salt deposits, causing their dissolution; the resulting brines have been the subject of exploitation from the Roman era to the present.

Minor structures with small displacement which are associated with the step faulting in the Tuzla salt basin are in the clastic rocks. The step faulting is a post-folding phenomenon and it has not particularly affected the internal structure of the salt deposits. This type of normal faulting is much better exhibited in the nearby Kreka coal basin where step faults may occur as a group in the mining area where each is throwing the coal from several centimeters to several decimeters.

3.1.4 *Shear zone*

The shear zone of dome deposits is discussed on the basis of Kupfer's work as Kelsall and Nelson described in their paper and as is given below.[3]

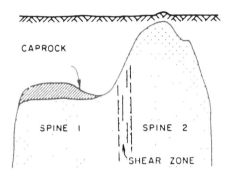

Figure 3.1.6. Internal shear zone of Jefferson Island dome (after Dekker)

Many domes contain small proportions of shale and sandstone derived either from material which was originally deposited with the salt or from material that was incorporated into the dome during its upward displacement. Close to the edge of a dome, sedimentary material might be incorporated by shearing as slices of shale sheath are caught between slices of salt and then further broken up during subsequent upward movement. Within a dome the sedimentary material might be shale sheath which is caught between adjacent spines of salt which were intruded independently.

Kupfer referred to the sheared zone at the edge of a dome as 'external shear zone' which might be identified by shearing within the salt (i.e. banding with constant strike parallel to the edge of the dome with little evidence of folding) as well as by high proportions of impurities. In the Weeks Island mine, Kupfer found evidence of distinctive shearing extending 150 m into the dome from the edge of the salt. In the Belle Isle dome, a hole drilled 100 m from the edge of the salt encountered 40% to 50% shale mixed with salt. It is common practice in the Gulf region mines to avoid mining within 100 m of the edge of the dome. This is evidently sound practice, but it excludes direct observation of the nature of the salt close to the edge of the dome.

Shear zones within the salt mass were designated by Kupfer as 'boundary shear zones' in cases where sedimentary material was found, and 'internal shear zones' in cases where there was evidence of pronounced shearing but no sedimentary material (Figure 3.1.6). Boundary shear zones have been observed by Kupfer in the Avery Island, Belle Isle, Jefferson Island and Weeks Island mines, where they are commonly associated with anomalous features such as brine and oil seeps and gas pockets. Boundary shear zones may be from 3 m to greater than 100 m wide, and they may be very extensive in both the lateral and vertical directions. A striking example is the major shear zone in the Avery Island dome which has been eroded to form a mark depression in the top of salt. However the great majority of domes do not contain a major shear zone.

3.2 REVERSE FAULTS

A reverse fault in which the hanging wall is displaced upwards relative to the footwall is almost the opposite of a normal fault. Formed by compression, reverse faults might be extreme results of folding originating through the failure of the steeper limb of an asymmetrical fold.[1] They therefore commonly occur in heavily folded areas such as folded salt basins where the trends of reverse faults and folds are similar, as for example the

Carpathian province in Europe. As Williamson stated, the hade is generally greater than that of most normal faults being often between 40 and 60 degrees.

3.2.1 *Thrust faults*

If the hade is greater than 45 degrees the faults are called a thrust, and it is possible that many reverse faults flatten out at depth into such faults.[1]

The thrust faults originated during uplift orogenesis and due to induced enormous lateral pressure the sedimentary strata were folded and ruptured (Figure 3.2.1). The main characteristic of the thrust fault structure in the salt basin is that strata above the thrust plane move up at some distance, and the dip is so small that the overlying block is pushed almost laterally.[5]

Low angle thrusts or overthrusts are found only in intensively deformed evaporite basins. These thrust structures are an expression of compressive forces which shortened the salt-bearing strata along dip. The mechanics of shortening is represented by breaking, with one sheet of beds overriding another.[5]

As a result of the thrust faulting in the Carpathian province, underlying tuff strata as a marking horizon of the salt deposits also overlay them. In this province thrust faulted Eocene sediments are partially pushed over salt deposits which act as a plastic mass and have been very good gliding surfaces.[6]

Thrust faulting, particularly of the overfolds, has certain influences on mining technology and strata stability considerations.[7]

3.2.2 *Thrusting along bedding*

Thrusting along bedding is usually exhibited when the thrust faults intersect the salt strata and then tend to die out along their sedimentation planes (Figure 3.2.2).[8] The thrusting along bedding planes with ductile salt strata is not easily detectable because of its intensive healing processes. The magnitude of displacement along bedding planes will depend on the degree of plasticity of each individual salt member within evaporite strata. The differential shearing along bedding planes occurs when the evaporite profile is composed by the layers of various mechanical properties and behaviour.[9]

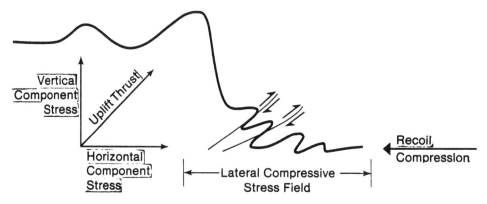

Figure 3.2.1. Model of thrust faulting of folded sedimentary strata

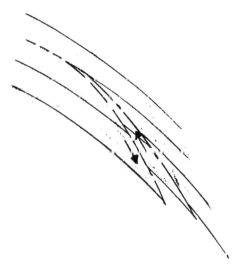

Figure 3.2.2. Thrusting along stratification planes

Regardless of the intensity of thrusting either at low angle to stratification planes or along them the majority of salt mass exhibit very compact state. The tectonic movement and its related pressure cause increased compactivity of the salt mass by recrystallisation and densification of salt grains along flow planes. Also the strength and homogeneity increase of all mineral facies are present in the deposit.

3.2.3 *Integrated reverse faulting*

This phenomenon has been studied and explained by Poborski and Slizowski for the Klodawa salt deposit in Poland. They give principal consideration to the uparched and dislocated top of the lowest structural stage, as illustrated in Figure 2.4.5. Some modifications of the described structural pattern should be accepted depending on the tectonic scheme in the deep substratum as developed by epeirogenic movements and subsequent kind of faulting. It is deduced in the event of very elongated salt domes running parallel to the major structural trends, that dislocation zones exist in the deep substratum. A good example of this feature is the Klodawa dome, and its tectogenesis having been described in the paper presented at the Fourth Symposium on Salt. There authors note reverse faulting and thrusting of the lowest salts (Z1) upon the older zones (Z2) as the case in the early halokinetic period.[7]

The integrated reverse faulting and thrusting can produce a very complex tectonic structure due to ductile nature of the salt and its intense flow abilities. Such a structure could represent some difficulties during underground mining operations.

3.3 TECTONIC BRECCIA

The tectonic breccia is very well exhibited in some sheared and faulted evaporite-bearing strata, particularly of rock salt deposits. The rock salt breccia occurs in many salt deposits over the world. The character of the discontinuities of breccia controls the property and behaviour of the salt deposits, which are not in favour for underground mining, particularly with the aspect of mine structure stability. The salt breccia also does not facilitate the solution mining, which requires a continuity of salt beds. The description of the tectonic breccia is given by three separate topics as discussed below.

3.3.1 *General considerations*

Tectonic or crush breccia demonstrates the mechanics of strong stress in the fault area. In general, two complementary shear displacements are represented along which the salt beds were sliced in angular fragments, and between which the shearing produced a pulverized material.[5] The claystone-sandstone and gypsum/anhydrite (the clastic rocks during shearing) are pulverized rather than fragmented, and as a result they are predominant material of the crush breccia cement (Figure 3.3.1). The salt fragments exhibit stretching in the direction of the shear flow before crushing. The appreciable thickness of tectonic breccia indicates that the normal stress across the fault was low which permitted a considerable volume expansion during crush faulting. The shear stress, however, was quite strong due to pushing up uplifting forces associated with flexural anticlines and dome. The tectonic breccia, of course, does not show evidence of salt stratification which has been completely destroyed by powerful movement, breakage and fragmentation.

Figure 3.3.1. Tectonic contact of salt breccia with marl beds (Wieliczka mine, Poland)

3.3.2 Breccia of flexural anticline

The flexural anticline breccia is formed at one side of the deposit, actually on the side where further folding had been stopped due to resistance of the retaining rock strata. The dome might exhibit the tectonic breccia on both sides of the salt body.

The analysis of tectonic breccia of the complex anticline at the rock salt mine Tusanj indicates that it could be the product of orogenic movement of the Rodan's phase. The lateral tectonic stresses folded and refolded the rock salt beds to the point where they confronted the Jala anticline, which acted as a retaining wall, along which the salt series has been pushed up (Figure 3.3.2). The erected tectonic breccia exhibited a maximum fragmentation at the contact of rock salt and anhydrite.[4]

3.3.3 Breccia of dome

The structural mechanics of the crushed breccia of the rock salt dome at the Pride rock salt mine (Romania) also has been studied (Figure 3.3.3). The development of the wide breccia zones up to several hundreds of meters in thickness and its location around the periphery of the salt dome indicate the development of a very strong shear stress in the direction of displacement and a very weak normal stress during salt dome uplift. The tectonic breccia of

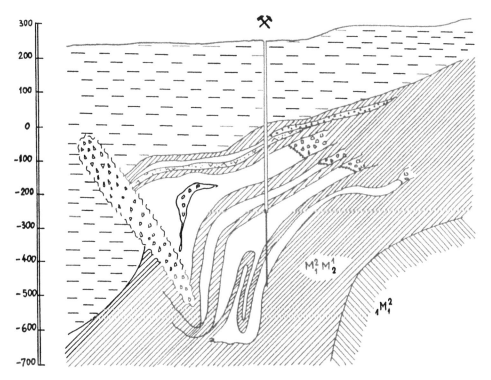

Figure 3.3.2. Crushed zone with breccia at the front of flexural folding of anticline (Tusanj rock salt mine, former Yugoslavia)

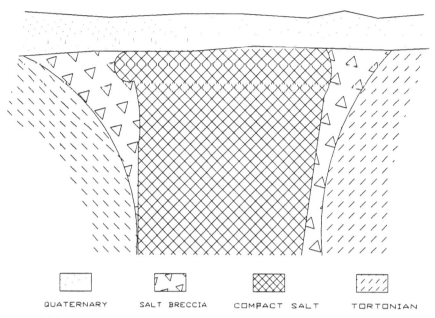

QUATERNARY   SALT BRECCIA   COMPACT SALT   TORTONIAN

Figure 3.3.3. Salt dome with tectonic breccia of rock salt (Pride rock salt mine, Romania

the rock salt domes of the Carpathian evaporite province consists of 30-60% rock salt; typically it occurs in smaller or bigger transparent cubical crystals cemented by clay/silt material.[6]

The tectonic breccia occurs always on the peripheral part of salt deposit because internally, salt bodies have undergone intensive plastic deformation, but when they come in lateral contact with clastic sediments, they are crushed together with them.

The mining of the salt bodies is strictly controlled in regard to tectonic faulting and brecciation. The complete control is achieved by mining the panels and blocks away out of the domain of tectonic clastic structures which could also be associated with the ground-water.

3.4 STRUCTURAL DEFECTS

The structural defects are the product of the orogenic deformations of the evaporite strata, particularly the clastic one. The structural defects are much more apparent in the clastic beds than ductile strata. Three basic types of structural defects are further commented.

3.4.1 *Fissures*

Fissures are developed in the salt of the extreme flow deformations. This tectonic feature is differently exhibited in the salt beds of high plasticity and rock beds of much lower plasticity. As a matter of fact, the adverse effects of the fissures in salt deposits are minimized due to their closing according to reheeling processes. On the contrary they could cause strength reinforcement of the salt due to the strain hardening effect as discussed in

Figure 3.4.1. Fissuring of the salt beds (Bochnia salt mine, Poland)

Chapter 9 (Section 9.3). The degree of fissuring in evaporite strata reflects its history of loading and unloading deformations. The localization of fissuring as well as jointing is accompanied by strata deformations (Figure 3.4.1).

In regard to the underground mining and strata stability the salt with reheeled and bonded fissures could be considered as compact evaporite rock mass. However, if the fissures within dislocation surfaces are not reheeled, then strength of salt deteriorates due to the porosity effect. Further consideration to this phenomenon is given in Chapter 4.

3.4.2 *Shear joints*

Shear joints usually represent the set of small dislocations more or less parallel to the existing structural axis of the salt deposits and they are the product of the compressive stress action. The joints are mostly developed in flat or gently dipping strata of harder evaporite beds as marl, gypsum and others. They are observed also in ductile evaporite beds as for example in trona deposits in Wyoming. Trona beds consist of minor joints parallel to major regional features. In trona mines significant structural defects have not been encountered. The adverse effects of the joints are not anticipated in regard to ground stability and strata control. The shear joints in the folded evaporite strata usually show some small displacements. Generally speaking, the joints better developed in elastic members of the strata than in highly plastic salt beds. However, in highly folded strata the jointing is equally expressed across all beds (Figure 3.4.2).

3.4.3 *Tensile joints*

Tensile joints occur in tectonically deformed salt-bearing strata, particularly in the regions with excessive extension deformation. The tensile joints are open fractures, on the contrary to shear joints with bonded surfaces by cohesion and interlocking shear stress. The tensile joints in the evaporite beds are usually filled or reheeled, and could be observed by their inprints, within highly squeezed plastic salts. It could be a fair statement that there is absence of open tensile joints in the virgin salt beds due to its high ductility and preference to flow deformations. However, in the salt beds which are in the active mining area the joints could be rejuvenated and are easily detectable by their opening (Figure 3.4.3).

The absence or presence of joints in salt mines in the majority of cases is not important for structural control of mine openings because they are not a constraining element of strata stability and ground support.

Figure 3.4.2. Compressive joints in highly folded salt strata (Klodawa mine, Poland)

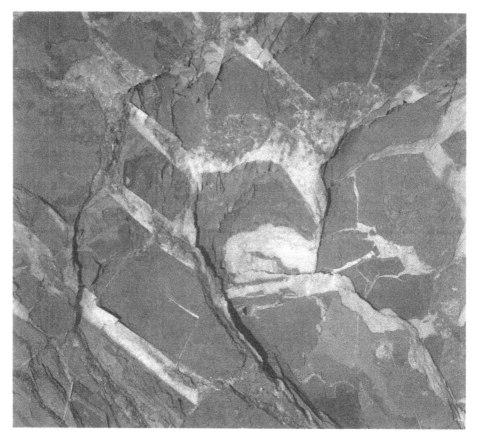

Figure 3.4.3. Tensile joints of highly fractured strata (Tusanj salt mine, former Yugoslavia)

REFERENCES

1. Williamson, I.A. 1967. Structural geology: faults and folds. In *Coal mining geology*, pp. 67-88. New York/Toronto: Oxford Univ. Press.
2. Dekker, L. 1985. Potash exploration at Malagawatch, Cape Breton, Nova Scotia. *CIM Bulletin* 78 (880): 27-32.
3. Kelsall, P.C. & J.W. Nelson 1983. Geologic and engineering characteristics of Gulf region salt domes applied to underground storage and mining. In *Proc. of 6th Int. Symp. on Salt*, The Salt Institute, Virginia, Vol. 1, pp. 519-542.
4. Stojkovic, J. 1984. New concept of tectonics of Tusanj rock salt mine, Tuzla. *Technol. Acta* (Tuzla), 12: 113-116 (in Yugoslav).
5. Press, F. & R. Siever 1982. *Earth*. San Francisco: Freeman, 313 pp.
6. Jeremic, M.L. 1965. Geological structures of the salt deosits of Romania. *Min. & Metall. Bulletin* (Belgrade), 3: 57-65 (in Yugoslav).
7. Poborski, J. & K. Slizowski 1983. Geological prediction of the mining feasibility within the salt domes of Upper Permian Zechstein, Central Poland. In *Proc. of 6th Int. Symp. on Salt*, The Salt Institute, Virginia, Vol. 1, pp. 303-310.
8. Jeremic, M.L. 1985. *Strata mechanics in coal mining*, pp. 210-214. Rotterdam/Boston: Balkema.
9. Hills, E.S. 1966. *Elements of structural geology*, pp. 282-286. London: Associate Book Publ.

CHAPTER 4

Composition of salt

The composition of salt is of great importance with regard to the academic aspects as well as practical engineering aspects such as choice of mining methods and ground control techniques. The composition of salt is represented by four topics: salt texture, homogeneous structure, heterogeneous structure and anisotropy of salt deposits, as further discussed.

4.1 TEXTURE OF SALT

Geologists' classifications of texture of salt deposits are of greater complexity than is required in engineering considerations. A description of an evaporite's texture and fabric facilitates understanding of those properties, most closely related to particle bonding, interlocking, imperfections and others. The texture of the evaporite deposits is represented by several principal types described below.

4.1.1 *Massive texture*

If the salt mass is homogeneous without stratification and flow-banding, it is considered to be of massive texture, Particularly, massive texture is exhibited by rock salt masses in domal deposits (Figure 4.1.1). In this case the salt is composed by fine grained halite with a variety of impurities. Massive salt deposits often contain anhydrite crystals both interlocked along grain boundaries or locked in halite crystals. Core samples from massive domal salt bodies indicate that strengths are moderate contours of poles to planes of diffuse failure, incomplete girdles with maxima which reach the plane normal to the core axis (Figure 4.1.2).[1]

The density of massive textured salt depends on the grain sizes.[2] A relationship between uniaxial compressive strength of salts and the number of grains per volume unit has been established. For example, with an increase of number of grains per volume unit, the strength of rock salt samples increased accordingly (Figure 4.1.3). Also the strength of salt with massive texture depends to some extent on the interlocking of halite grains as well as on the homogeneity of salts crystals.[3]

Salt deposits of massive texture are more favourable for safe excavation and strata control than other textural types.

Figure 4.1.1. Rock salt of
massive texture (Poland)

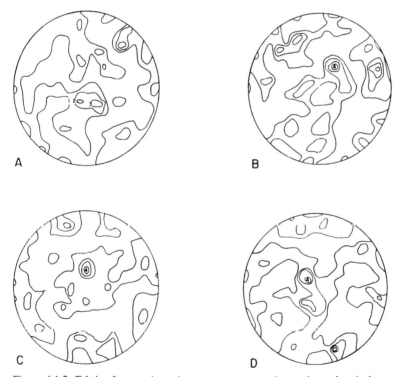

Figure 4.1.2. Fabrics for massive salts; contours on equal-area, lower-hemisphere projections are in
intervals of 1% per 1% area; the core axis is vertical in projections (after Hansen and Carter)
A. Jefferson Island domal salt; B. Weeks Island domal salt; C. Côte Blanche domal salt; D. Avery Island
domal salt

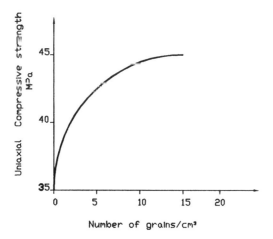

Figure 4.1.3. Relationship between grain density and compressive strength of rock salts

4.1.2 *Porous texture*

There are four types of porosity recognized in evaporite deposits: pore, fissure, fracture and solution cavities. The evaporite porosity depend on the microscopic cracks within minerals or microscopic cavities among minerals as well as macroscopic fractures and cavities.

1. *Pore porosity* is the main controlling factor of porous fabric of evaporite rocks. It depends on the grain sizes of the salts and their compaction for a given volume. The porosity of evaporites varies from low (> 10%) to high (> 20%) and is evaluated on the basis of grain density. The relationship between porosity and strength of evaporites is expressed by an asymptotic function, where its compressive strength decreases with an increase of porosity (Figure 4.1.4). The dome salt is considered with very low porosity (0.1-1.0%); the measurement of permeability of dome salt in situ showed that it is in the range between 0.0001 md to 0.01 md.

2. *Fissure porosity* differs from pore porosity and it is not threedimensional; it is characterized by short planar cracks of microscopic size (> 1 micron to sometimes > 10^3 microns). Fissure porosity of the salts facies may have little influence on porosity fabrics, because of bonding along the microscopic cracks which is similar to bonding in monolithic rocks. If moisture migrates along cracks, then presence of moisture deteriorates salts strength and stability of free faces. When the rock is deformed thin films of water coat the surface of newly formed fissures, causing the fissure faces to separate and facilitate further crack growth. Under these circumstances the permeability of rock is increased, and further increase is effected with more moisture content.

The degree of fissuring of anhydrite, gypsum, marl and others also influences rock behaviour, because of its porosity. For example, at the initial stage of loading cracks are compressed, which is exhibited by an asymptotic function (Figure 4.1.5). After compaction of the cracks, the further rock behaviour is linear. The concavity of the asymptotic function might be interpreted as a fissure spectrum.[4]

3. *Fracture porosity* is created by structural defects exhibited by faulting, fracturing and jointing of evaporite-bearing strata. A porous rock fracture will act as a drainage structure for water-bearing strata and at the same time as a water feeder for strata below, as briefly discussed in Chapter 8.

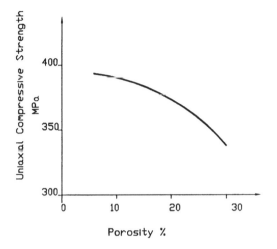

Figure 4.1.4. Asymptotic relation between porosity and strength of rock salt

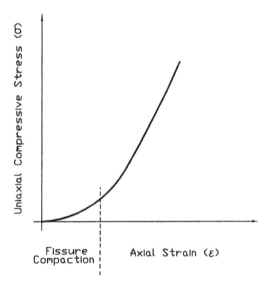

Figure 4.1.5. Model of relationship between σ-ε for fissured rock where an asymptotic function shows a 'fissure spectrum'

4. *Solution cavities* also are present in porous fabric of the evaporite deposits. Due to high solubility of the saline minerals, the volume of the pores may be increased and even transformed into smaller or greater cavities (Figure 4.1.6).[5, 6]

Porous fabric of the beds of evaporite-bearing strata, especially the salts facies, results in rock weakening because both compressive and tensile strengths of rocks are decreased. However, further strength deterioration is caused by the presence of water due to a negative water pressure and further increase of porosity.

Porous textured salt deposits impose a lot of limitation to the mining technology and strata mechanics, because of the possibility of gas outflow or outburst which may prohibit underground mining.[7] Under these circumstances most likely uncontrolled solution mining is the only possible means of extraction.

Figure 4.1.6. Cavitation of rock salts due to water solution and formation of high porous strata (Romania)

4.1.3 *Flow texture*

Flow texture can be observed in salt deposits which have undergone intensive tectonic deformations, such as folded bodies. The original random orientation of salt grains formed during evaporation processes, is changed during folding when the mineral grains are elongated in the direction of the differential shear flow (Figure 4.1.7).[3] Also, the fissures are oriented in the same direction as the mineral grains, e.g. in the direction of fluidal flow. It should be noted that during flow deformation there is increased density and decreased porosity of the salt mass. Also, the strength of rock mass had been increased as discussed in other paragraphs of some chapters of this text. It is interesting to note that during salt grain reorientation there is a grain recrystallization, which usually results in formation of finer grain texture. The fluidity of individual salt facies depends on their viscosity coefficient. In this regard, salt facies with a variety of coefficients of viscosity exhibit differential fluidal fabrics, which result in the formation of a highly anisotropic mass (Figure 4.1.8).

Flow fabric, with the grains flattening in the direction of flow, could cause weak foliations controlled by the geometry of flow. However, in a majority of cases this phenomenon is followed by annealing crystallization and related compaction.

Flow texture facilitates underground mining and strata control. For this texture as well as deposits of massive texture, the mining could be carried out by excavation of large rooms, where ground sustains stresses without any artificial support or reinforcement.

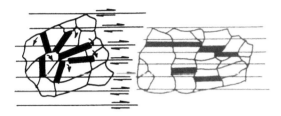

Figure 4.1.7. Original grain orientation is reoriented in the direction of
shear flow deformation

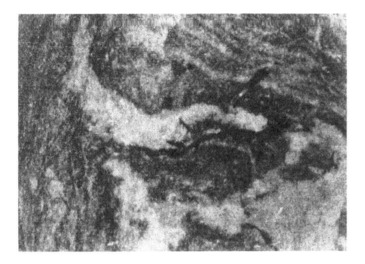

Figure 4.1.8. Fluidal
texture of complex salts
mass, exhibition differ-
ential flows fabrics
(Romania)

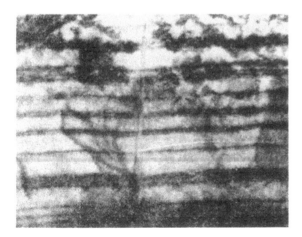

Figure 4.1.9. Rock salt of banded
texture (former Yugoslavia)

4.1.4 *Banded texture*

Banded texture is formed by individual laminae or bands that may be composed of the same
salt minerals, but differing in colour and proportion of fabrics. The banded fabrics are also
composed of layers of different minerals (Figure 4.1.9). The banded textures are very
common fabrics of the evaporite deposits which still maintain some degree of their original
sedimentary features.

The banded texture usually exhibits a greater content of moisture than massive and flow textures, due to higher porosity. Moreover, the impurity content, particularly laminae of clayey material, is often present. Rock salt is usually interbedded with anhydrite or gypsum which also occurs in crystal form in association with halite minerals. The thickness of the interbeds could vary in a wide range, from several millimeters to several meters.

4.2 HOMOGENEOUS STRUCTURE

A homogeneous internal structure of salt deposits is not typical and is very rarely exhibited in the mass of whole deposits, but only in certain portions. In this regard two types of homogeneity of salt deposits could be distinguished, as discussed below.

4.2.1 *Pure homogeneity*

This type of homogeneity structure is considered only for a certain part of deposits, which is controlled by salt bed (banded texture) or by salt lense (massive texture). This portion of salt deposits, which could be locally observed in underground mines, are composed of pure halite, where anhydrite is absent and only trace quantities of other insoluble material exist. Generally a major factor affecting formation of pure homogeneity salt structure is the presence of shear zones in their vicinity.

The Gulf region domes (USA), the pure salt beds of a thickness of over 20 m have been locally observed. The massive salt has no visible banding of anhydrite of any impurities, because their content is below 0.5%. Commonly massive pure halite is more coarsely crystalline with giant crystals, up to 1 m in size, found in some cases. In other areas, a large portion of coarsely crystalline salt may display poikiloblastic texture with all the crystals in a uniform crystallographic orientation. It is generally accepted that the coarsely crystalline salt results from dissolution and recrystallization during or subsequent to dome emplacement.[7] Also, in many salt regions over the world the lenses of pure coarsely crystalline salt have been found (Figure 4.2.1).

It would be a fair statement that pure homogeneous salt islands in the heterogeneous salt deposits are the product of secondary processes, either mechanical (tectonic) or chemical (solution and recrystallization).

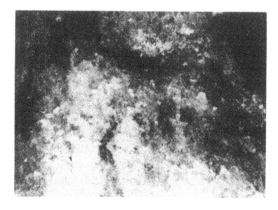

Figure 4.2.1. Transitional contact between halite crystalline pure salt and dark salt highly contaminated with impurities

4.2.2 *Relative homogeneity*

Relatively homogeneous evaporite deposits have a certain number of physical characteristics such as great thickness of the salt bodies, absence of frequent stratification planes and
mono-mineral composition.[6] These deposits have mainly lensoid and domal morphology.
Relatively homogeneous evaporite deposits could be considered to have more or less
uniform mechanical properties and behaviour.

Of particular interest are homogeneous rock salt bodies. These deposits are composed of
halite, where the content of NaCl is over 98% and the facies development or rock salt was
very simple and mono-mineralic. Homogeneity of such deposits may be achieved by
orogenic deformation, as for example formation of lenses of rock salt. In the stratigraphic
column of the Wapno salt deposit (Poland) there is a lower, older white rock salt of certain
purity (Figure 4.2.2). This salt body has a regular morphology and very nearly mono-
mineral composition. Z. Werner calls this deposit a 'magazine of salt' because the rock salt
mined out is directly delivered to the market. The salt does not need any beneficiation,
because of its bright white colour and high quality. The deposit has been named 'Wapno',
which means in English 'lime' , due to its white colour.[9]

Generally speaking, the increased homogeneity of a rock salt mass is due to greater
intergranular friction which results in greater density and strength.[2]

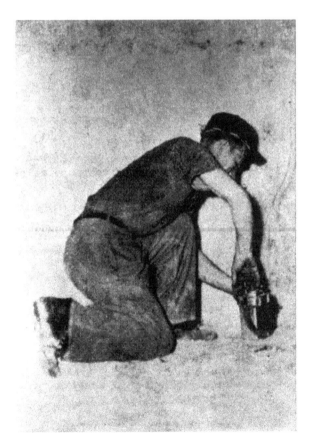

Figure 4.2.2. White rock salt composed by halite mass (Wapno mine,
Poland)

4.3 HETEROGENEOUS STRUCTURE

In most evaporite rock masses heterogeneity is due to differences in mineral composition and its deposition. The greatest degree of heterogeneity is noted in evaporite strata consisting of orogenically deformed various salts layers.

The geological mapping of underground mine workings and geological logging of core samples are insufficient to make an identification of the internal structure of evaporite masses. Numerical indices characterizing the degree of internal structural heterogeneities of the salt deposits have not been developed as they have for coal mining.

Four basic heterogeneous internal structures of the salt deposits are proposed which relate to various geological events.

4.3.1 *Domal heterogeneity*

Evidence regarding the internal lithology and structure of salt domes suggests that they could be very heterogeneous. The identification of dome heterogeneity is difficult, and in the majority of the cases the solution mining operation has been used to evaluate it. On the basis of chemical analysis of brines and calculation, the volumes of insolubles accumulated in caverns could be inferred degree of heterogeneity. In addition, by sonar surveying it is possible to define the volume and shape of the caverns.[7]

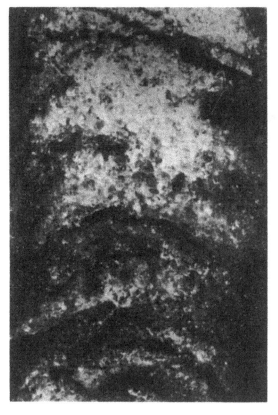

Figure 4.3.1. Domal salt heterogeneity

The main characteristics of salt heterogeneity is the mixture of halite and anhydrite. Pure salt exposed in the mines is white and contaminated salt with anhydrite is dark. Darker salts owe their colour to dissemination of anhydrite and internal reflection.[7] A proportion of 5% anhydrite to salt is sufficient to impact a grey colour, and salt with 15% to 25% of anhydrite may appear almost black. The mixture of the halite and anhydrite could be in various proportions at short distance, what will result in extreme heterogeneity (Figure 4.3.1). The content of anhydrite in salt dome could vary from 1% to 80%. The portion of salt domes, when content of impurities is over 10% is not considered for underground mining, but for solution mining.[8]

Also domal heterogeneity could be exhibited by mixture of the salt with other evaporites such as gypsum, marl, limestone and others. In many instances this heterogeneity is due to tectonic events rather than to sedimentary processes.[9]

The European Zechstein's domes are exhibiting heterogeneity in regard to a mixture of various salt minerals, particularly halite and potassium salts. This phenomenon will be discussed in Subsection 4.3.3.

4.3.2 *Sedimentary heterogeneity*

Sedimentary heterogeneity of evaporite deposits is due to the deposition of different minerals and the presence of bedding planes. The main factor in the formation of sedimentary heterogeneity is a cyclic sedimentation as a result of the interchange of marine regression and marine transgression and fractional evaporation of salt minerals from the shallow sea or closed basins.

The definition of sedimentary heterogeneity is related to the original sedimentary structure and the thickness of sedimentary beds which could be from several millimeters to

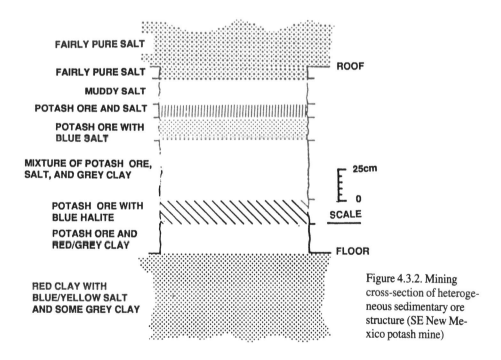

Figure 4.3.2. Mining cross-section of heterogeneous sedimentary ore structure (SE New Mexico potash mine)

several meters. Thick sedimentary beds are classified as relatively homogeneous masses. The cause of sedimentary heterogeneity is the frequency of pulsation, which could vary considerably for various evaporite beds.[10]

The sedimentary facial heterogeneity exhibited by interchange of evaporite layers results in different mechanical properties of the salt, particularly in the case of potassium mineral deposition, as illustrated in Figure 4.3.2. The effect of sedimentary heterogeneity can become quite complex when a spectrum of different salt facies is present.[11] For example, a halite salt layer deforms plastically if it is sandwiched between more rigid strata such as marl or anhydrite, but acts as a rigid bed if it is sandwiched in masses of more plastic potassium salts.[11]

4.3.3 *Mineral heterogenity*

With an aspect of mineral heterogeneity it is primarily important a spatial relationship between halite salts and potassium salts. In Central Europe, two types of mineral heterogeneities may be classified. Firstly, within the complex of flexural salt bodies of the Carpathian province and secondly, in the salt domes of the Zechstein's series.

The main characteristic of mineral heterogeneity of salt deposits of the Carpathian province is that potassium salts occur in association with magnesium salts but also in conjunction with sodium salt deposits. The multi-mineral development is of a complex nature and is usually located at the boundary of a sedimentary basin. The salt deposits are often intensely folded and partially brecciated. The development of potassium salts in a majority of cases occurs like a frame around the rock salt body, as illustrated in Figure 4.3.3. The understanding of these deposits has been established by geological mapping of underground mine workings on each mine level. The mapped data delineated the lateral extension and dip of individual salt facies and the relationship amongst them. Very complex heterogeneity of salt deposits interrupts continuous mining of the potash ore.

In contrast to the salt deposits described above the salt domes of the European Zechstein's basin contain a potassium mineralization in the inner part of rock salt bodies (Figure 4.3.4). For example, the Klodava salt dome is composed of several facies as listed

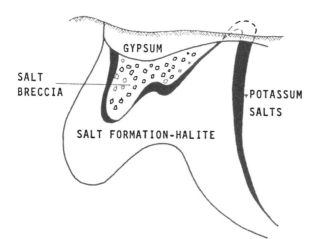

Figure 4.3.3. Schematic profile of rock salt and potassium salts flexural body at Tazlau, Romania

below:

- older rock salt;
- older potassium salts (hard anhydrite-kieserite mass);
- younger potassium salts (kieserite-carnallite mass);
- Younger salts.

It should be noted that within the same facies there is some differentiation due to the age of their sedimentation. This phenomenon further complicates mineral heterogeneity of the salt deposit, because a very complex structure exists at the scale of mining workings. A rock salt mass with a potash lense containing rock salt intercalations is illustrated in Figure 4.3.5.

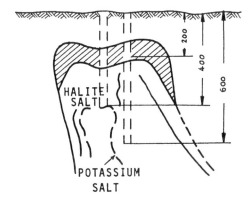

Figure 4.3.4. Schematic profile of rocksalt and potassium salt dome at Klodawa, Poland

Figure 4.3.5. Lense of potassium salts within halite mass

The orogenic mineral heterogeneity of similar composition differs from sedimentary heterogeneity, because it is not controlled by sedimentation planes and the structure of individual layers.[12] The high degree of irregularities in the distribution of individual facies sometimes prevents determination of the controlling structural factors and related difficulties in characterization of the heterogeneity. The orogenic mineral heterogeneity has a great influence on the mechanical behaviour of salt and it can change mining conditions and strata stability over short distances.

4.3.4 *Structural heterogeneity*

Structural heterogeneities of evaporite strata are the product of tectonic movements of the earth's crust in the region of their deposition.

The folding and faulting of the evaporite strata, relate to first order heterogeneities by dividing rock mass into large extraction blocks. This type of structural heterogeneity is of lesser influence on the mineability and stability of mine workings.

The inconsistency of original sedimentary strata, however, relates to the structural discontinuities. In this case the character of structural heterogeneity is controlled by the pattern of joints and other discontinuities.

The structural heterogeneities in evaporite strata could vary significantly due to differences in clasticity of individual layers. For example, gypsum, anhydrite, marl and others might have developed an apparent internal blocky structure (Figure 4.3.6). Halite salt and potassium salts, however, might consist beside an intensive folded structure also an

Figure 4.3.6. Structural heterogeneity of gypsum underground opening (Southern Ontario, Canada)

apparent internal fissuring. The dimensions of structural heterogeneities in those two cases could vary significantly. Under these circumstances there are two orders of heterogeneities, first in the scale from tens of centimeters to tens of meters, second in the scale from tens of microns to tens of millimeters.[13]

The structural heterogeneities of various orders have an influence on the stability of mine workings, mainly being responsible for rock falls and for strength deterioriation of given beds.

4.4 ANISOTROPY OF SALT DEPOSITS

The mechanical anisotropy of the evaporite strata will be considered with respect to its orientation relative to the planes of stratification. These anisotroic effects are the consequence of variations of individual layers in cohesive strength and resistance to deformation. The anisotropy of the evaporite deposits controls the mining layout and mechanics of strata control.

4.4.1 *Anisotropy perpendicular to strata*

An unaltered sedimentary structure is represented by a column of various beds, which could be grouped in function of their chemical composition or facial development as discussed in this book. The layers with the same anisotropic characteristics could occur in the geological column in several sequences, which is the case for example with rock salt layers. Under these circumstances the anisotropic condition will recur. The repetition of mineable salt beds in the geological column could be of interest for multiple bed mining similar to multiple coal seam mining. However, due to abundance of the evaporite beds and their impurities, the multi-bed mining method is seldom implemented and by this act it avoids complicated mining technology and strata control methods.

Mechanical anisotropy of sedimentary structures could be analyzed by three phenomena:

1. The strength of an individual bed could vary considerably, particularly in the case of different mineral composition. The variation of strength perpendicular to the sedimentary layers could effect the stability of the mine opening, particularly when very weak beds are present. Under these circumstances protective layers have to be left at the floor of the mine opening, as illustrated in Figure 4.4.1. Particular difficulties could be encountered if the sedimentary structure is composed of a large number of thin beds of very low strength.[10]

2. The stress distribution should be identified on the basis of underground stress measurements. Also, it could be evaluated during diamond drilling on the basis of intensity of core discing. For example, Hebblewhite et al. used the relative core discing frequency to determine the stress distribution and concentration with complex evaporite strata above a mine opening.[12] The use of discing for stress evaluation was possible due to appreciable friability of the strata and anticipating that a high stress field exists. As illustrated in Figure 4.4.2, the stress distribution perpendicular to sedimentary structure varies significantly, not only due to induced stress but also due to high anisotropic properties of evaporite strata.[9]

3. The time-dependent deformation of the sedimentary layers is greatly influenced by creep properties of individual layers. This usually results in a differential flow of individual layers into mine openings. In the case of mono-mineralic beds the anisotropic behaviour is a minimum, as will be discussed later.

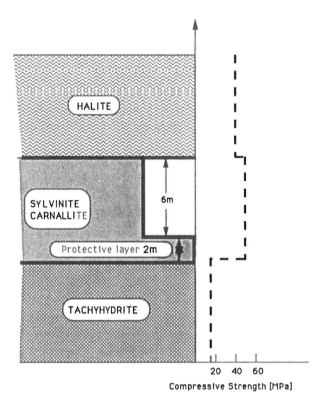

Figure 4.4.1. The transverse strength of salt beds of different anisotropic properties (Sergipe potash deposit, Brazil)

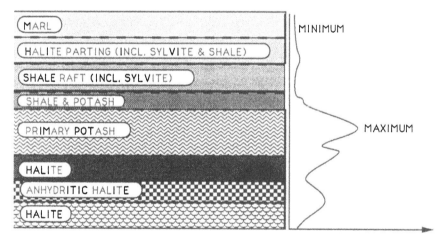

Figure 4.4.2. Stress distribution above mine opening, perpendicular to sedimentary structure (Boulby potash mine, North Yorkshire, Great Britain)

It is necessary to point out that for flat sedimentary strata the compressive strength perpendicular to bedding planes is an essential design parameter.

4.4.2 *Anisotropy parallel to strata*

Anisotropy parallel to sedimentary layers for flat evaporite deposits is of little influence, because of slow change in the mineral composition of salt layers. However, in the case of the vertical sedimentary structures the anisotropy along sedimentary units could be of particular importance if the evaporite beds have a finely banded texture.

In the case of loading of vertical salt beds two mechanical phenomena are shown. Dilatation of the beds due to separation forces and fracturing of individual beds caused by buckling forces (Figure 4.4.3).[14]

Anisotropy of vertical sedimentary strata which were brought into this position by orogenic movement, has a different influence on stability of the mine opening. In this case tensile strength between sedimentation surfaces and bending strength of an individual bed is the controlling factor rather than compressive strength of a sedimentary layer itself.

4.4.3 *Anisotropy of inclined strata*

Sedimentary layers which are folded during orogenic movement could have a wide range of angles of inclination from the extremes of flat to vertical.

Many investigators studied the relationship between the angle of inclination of sedimentary layers and the uniaxial compressive strength for various rock types. The common conclusion is that there is a functional relationship between compressive strength and angle of inclination of sedimentaty beds to the principal stress direction.[13]

The investigations indicated that the strength of rock salt decreases with an increase of angle of inclination of layers from horizontal until bedding reaches approximately 45°. After this angle, the strength starts to increase reaching a maximum at an angle of inclination of 90° to horizontal. It should be noted that the strength of rock salt with a load perpendicular to bedding is almost 50% higher than for the load parallel to bedding planes.

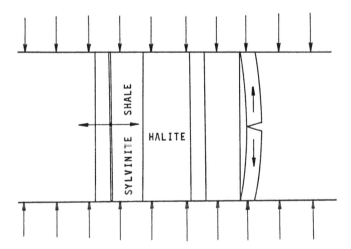

Figure 4.4.3. Loading of mine structure parallel to bedding and related fracture mechanics

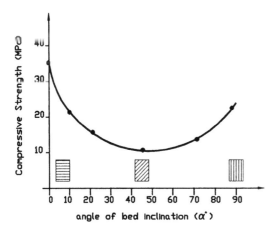

Figure 4.4.4. Compressive strength of strata in function of their angle of inclination

The minimum rock salt strength is at an angle of inclination of beds at 45° where it decreases to almost 55% of its maximal strength (Figure 4.4.4).

4.4.4 *Anisotropy of massive evaporite*

In greatly deformed evaporite masses sedimentary planes are highly contorted. The anisotropy which is related to the apreciable differences in the mechanical properties of the interfaces between layers cannot be observed. In this case consideration is given to mechanical anisotrpy governed by different distributions of minerals rather than stratification structure.[12] Design of mine openings in this situation must consider the effects of differences in bulk mineralogical composition rather than differences among layers. Compressive strength of the former should be used as the parameter for design.

REFERENCES

1. Hansen, F.D. & N.L. Carter 1980. Creep of rock salt at elevated temperature, pp. 217-227. In *The State of the Art in Rock Mechanics*, 21st Symp. on Rock Mech., May 28-30, 1980, Univ. of Missouri-Rolla.
2. Dreyer, W. 1973. *The science of rock mechanics – Part 1*, pp. 150-190. Clausthal: Trans. Techn. Publ.
3. Bujalov, N.I. 1957. *Structural Geology*, pp. 65-72 (in Russian). Moscow: Gostoptecizdat.
4. Goodman, E.R. 1976. *Methods of geological engineering in discontinuous rocks*, pp. 16-36. San Francisco: West. Publ. Co.
5. Stojkovic, J. et al. 1987. Salt deposits of Tuzla basin and prospect of future geological exploration. *J. Min. & Geol.* (Tuzla Univ., Tuzla), 15/16: 9-18 (in Yugoslav).
6. Jeremic, M.L. 1965. Geological structures of the salt deposits in Romania. *Min. & Metall. Bulletin* (Belgrade), 3: 57-65 (in Yugoslav).
7. Kelsall, P.C. & J.W. Nelson 1983. Geologic and engineering characteristics of Gulf region salt domes applied to underground storage and mining. In. *Proc. of 6th Int. Symp. on Salt*, The Salt Institute, Virginia, Vol. 1, pp. 519-542.
8. Jeremic, M.L. 1965. Methods of development and exploitation of salt deposits in Poland. *Min. & Metall. Bulletin* (Belgrade), 32 (2): 264-274 (in Yugoslav).

9. Jensen, M.L. & A.M. Bateman 1981. *Economic mineral deposits* (3rd. ed.) pp. 199-212. New York: Wiley.

10. Fairhurst, C.M. et al. 1979. Rock mechanics studies of proposed underground mining of potash in Sergipe, Brazil. In *Proc. of the 4th Int. Congr. on Rock Mech.*, Montreux, Switzerland, Vol. 1, pp. 131-135.

11. Hebblewhite, B.K. et al. 1979. The design of underground mining layout. In *Proc. of 4th Int. Congr. on Rock Mech.*, Montreux, Switzerland, Vol. 2, pp. 219-226.

12. Jeremic, M.L. 1964. Salt deposits of evaporite bearing strata of Poland. *Min. & Metall. Bulletin* (Belgrade), 3: 60-66, 4: 79-81 (in Yugoslav).

13. Trchaninov, I.A. et al. 1979. *Principals of rock mechanics* (translated from Russian), Rockville, Maryland, USA: Terraspace.

14. Jeremic, M.L. 1983. *Strata mechanics in coal mining*, p. 41. Rotterdam: Balkema.

CHAPTER 5

Form of salt deposits

The shape of salt deposits depends on the specific conditions of its emplacement in geological environment.[1] Generally speaking, four geological environments could be considered, namely: sedimentary with tabular structure, flexural structure controlled by tectonic, domes originated by diapiric processes and vein-like structure controlled by internal tectonics of deformed salt bodies.

5.1 BLANKET FORM

The principal geological unit of all sedimentary deposits is a bed, stratum, seam or layer. Generally speaking, the layers of evaporite strata reflect the conditions of their formation and sedimentation as discussed in the first chapter of this book. The salt deposits in this case have a uniform bedding of the enclosing sedimentary rocks.

Tabular beds of evaporite strata exhibit all morphological characteristics of sedimentary rocks, as for example regularity in attitude, variation in thickness lensing formation as well as narrowing and pinching out (Figure 5.1.1).[2] The nature of sedimentary formation controls the thickness of salt beds. The thickness of salt beds is also affected by palaeo-solution and secondary deposition, as discussed in the first chapter of this book.[3]

5.1.1 Tabular shape

Sedimentary structures in the same salt basin have been preserved exhibiting certain parallelity between roof strata and floor strata of salt beds. This type of salt deposit is most convenient for the dry mining, because the strata control is not complicated.

From a mining geological point of view, beside the character of the roof strata and floor strata also the thickness of salt deposit is an important parameter.[4] In regard to the mining the following classification of salt bed thicknesses is given:
- very thin deposits (< 0.5 m), not mineable;
- thin deposits (0.5-1.0 m), mineable by solution mining;
- moderate thick deposits (1.0-5.0 m), single level u/g mining;
- thick deposits (5.0-15.0 m) bi-level u/g mining, or single large room mining;
- very thick deposits (< 15.0 m), by-level u/g mining, multi-level u/g mining, or single high room mining.

Of course, all mineable thicknesses by underground mining are also mineable by solution mining.

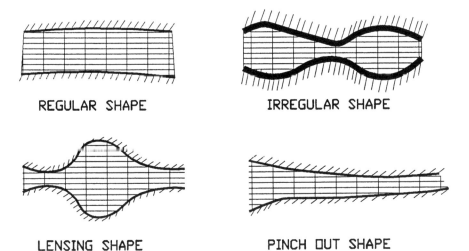

REGULAR SHAPE   IRREGULAR SHAPE

LENSING SHAPE   PINCH OUT SHAPE

Figure 5.1.1. Blanket forms of salt beds

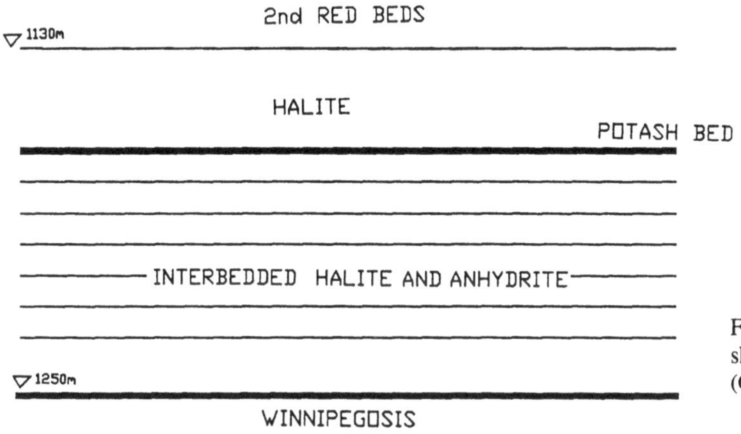

2nd RED BEDS

▽ 1130m

HALITE

POTASH BED

INTERBEDDED HALITE AND ANHYDRITE

▽ 1250m

WINNIPEGOSIS

Figure 5.1.2. Tabular
shape of salt deposit
(Canadian prairies)

The preceding classification of seam thicknesses suggests that mininig methods vary with salt strata thickness. In fact, the application of similar mining methods to salt beds of various thicknesses will produce different overall mine stabilities and related profitability.

The tabular shape of salt might be considered for a potash-enriched halite bed of prairie evaporite (Canada). This bed is located in the upper portion of the salt and anhydrite bedded complex of thickness of 120 m (Figure 5.1.2). The potash-enriched salt bed could be considered as a sheet, approximately 2.5 m thick, with a lateral extension of 10 000 m.

5.1.2 *Lensing shape*

The lensing form of tabular deposits is represented by the lateral thinning and thickening of the salt strata in one or more directions.[5] Under these circumstances the mining may be

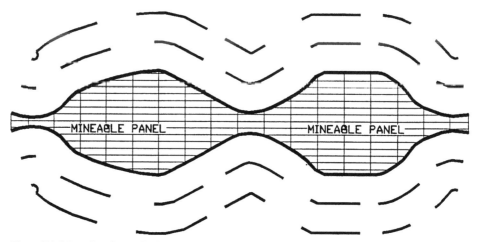

Figure 5.1.3. Lensing shape of salt bed

located only in the zones of adequate thickness, to achieve a satisfactory mineability and strata stability (Figure 5.1.3).

The lateral lensing shape of the salt beds may be related to post-depositional alteration. This form of deposit could be caused by the mechanical alteration due to weight of overburden which induced shear flow along bedding planes of the ductile salt material. Also, there is a possibility of the decantation of palaeo-solution in the basin with an undulated floor and with numerous depressions which had no natural outlet for surface drainage.[6]

The lensing shape of a salt body is not characteristic, such as other forms of salt deposits.

5.1.3 *Elongated shape*

The geometry of the elongated shape of salt bodies is restricted by the basins depositional environment, as discussed in the first chapter. The basin-like nature of the salt deposits is well exhibited by the Rybnik salt deposits in Poland, which were discovered in 1950 during exploration drilling for coking coal in carboniferous sediments (Figure 5.1.4). At this locality evaporite strata are 200-300 m thick with salt-bearing series of 50-80 m thick. Salt is not being mined there, because its exploitation above an active underground coal mine will jeopardize the existing equilibrium of mining operations.[7]

The elongated form unaffected by orogenic movement is a common type of salt deposits in the Carpathian salt province and the Dinaride salt province. These deposits are convenient for dry mining and strata control, but unfortunately due to high frequency of interchange of pure halite beds, and halite beds with anhydrite and other impurities decrease an overall quality of the salt ore. In this case usually the selective solution mining is recommended.

The salt bodies of the elongated shape have a thickness more or less similar to or greater than the blanket shape deposits, but their lateral extension is restricted up to several thousands of meters.

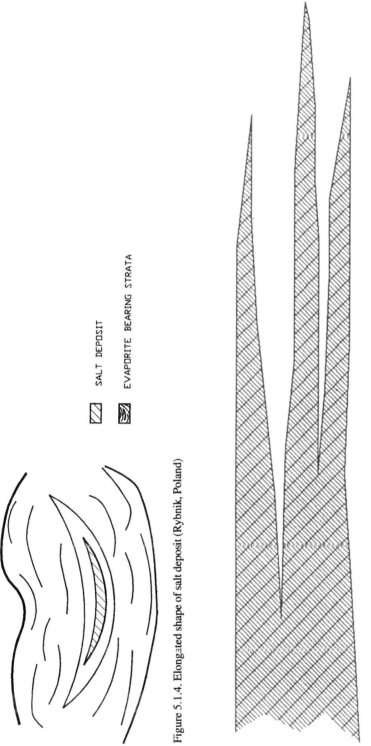

SALT DEPOSIT

EVAPORITE BEARING STRATA

Figure 5.1.4. Elongated shape of salt deposit (Rybnik, Poland)

Figure 5.1.5. Model of fingering shape of salt beds

5.1.4 *Fingering shape*

A fingering shape of salt beds is due to their pinch out, which is controlled by evaporation processes and the sedimentation mechanism of the basin. Because of specific sequences of marine regressive and transgressive phases the sedimentation of salt beds was interchanged with sedimentation of clastic or other evaporite sediments, as discussed in the first chapter of this book. As a result of these processes some salt deposits profiles are represented by pinching out, splitting and fingering of the beds (Figure 5.1.5).

A fingering shape of salt beds is of particular interest when the mining is carried out towards the periphery of salt deposits and when protective strata above mining workings gradually become thinner and thinner, that at certain points cannot support weak strata above (shale) or water-bearing horizons (water accumulation). Under these circumstances great difficulties encountered in the mine, either by roof falls or water inflow or both. The adverse effects of the pinch-out shape of salt beds have been experienced in some potash mines in the Saskatchewan area, particularly in regard to accelerated closure of the roof span and water inflow in the mine.

5.2 FLEXURAL FORM

The flexural form of salt deposits occurs in two types of geometries, namely lens form and complex flexure. The lens form is represented by several simple shapes, which differ from the complex flexural shape.

5.2.1 *Lenticular shape*

The salt deposits which occur in the lenticular shape could be of various sizes, depending upon the mechanics of their formation.[4] Lenticular salt bodies usually have irregular thickness and distribution as well as gently sloping. The salt is massed in a peculiar oblong shape with long and rounded tops with an abrupt pinch-out. In the whole they are representing a smooth and regular geometrical shape (Figure 5.2.1).

Lenticular salt deposits are formed by simple tectonic processes, where shearing along sedimentation planes and salt flow in the direction of the centre of the deposits took place. This resulted in shortening of the original blanket form of the salt body, because of the salt flow from boundary to centre, where the deposit was thickened. Due to this process the original form of the deposit had been changed but also the internal structure, due to the

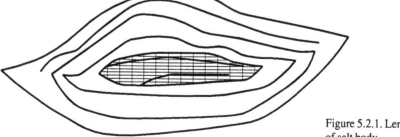

Figure 5.2.1. Lenticular shape of salt body

differential movement of the layers with various mechanical properties. Generally, the salt purification and concentration in the central parts of the lenticular body should be expected. This type of salt body shape has been observed in the Dinaride salt province (former Yugoslavia).

5.2.2 *Convex lens shape*

The convex lens shape originated during the moderate tectonic movement which produced open folds of regular anticlines and synclines. During folding the salt beds flow in the direction of anticline and syncline structure forming concave lenses which are facing each other. By this mechanics the original salt deosit had been partitioned in separate concave lenses, where salt masses were flown from limbs of folds, which consequently do not contain salt, only its traces. The thickness of salt lenses is multiplied several times of its original sedimentary thickness and usually is over 15 m (Figure 5.2.2).

The convex lens of salt deposits has been shown in the first chapter on general geology, for the Maha Sarakham evaporite formation which is characteristic for the Bamnet Narong salt province (North-eastern Thailand).

5.2.3 *Irregular lens shape*

An irregular lens shape of the salt body is the result of the intensive tectonic movements, which pushed the salt mass against the retaining rock strata. This type of lens shape is often accompanied by salt breccia, which was formed by crushing salt against resisting rock mass. The internal structure of the irregular lens is also irregular due to dragging of the salt mass in a lensoid form. Thick lens bodies of irregular geometry could be formed as in the case of the Trnovac salt deposit, Tuzla basin (Figure 5.2.3). The exploitation of these deposits may be by uncontrolled solution mining, where salt recovery is very small.[5] This morphological type of lens salt deposit is most economical for mining exploitation due to its great thickness. The thickness could be several hundreds of meters because it is controlled by deformed structural geometry.

The irregular lens shape associates with complex flexural geometry of salt deposits in the Carpathian and Dinaride salt provinces in Europe.

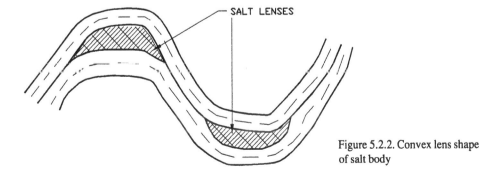

SALT LENSES

Figure 5.2.2. Convex lens shape of salt body

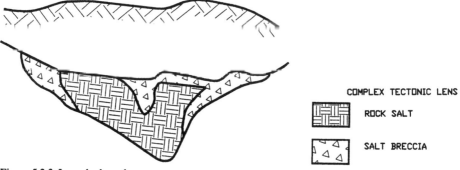

COMPLEX TECTONIC LENS

ROCK SALT

SALT BRECCIA

Figure 5.2.3. Irregular lens shape

Figure 5.2.4. Complex fold shape

5.2.4 *Complex flexural shape*

The salt deposits of complex flexural form are the product of intensive tectonic movement. It exhibits two geometries of folding: external fold shape and internal fold shape, which usualy do not totally coincide with each other (Figure 5.2.4). The morphology of the complex flexural structure of salt deposits is to a certain degree discussed in the previous chapter, and further discussion in this regard is not necessary. It should be pointed out, however, that flexural deposits represent a large body of rock salt, sometimes with a multi-salt facies. Mining in this type of deposit could be by large room of appreciable height or by multi-level mining.[6] The internal structure of the deposit is more important than the structure of its boundary because the mine layout has to follow the distribution of salt minerals.

The Carpathian province in Poland contains many salt deposits with a complex flexural morphology. These deposits are intensely folded and refolded and compacted over a relatively short distance. Flexural salt deposits of the Carpathian province usually have a relatively regular external structure. Internal structures, however, are more complex, particularly in regard to distribution of halite. For example, some folded salt deposits in Poland contain an area with relatively pure salts and an area with dirty and very dirty salts, which are called zuber salts. The layout of underground mining workings is not governed by flexural morphology of deposits but by structural distribution of pure halite.[7]

5.3 DOME FORM

The dome structural morphology of the salt deposits is unusual and intriguing with the mining geological point of view. The domal salt deposits are known in Mexico, USA, Germany, Poland, Spain, Romania, former USSR and Midlle East. It is estimated that mineable rock salt from know domes would supply the world demand for 300 000 years.[8]

To obtain a better understanding of the salt dome form, it is necessary to discuss some aspects in this regard, namely the dome formation, gypsum-anhydrite caps of dome, identification of dome shape and finally geometrization of dome structure itself.

5.3.1 *Dome formation*

The dome formation and dome growth are already discussed in Chapter 2, but rather from a geological point of view. In this subsection consideration is given mostly to the aspect of dome morphology.

The dome represents a gigantic plug, that has risen upwards diapirically because of its low density into the overlying strata. The bending of host rocks around flanks has been discussed in the previous chapter. The tops of domes may protrude through to the surface or terminate several thousands of meters below the surface.

Originally the sedimentary salt deposits, most likely of appreciable thickness, were at depths of several kilometers. Under hydrostatic pressure the very plastic salt of lower density (2.4 g/cm^3) starts to flow upward through sediments of greater density. Salt flow is controlled through the fault intersections resulting in a domal structure.

The Gulf coast salt domes in the USA have been studied by Kupfer in regard to their formation, size and shape. As a result of this investigation he suggested a concept of salt dome growth in this region, as illustrated in Figure 5.3.1. It should be noted that with dome growth its internal structure also grows, but rather with some complexity.[9]

5.3.2 *Gypsum-anhydrite caps*

Salt domes in a majority of cases have a caprock of gypsum or anhydrite, as shown in the salt domes of the Kujava marine region, Poland (Figure 5.3.2). The salt dome at Solno in this region, has gypsum caprock extending down its northern flank. The thickness of the caprock of the Kujawa marine region is up to 100 m, but in some other salt provinces it could be as much as 300 m. Also the depth to the top of salt domes in this region is between

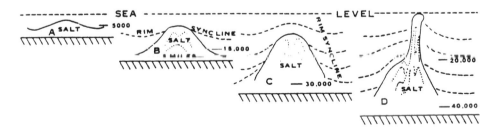

Figure 5.3.1. Concept of dome growth (after Kupfer)

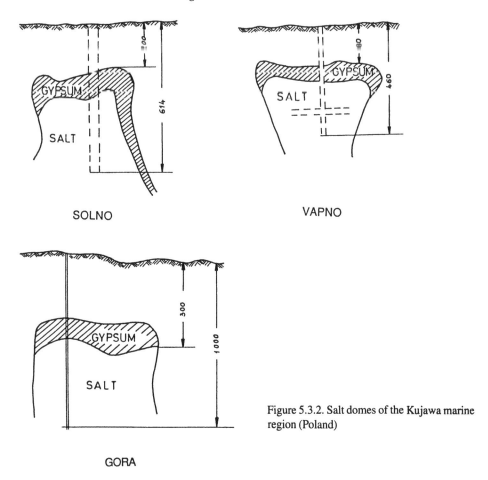

SOLNO

VAPNO

Figure 5.3.2. Salt domes of the Kujawa marine region (Poland)

GORA

100 and 300 m. The rock salt mass within domes usually is metamorphosed, recrystallized, and it may be intercalated with other salt facies.

W.M. Schwerdtner studied the gypsum-anhydrite caps of Arctic salt domes.[10] He commented that like the classical salt domes of Northern Germany and the Gulf coast region, the Arctic salt domes carry a cap of $CaSO_4$-$CaCO_3$. The composite cap of the classical domes is clearly zoned, mainly chemical in origin, and made up of liquified and recrystallized residues of dissolved salt and potash rocks. The stratified cap of the Arctic domes, on the other hand, is mainly structural in origin. Being composed chiefly of metasedimentary anhydrite, the cap owes its existence to profound changes in density and equivalent viscosity of the $CaSO_4$ material. Presumably, the primary gypsum lost its crystal water, upon shallow burial, and became a buoyant crystal mush which gradually turned into dense non-porous anhydrite.

Once formed, the two types of caps assumed similar mechanical roles; they behaved as competent masses on top of the rising salt domes and pierced the clastic sedimentary strata ahead of the rock salt (Figure 5.3.3).

For example, the structural features of a salt dome in Louisiana (USA) clearly illustrate the mechanical behaviour of the gypsum cap. The cap has been bent by the advancing salt

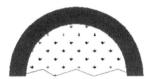

Figure 5.3.3. Schematic development of gypsum-anhydrite caps in the Sverdreys basin (modified after Schwerdtner)

while undergoing discontinuous internal deformations. However, the cap of the Arctic domes was bent at an early stage but partly unbent on while being compressed in a vertical direction. A salt-dissolution cap acts like a semi-brittle battering-ram which passively pierces the overlying sediments while being driven by a growing dome of buoyant rock salt. The final stage is bulging and partial breakup of the gypsum-anhydrite cap above mature salt diapir. The anhydrite breakup structure is represented by extension fractures propagated upwards.

5.3.3 *Identification of the dome shape*

The size and geometry of salt domes might be derived from gravity data. Actually the best identification of dome shape is achieved by a gravity model.

Gravity techniques for modelling the shape of salt domes have been used for the last fifty years. In the last ten years sophisticated computer-aided modelling of salt domes has been developed and successfully applied to detail the shape of salt domes.[11]

R. M. White and C. A. Spiers showed the relationship between Gulf coast salt domes and their associated gravity anomalies, as illustrated in Figure 5.3.4. The density of salt dome stock is constant and it averages 2.2 gm/cc, regardless of depth, because of salt, being non-porous, is not compressible. The surrounding sediment density varied from as low as 1.7 gm/cc at the surface to as high as 2.6 gm/cc at the depth where it is highly compacted.

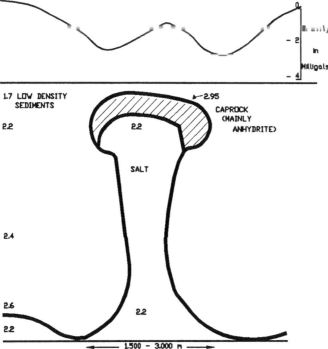

Figure 5.3.4. Model of dome shape identified by gravity anomaly (Gulf coast, USA)

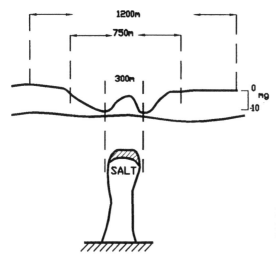

Figure 5.3.5. Model of shallow dome shape identified by gravity anomaly (Gulf coast, USA)

The deep salt is a strong density anomaly with a negative gravity anomaly. The shallow salt and caprock are a positive density anomaly. Caprocks composed mainly of anhydrite are common for salt domes in the Gulf coast region.

The minimum gravity data density suggests for shallow interior basin salt domes that the shape of the gravity anomaly due to the dome (Figure 5.3.5).[11] Further discussion of gravity surveying is given in Chapter 6.

Identification of the dome shape at the Gulf coast by detailed gravity surveying is supplemented by shallow high resolution reflection seismic surveys, surface electrical resistivity, and down-hole geophysical logging to add a third dimension to all of the surface techniques. In addition all geological data available from drilling have been utilized during the process of dome shape delineation.

5.3.4 *Structural characterization of domes*

The general structures of salt domes are well known from extensive exploration working particularly in regard to oil and gas. The major elements of the dome are the salt stock, the caprock and the overlying domed and adjacent upturned and faulted sediments. From the mining point of view, the major features of interest are the salt stock itself and the caprock through which the exploration drill holes are drilled and the shafts are sunk. The sediments adjacent to the dome are not of direct mining interest provided that the edge of the salt can be defined with reasonable accuracy.[12]

The general structures of several domes are illustrated in Figure 5.3.6. The illustrated domes show variation in size, shape and complexity both the salt stock and the caprock. The tops of the domes have a diameter range between 800 m and 4600 m, with an average of approximately 3200 m. The majority of domes increase in diameter with depth, although mushroom-like overhangs at the top of salt and associated reductions in diameter in the

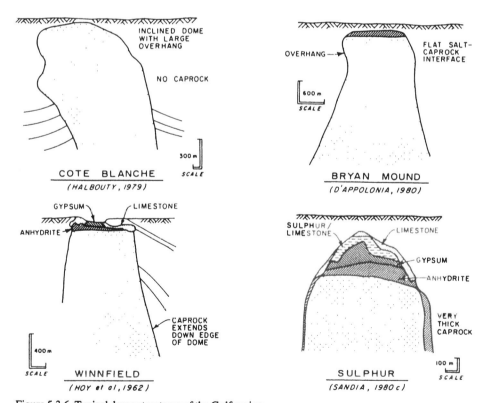

Figure 5.3.6. Typical dome structures of the Gulf region

depth range down to several hundreds of meters, as Kelsall and Nelson stated.[12] The depth of the salt dome top is between a few tenths to many thousands of meters. In the Gulf region there are 268 known onshore domes and half of them have a depth up to 1000 m and they might be the object of mining consideration. Salt domes extend downward to 3000 m or more. The salt domes at a depth greater than 2500 m are not the object of any engineering consideration.

In addition to the discussion of caprock structure given in Subsection 5.3.2, the Gulf domes suggest that caprock might be absent or very thin or developed in patches.[13] However, the thickness of very thick caprock in this region is up to 450 m. The shape of the caprock is uniform or with thickening over the centre of the dome.

The structure of the dome edges is controlled by tectonic contact with displaced sediments, as discussed in Chapter 3.

5.4 VEIN-LIKE FORM

The principles for determination of the vein form of salt bodies are the same as in hard rock mining, regardless of their mineral composition and surrounding. However, certain specific measures in regard to salt extraction should be taken into consideration, which are also in effect for vein mining.

Firstly, the precaution against water inflow into the mine from the surrounding rocks is really the most important factor in the mine design and exploitation of vein-like deposits. Protection against a water menace of this kind is usually undertaken by protection pillars of the uniform rock salt surrounding the exploitation areas. In such a case the salt miner has to work like a diver closed in a caisson on the bottom of the sea. He does not worry about the quantity or hydrostatic pressure of the surrounding water. However, he has to take great care of the water-tightness (imperviousness) and mechanical resistance of the caisson walls.[14]

Secondly, the mining exploitation is limited by the mine depth and the geothermal gradient, due to the ductile nature of the salt. The adopted criteria in function of factors above are based on the knowledge of experience in European salt mining. The depth of traditional dry mining or shaft mining is limited to 1200 m (-1000 m below sea level) and of solution mining about 2200 m (-2000 m below sea level).

These two phenomena described above make salt mining different from mining of other mineral deposits.

The salt veins could be classified in three principal groups, as discussed below.

5.4.1 *Regular vein shape*

Halite veins belong to the regular vein shape and are located within the flexurally folded deposits of the Carpathian salt province. The formation of halite veins is influenced by orogenic movement which caused complex folding of salt deposits. Differential movement of the salt beds separated the halite salt stratum and purified it by extrusion of anhydrite and other impurities. The final result was formation of pure halite veins enclosed by contaminated salt (so-called 'zuber'), as illustrated in Figure 5.4.1. The type of halite deposits is commonly known in Poland as the salt veins. Halite veins contain more than 98% NaCl. The contaminated salt or 'zuber' contains NaCl in a range of 15 and 80%. The contact between halite vein and enclosing salt is very sharp and well noticeable due to colour differences between white halite and dark-grey 'zuber'.

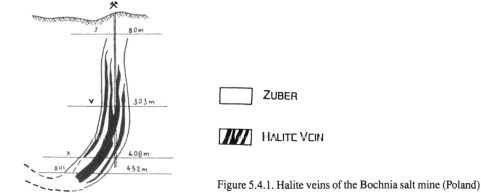

Figure 5.4.1. Halite veins of the Bochnia salt mine (Poland)

The inclination of halite vein could be from gently dipping to steeply dipping. The extension along dip is up to a couple of hundred meters. The thickness is uniform (Wieliczka mine and Baric mine). Vertical veins have a depth of several hundreds of meters and a very uniform thickness (Bochnia mine). The thickness of the halite veins is in the range between several meters and several tenths of meters. The structural features of the halite veins are considered very convenient for dry mining.[7]

5.4.2 *Irregular vein shape*

Potash veins belong to the irregular shape and are located within salt domes of the European Zechstein. The potash veins in the shape are very similar to some metal ore veins. The main factor of the veins' irregularity is an internal folding and refolding of potash beds within the salt domes, as illustrated in Figure 5.4.2. The shape of potash veins in many instances is controled by a monoclinal folding pattern on a general plane – a monoclinal fold, on local plane – drag folds. The degree of irregularity of the potash veins depends on the intensity of monoclinal folding.

The potash veins are pitching or steep pitching structures plunging toward a boundary of domes. The thickness of potash veins varies considerably and is controlled by local fold geometry. For example, fold core is thickening and fold limb is thinning of the vein. The depth of potash veins might be several thousands of meters. The present mining activities are confined to the upper portion of the potash veins.[15]

The dry mining of the irregular potash veins is carried out within domes of the Zechstein salt basin in Central Europe (Germany). These veins are the main source of potash ore in this country, which are mined out by mining methods implemented in hard rock mining.

5.4.3 *Thick vein shape*

Thick veins are also the product of the internal structure of the salt domes. Thick veins in Europe are the product of diapiric anticlines of the lower Zechstein's segment.[14] They consist of either halite salt (pure rock salt) or (potassium salts) potash ore or both.

In the Zechstein salt basin of Poland, the mineable thick salt veins are developed during the second structural stage and are represented by the large mass of older salts. The facial salt distribution is controlled by the structure of the anticlinale. The anticline core contains

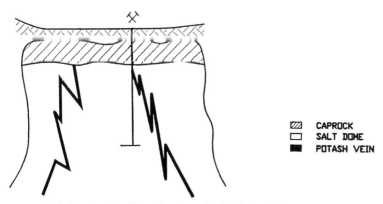

Figure 5.4.2. Potash vein of irregular shape, Zechstein's salt dome (Germany)

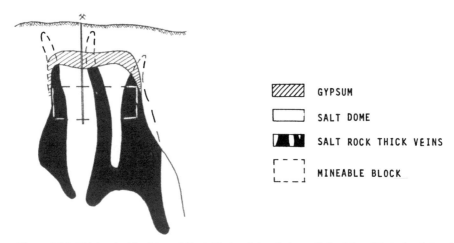

Figure 5.4.3. Thick vein (diapiric anticline), Zechstein's salt dome, Poland (modified after Poborski)

enormous concentration of older rock salt, but the limbs contain the older and younger potash salt, which is underlaid by younger salt beds.[16] From the mining point of view, the anticline limbs are approximated as thick potash veins with a tectonic hanging wall and a layered salt footwall.

The thick veins are dipping very steeply with a relatively regular shape (Figure 5.4.3). The mineable thickness is defined by a cut-off grade of potash and might be up to 50 m. The internal structure of the thick veins is to a certain degree complex due to irregular mineral distribution and the geometrical relationship among facies. The larger anticlined area of the salt domes corresponds to more simplified internal structures, which are more favourable for dry mining of thick veins.

REFERENCES

1. Hansen, F.D. & N.L. Carter 1980. Creep of rock salt at elevated temperature, pp. 217-227. In *The state of the art in rock mechanics*, 21st Symp. on Rock Mech., May 28-30, 1980, Univ. of Missouri-Rolla.
2. Bujalov, T.N. 1957. *Structural geology*, pp. 202-279 (in Russian). Moscow: Gostoptec-Izdat.
3. Dreyer, W. 1973. *The science of rock mechanics* – Part 1, pp. 102-104. Clausthal: Trans. Techn. Publ.4. Goodman, E.R. 1976. *Methods of geological engineering in discontinuous rocks*, pp. 16-36. San Francisco: West Publ. Co.
5. Stoikovic, J. et al. 1987. Salt deposits of Tuzla basin and prospect of future geological exploration. *I Min. & Geol.* (Tuzla Univ., Tuzla), 15/16: 9-18 (in Yugoslav).
6. Jeremic, M.L. 1965. Geological structures of the salt deposits in Romania. *Min. & Metall. Bulletin* (Belgrade), 3: 57-65 (in Yugoslav).
7. Jeremic, M.L. 1965. Methods of development and exploitation of salt deposits in Poland. *Min. & Metall. Bulletin* (Belgrade), 32 (2): 264-274 (in Yugoslav).
8. Jensen, M.L. & A.M. Bateman 1981. *Evaporation – Economic mineral deposits*, pp. 199-212. New York/Toronto: Wiley.
9. Kupfer, D. 1967. Mechanism of intrusion of Gulf coast salt. pp. 2-11. In *Proc. of Symp. on Geol. and Techn. of Gulf Coast Salt*, Louisiana State Univ.
10. Schwerdtner, W.M. 1983. Gypsum-anhydrite caps of Arctic salt domes, Queen Elizabeth Islands: Products of active and passive diapirism. In *Proc. of 6th Int. Symp. on Salt*, The Salt Institute, Virginia, Vol. 1, pp. 311-313.
11. White, R.M. & C.A. Spiers 1983. Characterization of salt domes for storage and waste disposal. In *Proc. of 6th Int. Symp. on Salt*, The Salt Institute, Virginia, Vol. 1, pp. 511-518.
12. Kelsall, P.C. & J.W. Nelson 1983. Geologic and engineering characteristics of Gulf region salt domes applied to underground storage and mining. In *Proc. of 6th Int. Symp. on Salt*, The Salt Institute, Virginia, Vol. 1, pp. 519-541.
13. Halbouty, M.T. 1979. *Salt domems, Gulf region, United States and Mexico* (2nd ed.). Houston, Texas: Gulf Publ. Co.
14. Poborski, J. & K. Slizowski 1983. Geological prediction of the mining feasibility within the salt domes of Upper Permian Zechstein, Central Poland. In *Proc. of 6th Int. Symp. on Salt*, The Salt Institute, Virginia, Vol. 1, pp. 303-310.
15. Willy, A.H. & A.H. Potthoff 1983. Potash mining in steep deposits (salt domes). In *Potash '83. Potash technology*, pp. 79-84. Toronto/New York: Pergamon Press.
16. Poborski, J.W. 1974. On the tectogenesis of some diapiric salt structures in Central Poland, Upper Permian, pp. 267-270. In *Proc. of 4th Symp. on Salt*. Northern Ohio Geological Society, Cleveland.

CHAPTER 6

Exploration and categorization of evaporite deposits

The generally accepted rule that exploration of mineral deposits continues until a last ton of ore is recovered does not apply to salt mining because the extent of exploration by drilling is limited. The exploration of deposits, however, should be to the point that ore reserves evaluation and evaporite rock categorization are complete.

6.1 SURFACE EXPLORATION

Before detailed exploration of deposits starts it is necessary to delineate potential areas. In this regard surface exploration is carried out, i.e. by the four principal methods discussed below.

6.1.1 *Regional exploration*

Regional exploration is a first step in the order of events to explore and determine salt deposits and evaporite strata. Regional exploration is usually carried out by reconnaissance mapping in order to determine the existence of potential sedimentary basins which contain evaporite-bearing strata. To achieve this aim several prospecting methods are usually applied to investigate a region, as listed below:

1. Regional mapping at large scales, such as 1:50 000 or greater. The mapping utilizes stratigraphic guides, palaeographic features and other geological phenomena.

2. Utilization of other indicators such as water wells and drill holes for oil and gas which have been particularly instrumental in the discovery not only of evaporite-bearing strata but also of economic salt deposits.

3. In recent times regional exploration uses the airborne geophysical prospecting and satellite reconnaissance data. This type of reconnaissance information may help to contour salt basin boundaries.

4. Computer techniques both in the evaluation of theoretical concepts and in the manipulation of raw data have been broadly applied and greatly expanded in exploration at this scale.

Regional exploration of evaporite-bearing basins could be of interest where knowledge of the geology and the potential of the salt deposits is very limited, i.e. in SE Asian countries. However, in countries with a history of salt mining, regional exploration is complete and the dimensions/location of sedimentary basins containing evaporite strata are known, e.g. in Romania (Figure 6.1.1).[1]

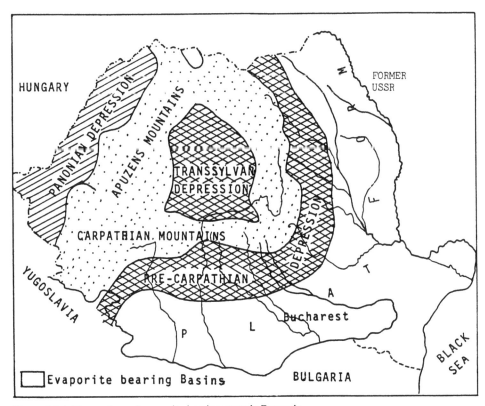

Figure 6.1.1. Depressions of evaporite-bearing strata in Romania

It should be added that regional exploration is carried out to make a field appraisal of known geological data, to choose targets for future surface exploration, e.g. to choose areas for detailed surface exploration. For example, palaeographic reconstruction of a corner of an evaporite basin where the last seawater bitterness evaporated, may reveal zones rich in potash. In this case regional prospecting should reach sensible conclusions regarding future exploration of selected promising target areas.

6.1.2 *Geological exploration*

Fundamental work to outline possible mineable reserves of salt deposits is carried out by geological mapping in several stages as exploration progresses. At this point consideration should be given to geological prospecting and geological mapping.

Geological prospecting is mostly directed toward mapping certain characteristics favourable to the occurrence of salt deposits such as:

1. the boundary between the evaporite-bearing strata and the strata below or above evaporites, also the lateral extent of the salt strata;

2. mapping of gypsum outcrops which are part of the salt-bearing strata;

3. ground surface depression which might be related to formation of sub-surface cavities caused by solution of the salt strata;

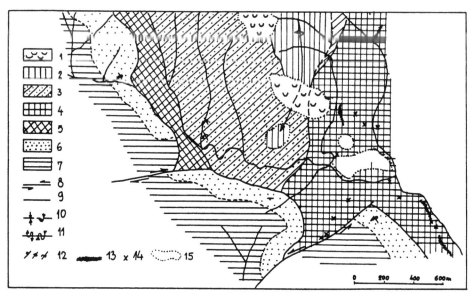

Figure 6.1.2. Part of geological map of the salt-bearing region Almas-Krakaoani (after O. Mirauta)
1. Quarternary; 2. Helvetian; 3. Upper Miocene – marl, sandstone and conglomerates; 4. Upper Miocene
– evaporite-bearing strata; 5. Upper Miocene – conglomerates and schists; 6. Oligocene; 7. Eocene; 8.
dislocations; 9. boundary of the formations; 10. axes of synclines; 11. axes of the turned-over synclines
and anticlines; 12. slopes; 13. outcrops of gypsum; 14. brine wells; 15. depressions

4. brine wells and ponds with saline water and recrystallized saline minerals due to brine evaporation.

The geological prospecting is carried out at a large scale: 1:25 000 or 1:20 000 (Figure 6.1.2). Its main aim is to propose target zones for further detailed exploration.[1]

The target zone for further exploration should be mapped in a more detailed manner, usually at the scale 1:5000. The main aim of this mapping is determination of stratigraphic zones of evaporite-bearing strata which exhibit specific lithological characteristics. It should be pointed out that detailed geological maps have to be kept up to date as further exploration progresses and mine development advances.

6.1.3 *Geochemical exploration*

Geochemical exploration has a long history for prospecting of salt deposits because brines which appear on the ground surface as wells, springs, etc. have been used as indicators of sub-surface salt deposits since ancient times.

In modern times geochemical prospecting has been improved as a result of new scientific knowledge, so that it can indicate not only occurrences of salt deposits but also their facial development.[2]

Two principal types of geochemical exploration are employed, which are briefly discussed as below.[3]

1. Biological exploration was in effect centuries ago when the phenomenon of plant growth above the salt deposits was recognised to show certain particularities. In these

circumstances some plants exhibit very extensive growth above salt deposits, but others do not grow at all in such areas. At present thirty plants are known to be useful biological indicators for detection of salt deposits; examples are:
- *Aster salicifolius Scholler,*
- *Salicornia herbaceae,*
- *Arenoria serpyllifolia* L. r. *viscida,*
- *Glaur maritima* (and others).

These plants belong to the group of halophyte flora.[1] The plants which grow above a potassium deposit are called 'flora potassium'. It is a well known fact that potash is used largely for the fertilizer market, and so it is quite a logical phenomenon that above such deposits plants grow with increased density and height and so can be used for the detection of these deposits.

2. Hydrochemical exploration is the most widely used method of prospecting for salt deposits because of the extensive presence of saline waters on the ground surface. The hydrochemical prospecting of salt basins is carried out by determination of certain chemical elements present in water samples. The important chemical elements and coefficients are given in Table 6.1.1 for two salt basins in two different countries. It should be noted that the important geochemical indicators are similar in both examples, regardless of the large distance between them. The hydrochemical anomalies indicated by geochemical coefficients are plotted by vertical profiles (Figure 6.1.3) and by plane isolines.[2]

Because water easily dissolves salt beds, it is understandable that this method is widely used for exploration of salt deposits. In closing it should be pointed out that the chemistry of the water in salt solutions might also be an important factor in mine stability analysis.

Table 6.1.1. Parallel analysis of the surface waters in the salt-bearing basins of Romania and former Yugoslavia (g/l)

Cl	Na	K	Mg	Ca	Na Σ jon	K Σ jon
Tazlau – Romania						
Brines						
25,430	16,410	0,061	0,084	1,014	38	-
8,430	4,927	0,039	0,017	0,128	35	-
0,381	0,234	0,003	0,005	0,145	35	-
0,751	0,409	0,101	0,101	0,132	-	7
Salty waters						
1,382	0,776	0,351	0,335	0,063	-	12
0,057	0,033	0,027	0,011	0,022	-	18
Sibosnica – former Yugoslavia						
Brines						
27,464	17,215	0,054	0,077	1,011	38	-
7,487	4,457	0,027	0,023	0,144	35	-
0,435	0,334	0,004	0,006	0,090	38	-
0,384	0,227	0,084	0,126	0,109	-	8
Salty waters						
0,418	0,241	0,196	0,121	0,046	-	19
0,082	0,046	0,044	0,012	0,013	-	20

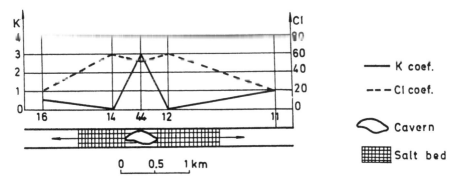

Figure 6.1.3. Hydrochemical profile above a modelled salt deposit with solution cavern (Tuzla salt deposit, former Yugoslavia)

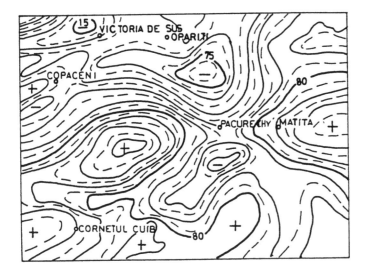

Figure 6.1.4. Buge's gravimetric map of Pacurechy region (Romania)

6.1.4 *Geophysical exploration*

Geophysical methods for the exploration of evaporite-bearing strata can be significant for detection of salt deposits. Geophysical surveying is often used in conjunction with other surface exploration methods as well as exploration drilling. Geophysical methods that are used for the exploration of salt deposits include gravity, electrical and seismic techniques. The gravity method is most extensively used because exploration results from subsequent drilling show that this technique produces accurate data for salt deposit delineation. In practice, drill holes are rarely spotted prior to gravity mapping.

Gravity measurements can detect salt deposits and also delineate their outline boundary and structural features. In Figure 6.1.4 a gravity survey map of the Pacurechy region (Romania) is presented. Gravimetrical minima in this figure indicate the approximate location of a salt deposit as well as its structural features which have the form of an asymmetrical anticlinal.[4] The very strong gravimetrical contrast between the salt deposit and the surrounding rock strata (which is 0.10 g/cm^3) indicates a complicated structure for the salt deposit. It is interesting to note that these evaporite-bearing strata are covered by younger sediments and were discovered by hydrochemical exploration.

It should be emphasized that gravity surveying permits a good estimate of the structure of the deposit; this permits identification of good locations for exploration drill holes.

6.2 EXPLORATION BY DRILLING

Exploration drilling from ground surface has been utilized for identification of the evaporite strata and interpretation of its stratigraphy and structure. Logging of drill holes is an important part of exploration drilling because it produces data for the evaluation of ore reserves as well as engineering rock categorization. All those topics are commented upon, including choice of technology of drilling and method of location of drill holes.

6.2.1 *Drilling techniques*

Three principal drilling techniques are applied for the exploration of salt deposits as briefly described below.[3]

1. *Diamond drilling* with coring is generally regarded as the most useful method of exploration because it offers visual inspection and testing of the evaporite-bearing strata (Figure 6.2.1). The success of exploration drilling depends on the percentage of core

Figure 6.2.1. Exploration by diamond drilling, Tuzla salt deposit (former Yugoslavia)

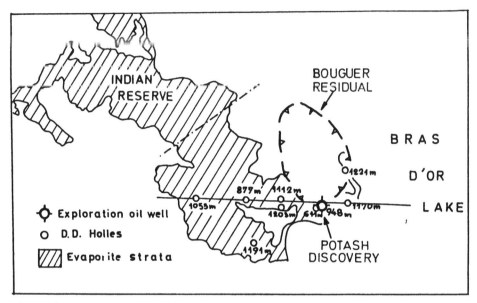

Figure 6.2.2. Plane of pattern of exploration drilling

recovery. If there is a poor core recovery, then the advantages of this method vanish. The fluid is the drilling medium and a core sample cut by a diamond bit is received in a barrel behind the bit. In deep holes a wire-line apparatus is used to remove and replace the core barrel inside the drill rods without taking the drilling tool from the hole. A special drilling fluid is used to prevent erosion of the salt beds. For potash exploration at Malagawatch, Nova Scotia, a truck mounted Longyear 44Rig for diamond drilling has been used. The first drilling tests, DDH M-5, was spudded 1250 m SW of the discovery location (Figure 6.2.2). To accelerate drilling, a second the Longyear 44Rig was contracted. Nine holes were drilled with the two diamond drilling machines; eight of them reached target at depths varying from 877 to 1221 m. The first hole was lost at 363 m depth and had to be redrilled. Total length of drilling was 9187 m of which 8131 m was cored with HQ and NQ size drill bits. Core recovery was excellent. The utlization of drill holes and core sample for logging, geological interpretations, and deposit evaluation is discussed in subsequent subsections of this chapter.

2. *Rotary drilling* is faster, cheaper and simpler to operate than diamond drilling but in this case only the rock chips of 0.5 cm to 1 cm are recovered from the 10.5 cm bit. Drill cuttings are flushed with compressed air and collected in a cyclone. The individual sample chips provide needed information for the exploration process. The geophysical logging of drill holes is required so that reasonable geological interpretation can be carried out.[1] Dry drilling is a technique in which air of a drilling fluid is used. It is a very convenient method for exploration of salt deposits which are very susceptible to water solution.

3. *Percussion drilling* is based on the principle of hammering of the drill rods accompanied by their partial rotation after each hit. The rock cutting can be caught by spoon and brought to surface. More recently a percussion drill with a compressed air hammer is used and air is utilized for lifting the cuttings to ground surface. Percussion drilling for salt

exploration is seldom done; it is commonly used for drilling holes to freeze ground to facilitate shaft sinking through water-bearing strata.

The exploration drilling of the salt deposits is used in conjunction with an extensive gravity survey both of which are carried out simultaneously. The drill holes and core samples produce valuable data for geological interpretation of deposits, ore reserves estimation, as well as rock strata categorization as it relates to rock mechanics.

6.2.2 *Methods of exploration drilling*

The methods of exploration drilling have been developed from practical experience over many years. It is estimated that in European salt basins over fifty thousand exploration holes have been drilled. The methodology of exploration drilling is carried out in several steps.

1. The first step for exploration of folded deposits is the location of the hole at the centre of the gravimetrical minimum. In the case of flat deposits with large lateral extensions, initial exploration drilling starts with several holes at centres spaced 5000 and 10 000 m apart.

2. If the result of the first hole is positive, then a subsequent number of drill holes is drilled either in the block pattern (Figure 6.2.2) or the profile pattern. The distance between drill holes along the profile is between 1000 and 2000 m.

3. If preceding exploration drilling has been successful, then detailed exploration of the deposit is initiated, as illustrated in Figure 6.2.3, where the distance between holes on longitudinal profiles is 1000 m, but on the cross profile is only 300 m apart. However, for exploration of the potash deposits the drill holes on both profiles are drilled 500 to 600 m apart.

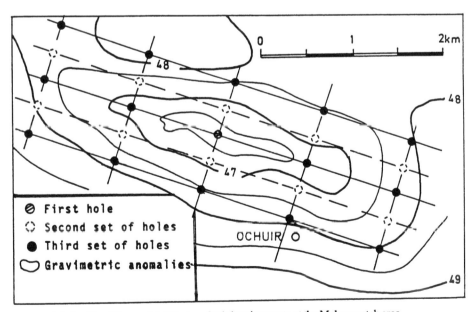

Figure 6.2.3. Profile of diamond drill holes of salt-bearing strata at the Malagawatch area

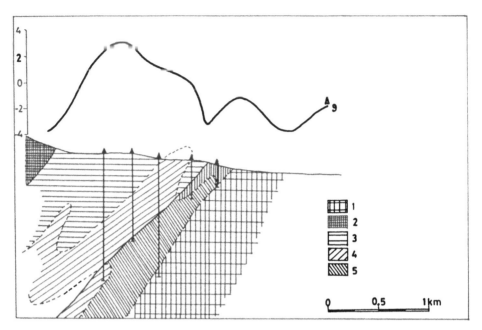

Figure 6.2.4. Location of drill holes by gravimetric profile; the middle hole is structural and the rest is subsequent
1. Sarmatian; 2. grey series; 3. red series; 4. conglomerate; 5. salt strata (Sarata, Romania)

It should be pointed out that distance between subsequent drill holes could vary from deposit to deposit, due to the characteristics of its geological structure. Also the distance between drill holes is dictated by categorization of ore reserves, as further discussed.

The first exploration hole is drilled with coring (diamond drilling) located at a centre of gravimetric anomaly. This hole is usually called 'structural', because it is drilled throughout salt deposits. If the geophysical survey indicates that the top of salt deposits is at a uniform height, then subsequent holes are drilled up to the boundary of the salt body. In this instance rotary drilling with water as drilling fluid is mainly used; a few holes with coring are necessary. These holes are ceased when they reach a top of salt deposits (Figure 6.2.4).

It would be a fair statement to make that the location of exploration drill holes is not carried out before geophysical surveying, because the latter technique produces very valuable data for interpretation of the contours of the deposit.

6.2.3 *Geological logging*

Geological logging of drill holes is carried out in the same manner as in the case of exploration of other mineral deposits: by drilling and coring. The fundamental aim of geological logging is to record the thicknesses and depths of individual beds and includes recording of facial development of the salt profile.[5] The representation of the drill log data might be done tabularly (Table 6.2.1) or graphically by columnar profiles, which is usually confined to the smaller lengths, of core samples as for example an ore section.[6]

Geological logging of evaporite strata is directed toward evaluation of the stratigraphy and structure of salt deposits. All data from each individual drill hole which can be utilized

Table 6.2.1. Log of typical drill hole at Goderich salt mine

Depth	Formation	Description
0-9.15	Overburden	Sand and gravel
9.5-88.0	Lucas (Detroit River Group)	Dolomite
88.0-119.0	Amherstburg (Detroit River Group)	Dolomite
119.0-191.5	Bois Blanc	Cherty limestone
	Bass Island	Dolomite
191.5-256.0	Salina (H unit)	Dolomite and shaly dolomite
256.0-289.5	Salina (G unit)	
289.5-326.0	Salina (F unit)	289.5-294.5 m: salt shaly dolomite; 306.5-314.0 m: salt
326.0-351.0	Salina (E unit)	Shaly dolomite
351.0-362.5	Salina (D unit)	351.0-362.5 m: salt
362.5-382.0	Salina (C unit)	Shaly dolomite
382.0-464.0	Salina (B unit)	Salt, interbedded shale, dolomite & anhydrite
464.0-580.0	Salina (A unit)	Mineable salt horizon, 507.5-532 m: dolomite & shaly, dolomite

for determination of the stratigraphy should be pieced together. For example, as L. Dekker[5] described at Malagawatch, Nova Scotia, exploration drilling indicated that the salt horizon consists of a uniform, massive grey halite with interbedded anhydrite and mudstone which grades upwards into a brecciated darker grey and orange-brown halite. Two potash zones occur in this interval. The presence of a thick anhydrite section below the halite is inferred from regional drilling for oil.[5]

The composition stratigraphy of Malagawatch salt-bearing strata is illustrated in Figure 6.2.5. There are at least three separate potash zones in the lower Windsor group, the lowermost one having the highest grade and thickness, but only where it is structurally thickened are economic grade and and thickness attained. The middle and upper zones are uneconomic, being too thin and of low grade. No drill core recorded all three zones. All potash zones are mineralogically simple, consisting of sylvinite, a mixture of sylvite and halite with traces of carnallite. The basal Windsor carbonate and the basal Windsor anhydrite were not penetrated by drilling but are inferred from stratigraphic correlation with other areas in Nova Scotia (Figure 6.2.6). There is an opinion that the lower Windsor group might be at least 2000 m thick.[6]

An example of structural deformation of salt-bearing strata, interpreted on the basis of data obtained by exploration drilling, is represented at Malagawatch and described by L. Dekker.[5] Initial attempts at a structural interpretation were stymied by a number of stratigraphic complications. A regional zone of salt removal and solution collapse extends from the surface to as deep as 415 m. Throughout this interval, correlation was difficult and sometimes impossible. The Malagawatch area exhibits folding and faulting structures. Large-scale overturned, tight folds with limb lengths measured in hundreds of meters are associated with similar small-scale folds with wave lengths and amplitudes in the order of several meters and located in individual structural blocks. The tectonic blocks are delineated by normal faults, as discussed in Chapter 3. The blocks separated by faults are displaced down with respect to each other as much as 600 m. Therefore, only the upper potash zone was reached and the main potash horizon was not intersected by drill holes in downward displaced blocks.[5]

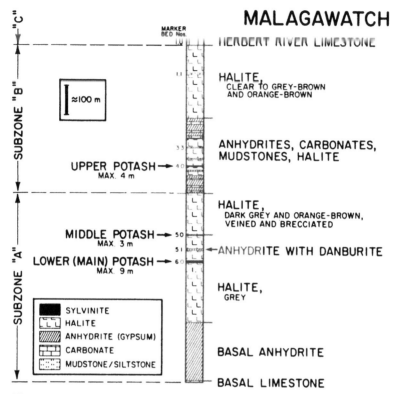

Figure 6.2.5. Composite stratigraphy of the lower Windsor group at Malagawatch (after Dekker)

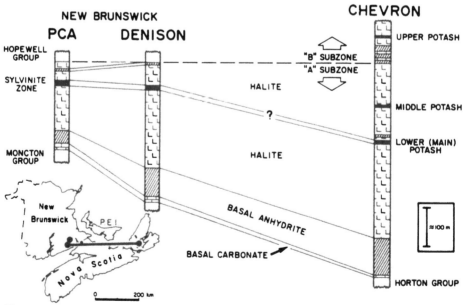

Figure 6.2.6. Tentative correlation of the lower Windsor group

The interpretation of stratigraphy and structure of salt-bearing strata at Malagawatch is based on the geophysical drill hole logging (see Subsection 6.2.4) and drill core logging; the latter exhibits petrographic data and facies of salt-bearing strata as well as ductile and brittle deformations of the rocks.

Finally, preparation of core samples obtained from drill holes is a part of geological logging. The recovered rock cylinders are usually cut to 25 to 35 cm in length, and then in half (longitudinally). The portion is then halved again to form a quarter which is crushed and ground for chemical analysis to obtain data for calculation of the ore grade which is an integral part of ore reserve evaluation. In addition a certain number of cylindrical core samples should be taken to be tested for their strength and behaviour; these data are part of the geotechnical logging which is briefly discussed in Subsection 6.2.5.

6.2.4 *Geophysical logging*

Geophysical logging of drill holes is an integral part of exploration of salt deposits. It is an important exploration tool, particularly in the instance when exploration holes are drilled without coring. Under these circumstances geophysical logging can produce data of fundamental importance in determining the stratigraphy and structure of the salt-bearing strata.

In the case of geophysical surveying all drill holes are logged by down-the-hole instrumentation using the following methods most often: gamma-gamma, gamma-neutron, sonic and 3-D. The geophysical logs are correlated with the logs of drill holes for which there is core and with the geolograph which records the rate of drilling (Figure 6.2.7).

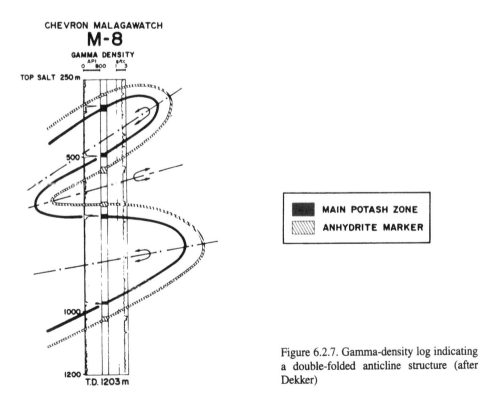

Figure 6.2.7. Gamma-density log indicating a double-folded anticline structure (after Dekker)

The most often used method of geophysical logging of drill holes is based on gamma density which indicates variations of rock properties as follows:
- compaction;
- porosity;
- saturation;
- texture;
- composition;
- weathering;
- fracturing;
- age of individual salt beds; etc.

The listed data can be effectively used for interpretation of geological structures; this is shown for the drill hole geophysically logged, as illustrated in Figure 6.2.7. In addition the data could also be used for determination of geotechnical parameters and hydrogeological characteristics of the evaporite strata.[1]

6.2.5 *Geotechnical logging*

Geotechnical logging uses geological and geophysical log data together with laboratory testing of the core samples to evaluate the strength properties and behavaiour of the mineable portion of the strata.

Geotechnical logging requires that core samples be taken for several meters above and below the mine opening.

On the basis of geotechnical logging classification of the individual evaporite rocks and salts can be carried out. Two methods are usually used:

1. Geomechanics classification (RMR), proposed by Z.T. Bieniawski,[7] with the following parameters:
- strength of intact rock material;
- drill core quality;
- condition of joints;
- spacing of joints;
- ground water conditions.

According to the magnitude of ratings, one of the five rock mass classes can be selected. The ratings range from 0 to 100.

2. Rock mass quality classification (Q), proposed by N. Barton,[8] with the following parameters:
- rock quality designation (RQD);
- joint set number (Jn);
- joint alteration number (Ja);
- joint roughness number (Jr);
- joint water reduction factor (Jw);
- stress reduction factor (SRF).

The rock mass quality is calculated as follows:

$$Q = (RQD/Jn) (Jr/Ja) (Jw/SRF)$$

The ratings range from 0.001 to 1000. The values are usually obtained by geotechnical logging of NX drill holes.

R.N. Singh and M. Eksi used both classifications to evaluate the bearing capacity of pillar structures in gypsum mines in Great Britain.[9] They came to the conclusion that the rock mass could be classified as 'very good rock and good rock', with a strength value rather low because gypsum and marl are in the weak rock group. They compared the RMR and Q systems by the following relationships.

$$RMR = 31 + 19 \, Q^{0.4} - 3 \, Q^{0.7}$$

They also made correlations between obtained RMR value and in situ strength:

the predicted in situ strength $= 0.45 \, e^{0.04 \, RMR}$

Geotechnical logging is of great importance for underground mine design and stability analysis; it is essential to obtain safe mine structures.

6.3 EVALUATION OF ORE RESERVES

The most important part of the exploration of salt deposits is the summarization of all data in a form which is convenient for evaluation of ore reserves. Actually, all exploration concepts and layout should be in agreement with requirements for ore reserve evaluation. The principal aspects for establishing reserves of salt deposits under question are briefly discussed below.

6.3.1 *Parameters of reserve computation*

The evaluation of the ore reserve is based on several parameters which have to be defined in advance, so that calculation of reserves of salt deposits can be carried out.

1. Ore grade cut off is a basic element for determination of the mineable section, particularly in the case of potash mining where potassium content may have an irregular distribution. The ore grade cut off for each deposit should be separately determined, usually using statistical analysis. In the case of Malagawatch's potash deposits at Nova Scotia the chosen ore grade cut off for core assay is 20% K_2O and the log calculated assay is 25% K_2O (Figure 6.3.1).[6]

2. The average grade is computed by determining an economical cut-off grade from data acquired from drill hole samples. The data given beside the sections in Figure 6.3.1 present (1) length in meters mineable thickness for K_2O versus thickness of insoluble material (waste rock), and (2) average K_2O for mineable thickness in (1) versus % insolubles for waste thickness in (1). On the basis of computed average grade a mineable thickness of the ore body is defined.

3. True thickness is used for ore reserve calculations; they are obtained by perpendicular projection of drilled thickness on the salt strata. As illustrated in Figure 6.3.1, it is seen that the true thickness of an ore section is appreciably smaller than that obtained by drilling. This is quite understandable for Malagawatch's potash deposit since it is extensively deformed by overturned folds (Figure 6.2.5).

4. The area of the deposit considered for ore reserve calculation is usually represented by regular geometrical contours. In the case of inclined and pitching salt deposits it is necessary to determine the true area which is along the dip of the ore body, not that projected to ground surface.

Determination of these parameters presents a certain number of difficulties due to limitations superimposed on exploration drilling.

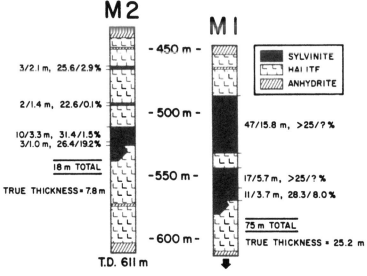

Figure 6.3.1. The grade and thicknesses of potash sections along exploration drill holes (Malagawatch, Nova Scotia)

6.3.2 Methods of reserve computation

Computation of ore reserves of the salt deposits is more or less predetermined by the pattern of drill holes executed during the exploration phase. Calculation of the grade and tonnage of a salt deposit is made by analysis of sample data framed in either rectangular or polygonal block or cross-sectional blocks, as briefly commented on below.[3]

1. Rectangular or polygonal blocks are usually used for ore reserve computation of the flat bedded deposits or massive domal deposits (Figure 6.3.2). The drill hole represents an individual block, and in the calculation the influence of each hole extends halfway to the adjacent hole. Actually it is assumed that the grade of an assay of a section of a drill hole approximates the grade of the ore around that hole.

2. Cross-sectional ore reserve computation is usually applied for tectonically deformed and pitching salt deposits. The sections are subdivided into areas or blocks (Figure 6.3.3). The sectional grade is assumed as a grade of the ore within a block for which tonnage is calculated.

Both methods of computing salt reserves contain the inherent problem of assigning an area of influence to an assay or calculated grade of ore section, particularly in the case when mineral composition of various facies is complicated. It should be stated that the geometrization of salt deposits by a simple contour block or cross-section is a largely approximated, particularly when the ore body has been folded and refolded.

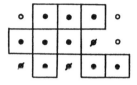

Uniform spacing

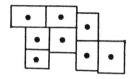

Random spacing

Figure 6.3.2. Rectangular blocks around ore grade point

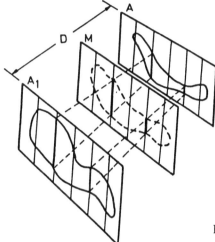

Figure 6.3.3. Cross-sectional profile along drilled holes

6.3.3 *Categorization of reserves*

Two principal factors will influence categorization of salt reserves:[10]

1. Morphological types of salt deposits, which occur between simple and complicated geological structures and which are discussed in geological-structural chapters of this book. The four morphological types are illustrated in Figure 6.3.4 and are named:

type I   – flat beds;
type II – lensoid beds;
type III – folded beds;
type IV – domal beds.

Each proposed type of salt deposit requires a specific spacing of drill holes for exploration and categorization of ore reserves.

2. Spacing of drill holes directly influences the accuracy of the ore reserve calculations. Of course, the same spacing of exploration drill holes for regular and simple beds and irregular and folded beds will produce a different degree of accuracy for ore reserve calculation. In Table 6.3.1 categories of ore reserves based on the frequency of drill holes per unit area or per unit length are presented. In addition the changes in the reserve categories are given when underground mine workings are implemented, which appreciably increase the degree of exploratory work.

In a majority of European and other countries the ore reserves for salt deposits are defined by the four categories; they are presented in Table 6.3.1 and are briefly commented on below.[10]

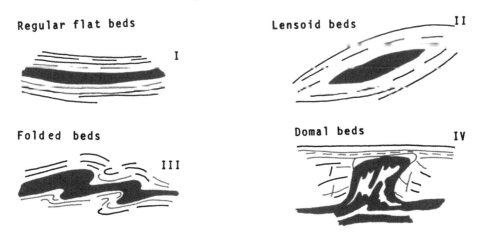

Figure 6.3.4. Morphological types of salt deposits in regard to the reserve calculation

Table 6.3.1. Categorization of salt reserves

Type of deposit	Surface exploration drilling		U/G exploration workings	
	C_2 Exploration drill holes per area (km^1)	C_1 Distance between exploration holes (m)	B Distance between workings (m)	A Distance between workings (m)
I Flat beds	4	3000 m	400-600	300-400
II Lensoid beds	3	2000 m	300-400	200-300
III Folded beds	2	1000 m	200-300	100-200
IV Domal beds	1	500 m	100-200	50-100

C_2 – called inferred reserves whose presence is indicated, based upon one drill hole;

C_1 – called possible reserves whose outline is indicated by a certain number of drill holes;

B – called probably reserves; quantity and quality are determined by drilling and underground mining workings;

A – called proven reserves; are determined only on the basis of mine working of appreciable density.

It should be noted that solution mining is implemented only on the basis of ore reserves estimation by drilling and geophysical surveying; this is not the case in a dry mining operation which could be implemented only on the basis of probable and proven reserves.

6.3.4 *Classification of ore reserves*

Some European countries group ore reserves into three classes in terms of cost and mineability.[3]

1. *Mineral reserves* correspond to total ore reserves in place, calculated and assumed. These reserves belong to underground geological-structural environments. Mineral reserves are divided into three sub-classes (Figure 6.3.5):

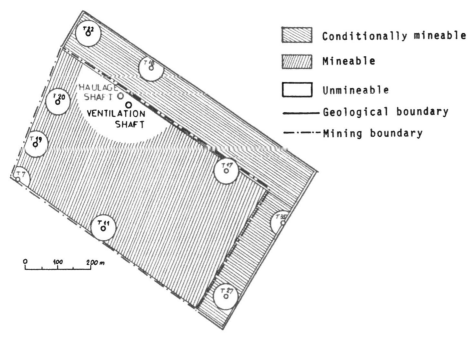

Figure 6.3.5. Classification of ore reserves of Tusanj salt mine, Tuzla basin, former Yugoslavia

a. Mineable reserves with a grade, tonnage and shape which are adequate for profitable mining operations.

b. Conditionally mineable reserves are represented by parts of deposits which have not been sufficiently explored so that their mineability can be identified. If further exploration gives positive results, then they could become mineable. In addition, if mining of the ore at the present time is not profitable, but with technological changes or market price improvement their exploitation could become profitable, then they can be reclassified as mineable reserves.

c. Unmineable reserves are represented by complex shapes and low quality of ore. Ore left in safety pillars and in some cases where hydrogeological conditions do not permit mining operation also falls into this classification.

2. *Industrial reserves* represent only the ore which is recovered and brought to surface as defined below:

industrial reserves = mineable reserves – unrecovered reserves

Only recoverable reserves can be identified with the industrial reserves. The tonnage of industrial reserves depends on the coefficient of ore recovery. For example, a potash deposit mined in France with longwall mining achieves a very high coefficient of recovery (approximately 90%), but very similar potash deposits in Saskatchewan mined by room-and-pillar mining achieve a low coefficient of recovery (approximately 35%). In the last case, room-and-pillar mining has to be applied to support a water-bearing horizon above the deposit and eliminate water inflow to the mine workings. The ratio between mineable and recoverable (e.g. industrial) reserves in the Saskatchewan deposit is low. For example, recoverable reserves of potash ore of Tombil potash mine were 314 million tonnes assaying 27.3% K_2O. Assuming an extraction ratio of 35% for the room-and-pillar mining, industrial reserves are estimated only at 110 million tonnes which will be recoverable for refining into muriate of potash.

3. *Commercial reserves* correspond to the tonnage of the ore which is delivered to the market. To obtain commercial reserves it is necessary to subtract from industrial reserves a loss of ore due to beneficiation and transportation processes. The beneficiation loss is related to the amount of waste partings and other impurities which have to be removed from the ore in order that it becomes commercially valuable. The transportation loss is due mainly to loading and unloading operations of the mineral commodity.

6.4 CHARACTERIZATION OF EVAPORITE STRATA

Exploration of salt deposits should produce data both for evaluation of ore reserves and for characterization of evaporite strata. The characterization of rock strata is carried out by a broad spectrum of factors including those of a geological character (petrology, shape and structure, homogeneity, texture, etc.), geotechnical character (joints, fissuring, porosity, cavitation, strength, etc.) and mining character (thickness of mineable profile, characteristics of mine roof and mine floor, etc.).

On the basis of these parameters rock characterization can be done for each facial unit of the evaporite strata. The discussion of rock characterization is presented using comments on individual facies with practical examples rather than by descriptive generalizations. The main attention in individual characterization of the rock units is given to the nature of their occurrence within evaporite strata, their physical and strength properties, solubility, as well as characterization with respect to ore mineability and ground stability.

6.4.1 *Carbonate rocks*

There is a wide range of mineralogical variations in carbonate sediments within evaporite-bearing strata. Two groups have been distinguished on the basis of engineering criteria rather than petrological criteria.

1. *Calcium-carbonate sediments* are represented by limestone, dolomite and dolostone. Only chemical carbonate sediments are considered here. The calcium-carbonate sediments of evaporite-bearing strata were formed by chemical precipitation during evaporation processes of sea water. The carbonate rocks, regardless of their origin, have more or less similar mechanical characteristics. The carbonate rocks may underlay or overlay salts and they may occur as interbedded strata with these deposits.

At the Cory potash mine in Saskatchewan the strength characterization for the same carbonate rock strata varies considerably.[11] This phenomenon may be due to the porosity of the carbonate rock, because it appears that it is the main controlling parameter in determining their strength. The most dense rock (Figure 6.4.1) has the highest strength, and the most porous rock has the lowest strength. D.J.Gendzwill showed that the strength of the Cory mine carbonate rock is inversely proportional to its porosity with statistical correlation coefficients between 0.899 and 0.990.

Exploration of carbonate sedimentary rocks showed that they are not uniform, because of the variation of rock porosity along strike and dip of the deposits. Also, the thickness of individual strata varies from thin, weak beds to thick sections of strong rock. The presence

Figure 6.4.1. Limestone specimen, Goderich, Ontario

of shaly impurities weakens carbonate sediments. Finally, joints could be the most important feature for characterization of carbonate sediments. It is estimated that values of strength and stiffness for the carbonate rocks in situ due to structural defects may be only 1/3-1/2 of that estimated by laboratory testing of core samples.

Due to the degree of porosity and frequency of structural defects (including shaly layers and joints), carbonate rocks can be divided into two classes. First are those rocks which are relatively homogeneous, dense and without shale impurities, strong and brittle, and which occur in thick beds with widely spaced joints. It is most likely that such a carbonate sediment overlying an underground mine opening would be self-sustaining and capable of storing enough energy to create rock burst prone zones. Second are those rocks which are heterogeneous, porous, interbedded with shaly layers, which occur in thin fractured beds and are most likely to have very limited self-sustaining abilities. Such units create some zones of weakness along which failure could propagate easily and rock falls could occur. It appears that such a characterization of carbonate rocks could be applied to a majority of evaporite basins.

2. Calcareous sediments are represented by calcareous sandstone (coarse grains), and calcareous clays and silts (fine grains). They are known as marls and marlstones, or impure limestones. They are mainly the product of processes of deposition in a terrigenous environment. Calcareous rocks can occur as substrata or over strata of rock salt and potash deposits. Of course the calcareous rocks occur as interbedded layers of these deposits because of changes in the depositional environment.

The calcareous rocks represent an intimate mixture of clay, silt, sand and particles of calcite or dolomite and contain up to 50% of calcium carbonate.[12]

Most calcareous rocks will have a composition lying between the homogeneous carbonate rock and heterogeneous carbonate rock categories. The wide span of mechanical properties of calcareous rocks is caused by the wide variety of compositions and frequency of geological discontinuities.

6.4.2 *Sulphate rocks*

Of engineering interest are calcium sulphate rocks consisting largely of anhydrite and gypsum. However, of less interest are magnesium sulphate rocks such as kieserite and others which occur as an association of minerals rather than mono-mineralic rock beds.

Anhydrite usually occurs in granular or fibrous crystalline massives. It is distinguished from gypsum by its higher density and hardness. Anhydrite can absorb moisture and change to gypsum; this is accompanied by an increase in volume and so in places, large masses of anhydrite have been thus altered. Anhydrite occurs in the same manner as gypsum and it is often associated with that mineral. Anhydrite beds underlay or are interbedded with salt deposits. Also they are known as the caprock of salt domes.[13]

Gypsum is widely distributed in sedimentary strata, often as a thick bed, and usually underlays rock salt deposits, having been crystallized there as one of the first facies during evaporation of salt waters. Gypsum is the most common sulphate mineral and extensive deposits are found in many localities throughout the world, mainly in large salt basins, e.g. in Michigan. Geologically, gypsum occurs in a sequence of marine evaporites in the evaporite facies basins, but also in the basins of other sedimentary strata.[13] Canada has large reserves of gypsum, which are located in various evaporite basins in the prairies, maritimes and Michigan (Figure 6.4.2).[14]

Laboratory investigations of gypsum from European deposits indicate that the strength characteristics of gypsum are favourable for underground excavations.[9] However, if gypsum is partially dissolved by water, its strength is altered to the point where it may cause significant instabilities. The anhydrite which is of much lesser importance for underground mining has strength characteristics similar to gypsum. The solubility of the gypsum and anhydrite would be the main controlling factor of their strength parameters.[9]

Figure 6.4.2. Gypsum specimen, Grand Rapids, Michigan

6.4.3 *Trona ore*

The beds of trona ore could be a mono-mineral where the trona occurs as a prismatic crystalline mass with some massive mono-clinal crystals (Figure 6.4.3).[15] The multi-mineral beds consist of trona in association with other saline carbonates and halite. The content of halite in trona deposits can be appreciable, since it forms a mixture with them.

Trona is a soluble mineral and has a significant and varying influence on the mechanical properties of rock if it is partially or totally saturated with water.

Dry trona ore has a strength characterization similar to other hydrous carbonates. The investigation of uniaxial compressive strength of trona from Wyoming indicated that it varies if mixed with halite.

The mineable trona beds in Wyoming are more or less of massive structure, and a remarkable uniformity of these trona beds is exhibited.[16] Trona beds with impurities,

Figure 6.4.3. Trona specimens showing a prismatic crystalline structure and a massive monoclinal structure

Figure 6.4.4. Trona-bearing strata, outcropping below a shale marker bed, and showing a fine lamination of sediments

so-called 'dirty trona', have a heterogeneous structure, and they usualy occur above homogeneous trona beds. The trona with impurities is not subject to mining and it is left in place as the roof layer.

The very prominent lamination of trona-bearing strata (Figure 6.4.4) and the high solubility of trona should be the main factors for its characterization in regard to choice of the mining system as well as the design of strata stability control.

6.4.4 *Halite salt*

Halite salt deposits have a wide range of morphological features as discussed previously. For this reason physical characterization of rock salt deposits is different for the case of flat, uniform and moderately thick layers versus large irregular bodies of heterogeneous composition. The rock salt has a grained structure composed of the halite and related salts (Figure 6.4.5). It varies from fine grained masses to very coarse grained masses. The geometry of rock salt deposits and their internal structure very much depends on the intensity of orogenic movement in their proximity.

The characterization of a rock salt mass and its quality designation is based on its density, grain size, and some structural defects, but not by joints because they rarely occur in rock salt beds.

Exploration in Romania[17] has shown that deposits deformed in a flexural anticline have changed their internal structure from fine and medium grained to very coarse grained. This has been accompanied by a density increase due to mechanical alteration and halite recrystallization. The rock salt is significantly compacted and has an increased bearing capacity due to the greater intergranular friction, particularly in the case of very coarse

Figure 6.4.5. Halite specimen, Griegsville, New York

grains. The ultimate compressive stress of the recrystallized rock mass may be increased as much as 35-50%, and might offer a stability of mine rooms with a height to 50 m. Generally speaking, it can be concluded that the mechanically altered rock salt always exhibits strength parameters in the higher ranges. Due to the solubility of rock salt its strength may deteriorate significantly and disappear completely in the presence of water.

6.4.5 *Potash ore*

Potash is composed of the various potassium minerals sylvinite, carnallite, polyhalite, and magnesium, sodium, calcium and other saline minerals as well (Figure 6.4.6). The characterization of the potash ore mostly depends on its particular mineral associations and the quantity of individual minerals present. For example, potash from Saskatchewan occurs in beds and the dominant mineral is carnallite which lowers its strength significantly. For example, the presence of about 10% of carnallite decreases the compressive strength of potash ore up to 50%.

Potash ore in Sergipe (Brazil) has been investigated by C. Fairhurst and associates[18] with regard to the design of an underground mine. They determined nine basic lithologies in this mine as follows:
- coarse-grained sylvinite;
- fine-grained silvinite;
- fine-grained carnallite with a small amount of halite;
- fine-grained carnallite with a large amount of halite;
- coarse-grained carnallite with a small amount of water;
- coarse-grained carnallite with a large amount of water;
- pure tachyhydrite;
- impure tachyhydrite;
- halite.

Quality designation of potash ore from Sergipe is affected by the presence of tachyhydrite because of its very low strength and the ease with which water can penetrate along the grain

Figure 6.4.6. Potash specimen, Rocanville mine, Saskatchewan

boundaries. For example, tachyhydrite has a solubility as high as 140 g/100 g water compared with halite with values of 40 g/100 g water. The presence of tachyhydrite within the mining profile suggests that such ore would be designated as weak and unstable. If tachyhydrite could be excluded and isolated by leaving a sylvinite layer as a protective bed below it, then the ore could be considered to have satisfactory strength.

REFERENCES

1. Jeremic, M.L. 1964. Possibility of utilization of Romanian experience for exploration and development of salt deposits in former Yugoslavia. *Technol. Acta* (Tuzla Univ.), 3-4: 15-27 (in Yugoslav).
2. Jeremic, M.M. 1965. Hydrochemical characteristics of salt basins in NE Bosnia. *Min. & Matall. Bulletin* (Belgrade), 5: 73-80 (in Yugoslav).
3. Jeremic, M.L. 1964. *Exploration of mineral deposits*, pp. 15-174, pp. 353-445. Faculty of Mining of the Sarajevo Univ., Sarajevo Univ. Press (in Yugoslav).
4. Airinei, S. et al. 1960. Exploration of salt domes by gravimetry surveying in Romania. *Revue de Géologie et de Géographie* (Bucuresti), 4 (2): 43-52.
5. Dekker, L. 1985. Potash exploration at Malagawatch, Cape Breton, Nova Scotia. *CIM Bulletin*, 78 (880): 27-32.
6. Muir, W.M. 1966. Salt mining at Goderich, p. 16. In *Proc. of General Meeting of the CIM*, Quebec City.
7. Bieniawski, Z.T. 1984. *Rock mechanics design in mining and tunneling*, pp. 190-214. Rotterdam: Balkema.
8. Barton, N. et al. 1974. *Engineering classification of rock masses for design of tunnel support*, pp. 4-22. Oslo: Norwegian Geotechnical Institution, publ. no. 106.
9. Singh, R.N. & M. Eksi 1987. Empirical design of pillars in gypsum mining using rock mass classification system. *J. Mines, Metals & Fuels* (India), 35 (1): 16-23.
10. Jeremic, M.L. & J. Moravek. Development and mining methods of salt deposits in Poland. *Min. & Metall. Bulletin* (Belgrade), 2: 267-273 (in Yugoslav).
11. Gendzwill, D.J. 1983. Elastic properties of carbonate rocks over the Cory mine, pp. 299-403. In *Proc. of Potash Technology Conf.*, October 3-5, 1983. Saskatoon, Saskatchewan.
12. Dreyer, W. 1973. *The science of rock mechanics* – Part 1: The strength properties of rocks, p. 500. Clausthal: Trans. Techn. Publ.
13. Dand, E.S. 1972. *Dand's manual of mineralogy* (17th ed., revised by C.S. Hurlbut Jr.), p. 609. New York: Wiley.
14. Fomlet, J.H. & G.C. Adams 1988. Gypsum in Canada – present status and future developments. *Min. Engng. AIME* 40 (2): 120-124.
15. Kostick, D.S. 1985. US soda ash industry – the next decade. *Min. Engng. AIME* 37 (10): 1205-1212.
16. Harris, R.E. 1985. Overview of the geology and production of Wyoming trona. *Min. Engng. AIME* 37 (10): 1204-1208.
17. Jeremic, M.L. 1965. Geological structures of the salt deposits in Romania. *Min. & Metall. Bulletin* (Belgrade), 3: 57-65 (in Yugoslav).
18. Fairhurst, C. et al. 1979. Rock mechanics studies of proposed underground mining of potash in Sergpipe, Brazil. In *Proc. of 4th Int. Congr. on Rock Mech.*, Montreux, Switzerland, Vol. 1, pp. 131-136.

CHAPTER 7

Access and mine development

Access and mine development is important in dry salt mining, since excavations are made in soft rock with water-bearing horizons. Several topics are discussed in this chapter. Pre-developing work has to provide enough information so that the design of the access and mine development will not only satisfy economic requirements but also ensure safe mining operations, particularly regarding possible water inflow into the mine openings. To ensure stable ground conditions, the technology of shaft sinking and lining, and ground consolidation has to be analyzed, as well as development links and mine access which differs for various types of mining methods.

7.1 PRE-DEVELOPMENT WORKINGS

Pre-development work to access ore reserves is important for successful underground exploitation of salt. Several aspects are briefly discussed.

7.1.1 Basic considerations

In evaluating salt deposits one should consider factors which are specific for the area.[1] There are a certain number of criteria for the location of mine openings, as listed below:
– a sufficiently high quality salt must be found to be competitive with that produced by other suppliers in the regional area;
– the deposit must be found at a reasonable depth;
– the deposit must be found where other necessary facilities are available such as road, rail and transportation;
– salt mining rights must be obtainable over an extensive area and those rights must be available at a reasonable cost.

All of these criteria exist at Goderich salt mine, Ontario. The task reduces itself to determining the quality of salt and to outlining sufficient tonnage to ensure several decades of mining operation.

In addition mineral lease boundaries should be considered because they are important for mine development and planning. For example, the mining operation of Potash Corporation of Saskatchewan Mining Limited, Cory Division, is situated on a sub-surface mineral lease just west of the city of Saskatoon (Figure 7.1.1). This lease contains an area of some 36 000 hectares. In a lease of this magnitude and assuming two potash-bearing zones (A and B) continuous over the entire area, there could be some 5×10^9 tonnes of ore in place.

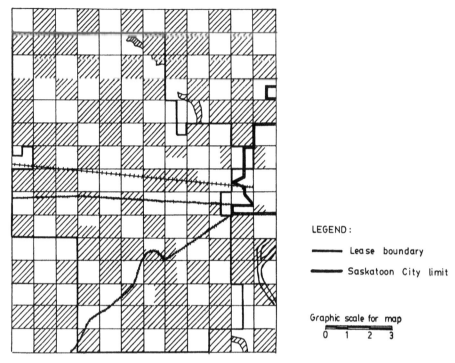

Figure 7.1.1. The boundary of sub-surface mineral lease (National Potash Company, Saskatchewan)

For an extraction ratio of 40% and assumed mining production of 3.2 million tonnes annually, the ore reserves could last between 400 and 500 years.

7.1.2 *Opening of salt deposits*

A basic consideration when designing the access to salt deposits is to avoid the difficulties which could be encountered caused by water intrusion either during opening and development or during mining. There are two methods of mine access, to a great deal depending on the depth, the morphological and geological conditions of the deposit as well as the configuration of the ground surface.

1. *Access by tunnels* could be used to decrease the risk of water penetration in the mine, if a water-bearing horizon is above it. However, if the salt deposits are located in flat or rolling hill areas, then tunnels cannot give access to the salt deposits. In this case, winzes are required to develop the ore below the tunnel. This type of access to mineable ore reserves is applied if the depth of the winze is up to 150 m. Dez salt mine in Romania where only one horizon of mining is applied by trapeze rooms is developed by a tunnel excavated 400 m through marl and 430 m through salt. From the tunnel a lower exploitation horizon is opened by a winze of 70 m depth (Figure 7.1.2). If the geological and geomorphological conditions are right, access to ore reserves by tunnel might have certain advantages, particularly with respect to increasing rate of production transportation, capital cost, duration of mine openings, etc. It should be mentioned that only a few salt mines are opened

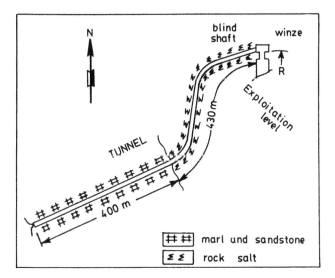

Figure 7.1.2. Access to salt reserves by tunnel and blind shaft (Dez salt mine)

Figure 7.1.3. Access to ore body by tunnel (Tusanj salt mine, Tuzla)

by a combination of tunnel and winze.[2] The opening of Tusanj salt mine, Tuzla, with tunnel access to the shallow section of the ore body is shown in Figure 7.1.3.

2. *Access by shafts* is the most common way to open salt deposits (Figure 7.1.4). Two shafts could be located either outside the ore body or within the ore body. The Carpathian province where deposits contracted by folding is an example of the shaft location outside

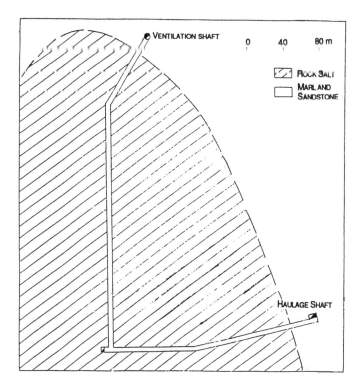

Figure 7.1.4. Access to salt reserves by pair of shafts (Projd salt mine, Romania)

the salt body. Whereas the Canadian prairies where deposits are undeformed and have infinite lateral extension the shaft location is inside the salt strata. One shaft is used as a production shaft, and the other shaft is used as a service shaft. The shafts could be closely spaced to provide parallel ventilation or placed some distance apart to employ diagonal ventilation. The decision to open the ore body by shafts is greatly influenced by hydrogeological conditions. For example, if water-bearing sediments are located over and under the salt deposits and circulating underground waters intersect the salt deposits, then access to the salt by shafts is inadvisable. All forms of dry underground mining are suspect and solution mining should be resorted to.

7.1.3 *Pilot hole drilling*

The required information for shaft-sinking is obtained from the pilot hole, drilled at the centre of each planned shaft and to the full depth of each shaft.[3]

The pilot holes for the production shafts were drilled and cored through prairie evaporites (Saskatchewan). In Table 7.1.1 a description of the strata over the proposed shaft is given from the surface to the bottom at 1032 m. The core of the pilot holes was carefully logged and photographed using the core colour photo method. This datum was used during shaft-sinking because it reveals zone transitions, fractures, shale partings, porosity, and other critical information.[3]

In prairie evaporites the porosity and permeability of the core samples obtained from water-bearing formations were determined. Special laboratory tests were conducted to obtain information about the groutability of the water-bearing zone. The flow potentials of

Table 7.1.1. Strata description at proposed shaft sites

Depth from collar (m)	Geology of strata
0-88	Glacial till sand, clay & boulders with porous water-bearing sands & gravels
88-392	Upper and Lower Cretaceous shale containing bentonitic zones that require close ground support
392-437	Blairmore formation unconsolidated sands, clays and shales. Sands are water-bearing. Formation pressure: 4.82 MPa
437-514	Lodgepole formation cherty limestone grading into argillaceous limestone at base
514-522	Bakken formation weak shale grading to siltstone; entire zone requires close support
522-566	3 Forks formation interbedded shales with limestone
566-589	Nisku formation competent fractured limestone with porous water-bearing zone. Formation pressure: 5.81 MPa
589-755	Duperow formation fractured limestone and dolomites; some weak shale beds
755-862	Souris River formation competant limestone, argillite and dolomites with water bearing zone. Formation pressure: 7.58 MPa
862-872	First red bed argillite, shales & anhydrites
872-916	Dawson Bay formation competant dolomites & limestones
916-925	Second red bed dense, massive argillite
925-1032	Prairie evaporite formation

the water-bearing zone were obtained by drill stem test evaluation. A carefully selected logging program covered all the formations pertinent to shaft-sinking.[3]

The pilot hole of the shaft for 'Sylvite of Canadian Division' (Hudson Bay Mining and Smelting Co. Ltd.) was drilled below the mining level.[3] The pilot hole was drilled below the planned shaft bottom to a pre-determined depth to establish the presence of an undisturbed protective salt zone between the shaft bottom and the water-bearing Winnipegosis formation. The pilot holes below the Blairmore formation were abandoned and sealed by cement plugs. Every water-bearing formation was sealed by a mechanical plug as an extra precaution. During the freezing operation the pilot hole for each shaft was used as a relief hole to prevent pressure build up which could cause ground fracturing. The Blairmore section of the hole was cased, but left uncemented by using the stage-cementing technique to insure close contact between the pressure relief hole and the formation during freezing. The casing was perforated at the Blairmore formation. The static water level of the Blairmore formation was established before the commencement of shaft-sinking, using the pilot hole.[3]

The shaft-sinking personnel and sinking crew should be provided with all data obtained from the pilot hole. On the basis of these data, adequate techniques can be developed for rock excavation, ground stabilization, shaft support etc. Correct and maximum utilization of pilot hole information can eliminate costly down-time which could be in some instances 150-200 days per year when a shaft unexpectedly strikes a water-bearing horizon.

7.1.4 *Location of the shafts*

The location of the shafts has a significant influence on the overall operating costs during the life of every mining venture. Basic characteristics of a potash mining operation are the very large tonnages to be transported, and the rapidly increasing distances from the mining face to the production shaft, resulting in steadily increasing operating costs. If properly chosen, maintenance costs for the haulage and ventilation entries are minimized due to satisfactory ideal mining conditions.

The general area for shaft-sinking was selected where the criterion cost, derived from the depletion, ore haulage, ventilation and mantrip costs, was at a minimum, indicating the optimum shaft location.[3]

As de Korampay discussed in his paper,[3] a successful shaft-sinking operation depends to a great extent on properly locating the shaft, taking into account the optimal capacity. Shaft-sinking for potash development in Saskatchewan is a challenging venture because shaft-sinking conditions in the province are extremely difficult.

The interpretation of the logs of the exploration drill holes provides the necessary information to choose a proper geological cross-section, which should be used for selecting the shaft site within the area of optimum shaft location. For example, the shaft should be located on the top of an anticline so that water flows out of the shaft zone, not toward it.

The hydrogeological characteristics of water-bearing strata have to be assessed accurately before the shaft excavation penetrates them. The importance of this requirement can be seen from practical examples. In the seventies in Saskatchewan eight conventional potash mines utilizing fifteen vertical circular shafts were operating. Six of the fifteen shafts were flooded by water intrusion from the Blairmore and Dawson Bay formations. Therefore, great care was taken to delineate an area where the glacial till and the Blairmore formation were adequate for shaft-sinking and where the flow capacities of the water-bearing carbonate formations in the Devonian were minimal.[3]

The ground surface of the chosen location for the shaft has to be drained in advance to the extent of the salt deposit boundary. The water drainage is usually done by a system of dewatering channels, which eliminate the adverse effects of surface water run off.

7.2 GROUND CONSOLIDATION

Excavation through water-bearing ground usually requires in advance ground stabilization. When shaft-sinking for access to salt deposits is necessary, ground stabilization by dewatering of the water-bearing strata, using well points to the base of the strata to be dewatered, is not usually feasible as the water-bearing zones are too deep. Also, the high porosity of the water-bearing zones would result in a prohibitively high pumping rate for effective dewatering. Under these circumstances ground stabilization could be done by two methods; either by grouting or by freezing.

7.2.1 *Principles of cementation*

Cement grout is essentially an unstable suspension of cement in water. Migration of water to the top of the grout when the suspension comes to rest, is known as 'bleeding'.

The velocity at which the cement particle settles is given by:

$$v = \frac{2gr^2(\rho - p)}{9\eta}$$

where:

v = settling velocity;
g = acceleration due to gravity;

r = radius of the particle;
ρ = density of the particle;
p = density of the fluid;
η = viscosity of the fluid.

From this relationship, it can be seen that the settling velocity of a cement particle varies directly as the square of its radius. Therefore, agglomeration of cement particles should be prevented to reduce bleeding. This can be accomplished by using a dispersant.[4]

The settling velocity also varies inversely as the viscosity of the fluid. Therefore, by increasing the viscosity, bleeding will again be reduced. This is accomplished by using a suspending agent.

There is a lower limit to the size of the opening that any cement grout will penetrate. Unless excessive pressure is used, cement grouts will bridge openings that are slightly less than three diameters of the cement particles (Figure 7.2.1). In water-bearing strata, many of the openings are smaller than 0.25 mm and cannot be sealed with cement at all. It was this limitation that led to the introduction of chemical grouts.

When sodium silicate is neutralized or partly neutralized, polymerization becomes much more extensive in three dimensions. Water is locked into the web and a gel is formed. If the material used to neutralize the sodium silicate reacts slowly, then the mixture can be pumped into a cavity before the gel forms. If small proportions of dilute calcium chloride solution are added to the mixture, the resultant gel will contain amounts of calcium silicate and will exhibit more strength and stability than pure silica gels.[4]

Pumping salt-free water in advance of silicate grouting is used to improve penetration. The effectiveness of this method is limited by the nature of the movement of water in small openings. The volume of this movement in round capillaries is governed by the relation:

$$Q = \frac{P\pi a^4}{8\eta l}$$

where:

Q = volume rate;
P = applied pressure;
a = radius of the capillary;
l = length of the capillary;
η = viscosity of the grout.

From the equation given above could be seen that volume of flow into a fine opening varies directly with the fourth power of its width.

The chemical grouts are: acrylamide, chrome lignin and silicate grouts.

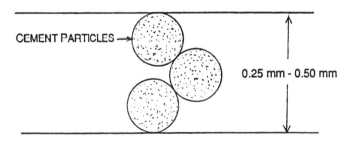

CEMENT PARTICLES →

0.25 mm - 0.50 mm

Figure 7.2.1. Bridging of opening slightly less in size than three diameters of particles

7.2.2 *Application of cementation*

Cementation for ground reinforcement[4, 5] can be done either prior to sinking, or during sinking, or during development work.

1. *Pre-cementation* can be used for certain strata conditions and it is carried out by grouting the ground that the shaft to be sunk through. The most successful application is for unconsolidated and water-bearing overburden and also where the upper rock formations are substantially leached. Unconsolidated ground requires a greater number of holes than does fissured rock. For example, pre-cementation has been successfully applied in the sinking of the Pugwash no. 2 shaft in Nova Scotia (Figure 7.2.2).

2. *Shaft bottom grouting* is carried out by drilling injection and cementation holes around the shaft (Figure 7.2.3). All holes are drilled radial and tangential to the shaft and inclined to ensure an adequate treatment zone around the shaft. Enough holes are placed to ensure that all fissures are grouted. After receiving a small dosage of lubricant chemical, the holes are injected with cement in order to fill up any open fissure. This avoids any excessive loss of chemical solution through open channels. If severe water conditions are encountered, a method of double injections are used. This process is primarily designed for chemical treatment.

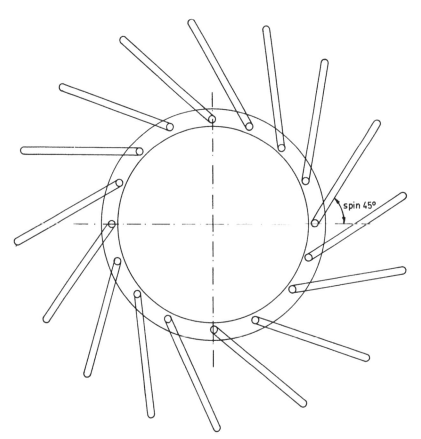

Figure 7.2.2. Arrangement of 50 m deep grouting holes for circular shaft

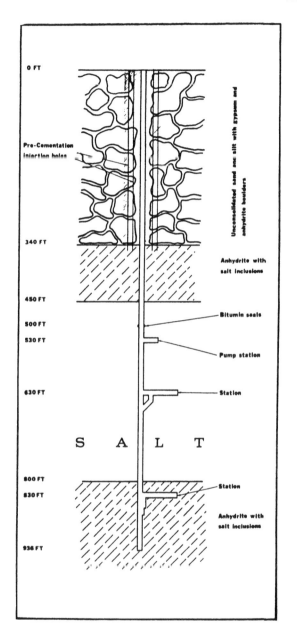

0 FT

Pre-Cementation
Injection holes

340 FT

450 FT

500 FT

530 FT

630 FT

800 FT

830 FT

936 FT

Uncemented sand ane silt with gypsum and
anhydrite boulders

Anhydrite with
salt inclusions

Bitumin seals

Pump station

Station

S   A   L   T

Station

Anhydrite with
salt inclusions

Figure 7.2.3. Diagrammatic section of
pre-cementation (no. 2 shaft, Pugwash, Nova Scotia)

3. *Drifting under grout cover* requires the drillers to probe ahead of the face to ascertain the existence of water-bearing fissures and to carry out injection. Drifting through water-bearing ground can be controlled by drilling two horizontal pilot holes inclined outwards at 6° from the centre line of the tunnel for a length of 75 to 100 m. If water is intersected, the holes are immediately injected if the water inflow is between 15 and 1000 l/minute. If water inflow is greater than 1000 l/minute, then the holes are plugged and the drift continues up to 10 m of inducted water feeder (the fissure).

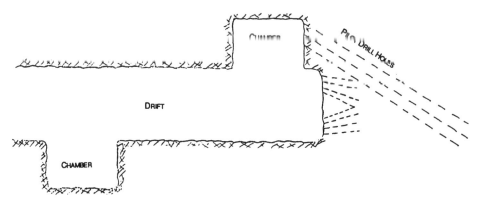

Figure 7.2.4. Horizontal pilot holes, drilled from chambers

At this point injection holes are drilled at an angle of 25° outwards from the axis of the drift. In some cases chambers are excavated on each side of the drift. Pilot holes are drilled from the chambers. The chambers are staggered every 40 m on alternate sides of the drift (Figure 7.2.4).

7.2.3 *Properties of ground affected freezing*

Before freezing can be attempted, it is essential to collect data of the ground characteristics and how they apply to freezing.

1. *On freezing, the physical and mechanical properties* of the ground are significantly changed. Freezing of ground or rocks causes a significant increase in cohesion. For example, the cohesion of the Blairmore formation at its original temperature (18°C) ranged between 0 and 205 KPa. However at -15°C, this increased to a range between 0.65 and 1.00 MPa. When freezing occurs, the water in the voids of the rocks freezes, interconnecting the rocks. This causes an increase in the strength of the rock. The unconfined compressive strength of the frozen rock depends upon the unconfined strength of the ice, the degree of saturation and on the angle of internal friction. Karl Terzaghi has derived the following formula for unconfined strength of frozen rock:

$$K = q(t) \tan (45 + \phi/2)$$

where:
K = unconfined strength of the frozen rock;
$q(t)$ = unconfined strength of ice as a function of temperature;
ϕ = angle of internal friction of the rock.

The effect that freezing has on ground can be seen in Table 7.2.1. Note how the strength increases with a decrease in temperature.

The compressive strength of the Blairmore sandstone rose from approximately 7 MPa at the original temperature to 17 MPa at –25°C.

2. *Thermal properties* needed for shaft design are:
– specific heat;
– thermoconductivity;
– diffusivity;

Table 7.2.1. Unconfined compressive strength of frozen formation

| Ground | Unconfined compressive strength at different temperatures in MPa | | | | | | |
	0 °C	–3 °C	–6 °C	–10 °C	–13 °C	–18 °C	–25 °C
Saturated sand	5.5	8.3	10.3	12.4	13.8	14.5	15.2
Clayed sand	1.7	4.8	6.9	7.9	8.6		
Sandy clay	1.8				6.2		
Clay	1.7	2.4	3.8	5.5			
Silt	1.2	1.7	2.8	3.4			

– freezing point.

Specific heat depends upon mineral composition. Thermoconductivity depends upon porosity, dry density, degree of saturation and temperature. At the freezing point, there is an abrupt change in the magnitude of the thermoconductivity. For ground in the frozen state, thermoconductivity is higher than for ground in an unfrozen state. Consequently, freezing of ground is relatively easier than thawing.

Diffusivity is the rate at which the temperature of a body adapts itself to changes in the temperature of the surrounding medium. It is higher in ice than in water. Therefore the average temperature of a water-bearing rock in the frozen state changes more rapidly than in an unfrozen state. The characteristics of this type of ground make it ideal for the freezing technique.[6]

7.2.4 Ground freezing

Two aspects of ground freezing are discussed; namely, the technology of the freezing process and the elements of the shaft freezing process.

1. *The ground freezing process* depends a great deal on the selection of an adequate freezing pipe. A poor selection of pipe might create problems with respect to heat transfer, brine flow, energy losses, costs, etc.

There are two basic ways of circulating the chilled brine in the freezing hole:
– the brine can pass down through the brine pipe and flow up the annulus between the freezing and the brine pipe, or
– the brine can be pumped down through the annulus and return up the brine pipe.
The first case is more in use.

The brine returns up the annular ring, extracting the heat from the surrounding strata (Figures 7.2.5 and 7.2.6). Twenty-four of these holes are usually drilled outside the perimeter of the excavation area. The circulating of a coolant through the system of freezing holes causes the building up of long pencils of frozen strata around each column. These increase in diameter until they begin to merge, thus 'closing the ice wall' (Figure 7.2.7). In order that the columns properly merge, it is imperative that the freezing holes be as vertical as possible. Holes must not wander into the excavation area and no two holes must be allowed to wander apart with depth. If this occurs it may jeopardize heat extraction which has a tolerance of 500 mm. Also, uniform spacing is equally important. From the knowledge of other freezing operations in Saskatchewan, it is known that a spacing of 1250 mm is optimal.[7]

The design of freezing operations should be divided into two basic stages:

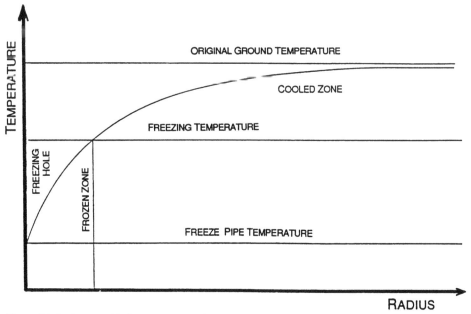

Figure 7.2.5. The area of influence around a freeze hole

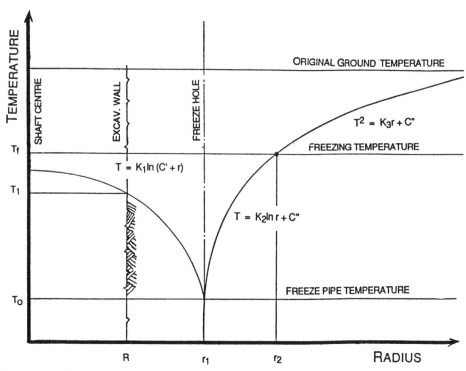

Figure 7.2.6. Curves showing temperature distribution in the ice wall

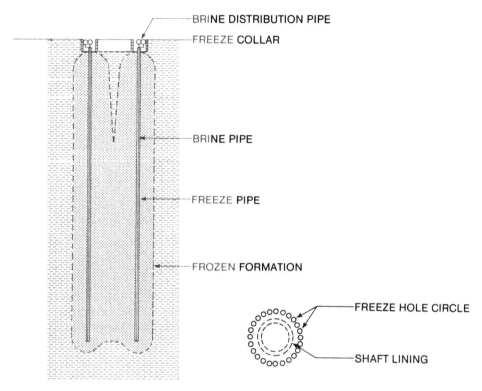

Figure 7.2.7. Formation of an ice wall around a shaft

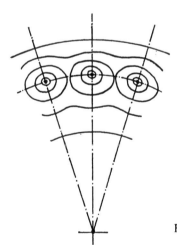

Figure 7.2.8. Temperature isolines around freezing holes

a) Freeze around single freezing holes until the single columns will connect with each other around the future opening (Figure 7.2.8).

b) Growth of the frozen ring to the inside and outside (Figure 7.2.9). The frozen ground inevitably extends into the shaft excavation area, thus eliminating sloughing of walls during excavation.

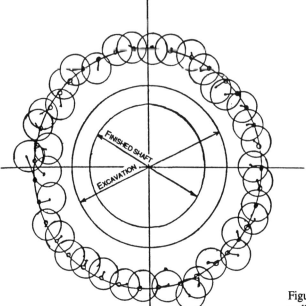

Figure 7.2.9. Typical plan of an ice wall showing complete closure

Using the 1250 mm spacing between the freezing holes, the closing time between cylinders (freeze holes) occurs in 27-30 days. However, about 90 days are required for the closure of the ice wall within the excavation radius.

Twin shafts should be developed because freezing technology is optimally utilized. The ideal way to accomplish this is to start with one, and utilize an adequate delay before starting the second. This accelerates the pre-freezing time. Since the first shaft will be completed first, all freezing units can concentrate on the remainder of the second shaft.

2. *Elements of frozen ground* related to shaft-sinking are briefly commented on the basis of Thyson Mining Construction of Canada Ltd. suggestions which could be used as principal guides as follows.[7, 8]

a) Circular shaft-sinking using the freezing technique provides the consolidation (freezing) of the weak and water-bearing sediments.

b) The frozen ground will take the form of a circular zone inside which the shaft is sunk and lined.

c) During the shaft-sinking process, within a frozen cylinder, the resultant ice wall functions as a temporary support. The support is given against the hydrostatic and ground pressures until the final lining is installed.

d) The sinking and lining operations can be simultaneous. Concrete or concrete blocks with steel or plastic sheeting, offer a water-tight shaft. Also, sinking and cast iron tubing operations could be applied. The cast iron tubing ring may be installed immediately following excavation. However, it is not fool-proof against water inflow.

The key factor in ground stability is to design the ice wall, which has to be continuous (no gaps), sufficiently thick and of adequate strength, so that it will support the outside pressure which will be exerted on it. Improper determination of the ice wall thickness could prove disastrous once the excavation of the shaft or drift has started. Mohr's concept for the stress

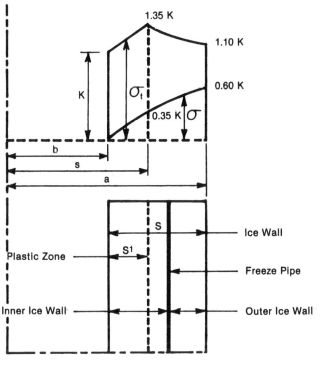

S = ICE WALL THICKNESS.

K = ALLOWABLE COMPRESSIVE STRENGTH OF
FROZEN GROUND.

σ_t = TANGENTIAL STRESS.

σ_r = RADIAL STRESS.

a = RADIUS TO OUTSIDE OF ICE WALL.

b = RADIUS TO INSIDE OF ICE WALL (EXCAVATION
LIMIT - DA)

s = RADIUS OF PLASTIC LIMIT.

S¹ = PLASTIC ZONE.

Figure 7.2.10. Stress distribution in the plastic and elastic zones of an ice wall

distribution in the plastic and elastic zones of the ice wall is given in Figure 7.2.10.[9]

An equation for determining the ice wall thickness (Domke's equation) is given below:

$$S = 0.5 \times D_A \, 0.29 \times \frac{P}{K} + 2.3 \frac{P^2}{K}$$

where:

  S = ice wall thickness;
  D_A = excavation diameter;
  P = pressure on ice wall;
  K = compressive strength of frozen ice wall (allowable).

The depth of the freezing holes is determined by the depth of the deepest water-bearing horizon, or, in some cases, an impervious rock layer underlying a water-bearing stratum. It is imperative that the freezing holes be drilled as close to vertical as possible. If the holes are not vertical, the distance between the bottom of the pipes may become too great to allow the proper formation of an ice wall. Figure 7.2.11 illustrates the area of influence and the formation of the ice wall around the freezing holes.

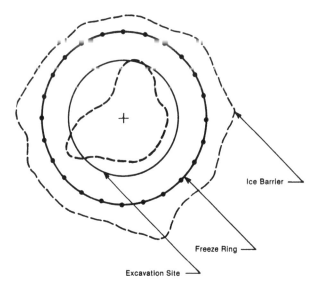

Ice Barrier

Freeze Ring

Excavation Site

Figure 7.2.11. Area of influence
around freezing holes

7.2.5 *Advantages and disadvantages of consolidation methods*

The advantages and disadvantages of the ground consolidation by cementation and by freezing are briefly discussed, but mostly as to the aspect of their disadvantages.

1. *Ground cementation* by grout mixture of variable viscosity which penetrates rocks of different porosity and structure is a fast and cheap method. However, many grout mixtures are attacked by the sulphates contained in the water-bearing formations, resulting in a loss of rock strength and volume changes which tend to induce hairline cracks in the grout curtain.[5] The grout must be placed at high pressures for effective sealing of the ground. This can result in migration of the grout through vertical fractures in the ground. This results in instability of the ground due to widening of the fractures and possible failure of the previously lined concrete portions of the shaft due to pressure above the design pressure.[5]

2. *Ground freezing* is successful in ground where the cementation method failed. The advantages of freezing is that the crystallization of the pore water increases the cohesive strength and unconfined compressive strength of weak or unconsolidated strata. Even poorly competent ground can be self-supporting during the time required for installation of the lining after shaft excavation has been performed. The main disadvantages of the freezing method are high costs and the close supervision required of the freezing hole drilling operation to obtain correct hole alignment and the desired ice-wall thickness at the appropriate shaft section. Mobilization for ground freezing must start in advance of the shaft excavation to insure full freezing of the section over depth all the time. Finally, concrete placed adjacent to the ice-wall will require a longer curing period to attain full strength, due to the reduced temperature.

7.3 SHAFT SINKING AND LINING

Both the production shaft and ventilation shaft have to be designed accordingly to ground conditions, position of water-bearing horizons and excavation depth.

7.3.1 *Sinking of the shafts*

The simultaneous sinking-and-lining technique is usually selected using a multi-deck Galloway stage. Usually a contractor is commissioned to sink shafts since he has the expertise for shafting and drifting operations. Sinking of the production shaft usually starts before the service shaft.

Drilling and blasting techniques employed for shaft-sinking are well documented. However, conventional blasting cannot be employed in frozen water-bearing horizons, because this could cause rupturing of the freezing pipes located outside the perimeter of the shaft. Even drilling and blasting technique using reduced charges is undesirable since it also tends to fracture the ice wall, thereby reducing its strength. For the frozen shaft section the excavation is usually executed by heavy paving breakers and a Cryderman shaft mucker. Conventional drill and blasting techniques continue after the water-bearing formation is sealed off. In some instances pneumatic hammers are used for ground excavation. This is a very slow technique because breakage of the ground is carried out manually by rock chipping.[3]

7.3.2 *Shaft-lining methods*

Shaft support is achieved using a required thickness of concrete lining in areas where little or no formation water is present. The only exception is in glacial till strata where a 'picotage' ring is used for sealing when tubbing is employed, or a chemical sealant is used when employing a composite concrete/steel plate liner. The picotage is a tubbing section with a heavy duty flange. Kiln dried wooden wedges and pegs are driven between the

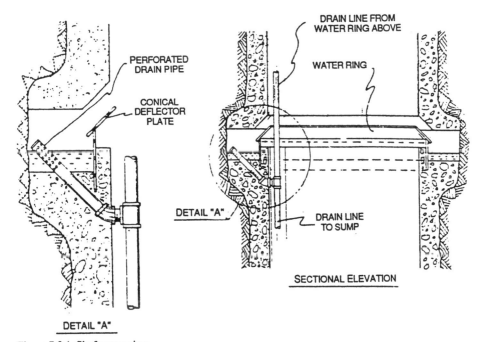

Figure 7.3.1. Shaft water ring

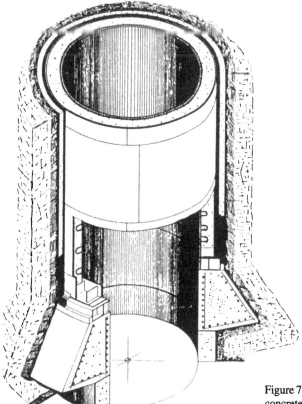

Figure 7.3.2. Double steel lining, resting on concrete foundations in the rock

picotage ring and the excavation perimeter until further placement is impossible. On saturation, the dried wood expands and forms an effective water seal. At intervals of 10 m below each section of water-tight lining, a water ring is installed to trap any leakage through the lining (Figure 7.3.1). Both the steel plate liner and the tubing sections require cathodic protection from corrosion, due to the presence of moisture and salt in the atmosphere. For this purpose, sacrificial zinc anodes are used. Concrete and steel plate liners are used to resist relatively low water pressures. The steel liner is used to preserve the water-tightness of the lining since shrinkage crack formation of the concrete during curing is anticipated (Figure 7.3.2).

In areas where formation water pressures are high, cast iron tubing is used for water-tightness and the necessary structural support. The tubing sections are fitted with lead gaskets to prevent any leakage.

The tubing and steel plate liner sections are continued 4.6 m above and below the water-bearing formations.

7.3.3 *Lining pressure*

Ground stress prior to shaft excavation depends a great deal on the presence of water in the strata. For example, one particular mine site in the Canadian prairies, for the various

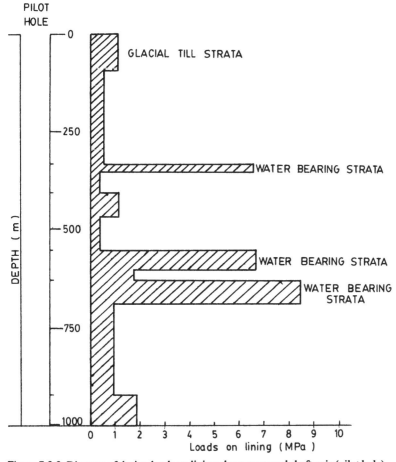

Figure 7.3.3. Diagram of design loads on lining along proposed shaft axis (pilot hole)

horizons in the geological profile design loads on lining, varies significantly due to the magnitude of water pressure (Figure 7.3.3). The formation water pressures were obtained from pilot hole measurements except in the case of glacial till strata, where pressures were determined using equations to calculate the stress on lining. The equation often used to calculate the stress on a lining is as follows:[7, 8]

$$P_o = P_w + P_g$$

$$P_w = \gamma_w \times \frac{h}{10}$$

$$P_g = \lambda_g \gamma_{gw} \times \frac{h}{10} + \left(\gamma_g \times \frac{H-h}{10} \right)$$

$$g_w = \gamma_g - (1 - n_p)\gamma_w$$

$$g = \frac{2(1 - \sin \phi)}{1 + \tan^2(45 + \frac{\phi}{2})}$$

where:

P_o = uniform horizontal stress;
P_w = hydrostatic pressure;
P_g = uniform horizontal ground pressure;
γ_{gw} = submerged unit weight of the solid formation mass;
λ_g = ground pressure index;
H = depth;
h = depth of water;
γ_w = unit weight of water;
ϕ = angle of internal friction;
γ_g = dry unit weight of the solid formation mass;
n_p = pore volume of the formation.

Constants are as follows (empyrical):

$\gamma_g = 1.85$
$\gamma_w = 1.00$
$n_p = 0.30$
$\lambda_g = 0.26$ (for $\phi = 30°$)

While excavation is being carried out, the shaft must be lined with a material which is capable of withstanding the pressures exerted on it by the surrounding formation, including the high pressures of hydrostatic head. Many different types of lining have been used in shafts of this type, the most common of which are a combination of reinforced concrete and cast tubing or steel lining. Regardless of the type of lining used, it is imperative that this lining be designed to withstand the pressure exerted on it.[7] Methods of installing the lining may vary depending upon the ground conditions encountered during excavation. After the support is installed, the ground is allowed to thaw.[8]

7.3.4 *Water inflow*

Water inflow in shafts has been summarized as follows by de Korampay, for the shaft of the Sylvite of Canada Division:[3] 'The accurate prediction of water inflow during shaft-sinking is essential in selecting the safest and surest method of sinking and lining'.

Reliable information for prediction water inflow in each shaft was obtained by conducting 26 drill stem tests in the production shaft pilot hole and 14 tests in the service shaft pilot hole. Log interpretation was used to determine the water inflow distribution within the drill stem tests interval.

The drill stem test evaluation indicated the presence of an extra water-bearing zone in the Souris River formation with respect to the service shaft. This zone was closed off by tubbing lining. The flow potential of the frozen Blairmore formation was 9100 lpm according to the DST interpretation. During sinking through the unfrozen portion of the ground, probe holes were drilled ahead of the bottom of the excavation. The water inflow from the formation was measured by AMC Construction Ltd, using the probe holes. The measured flow rates were corrected according to the diameter and drilling pattern of the probe holes. The results are as follows:

predicted flow rate by DST:
production shaft – 201.0 *l*pm;
service shaft – 277.5 *l*pm.
Actual measured flow rate:
production shaft – 184.5 *l*pm;
service shaft – 239.5 *l*pm.
The water inflows into the shafts were reduced by installing tube lining and at the end of shaft-sinking the reductions were:
production shaft – from 184.5 *l*pm to 14.5 *l*pm;
service shaft – from 239.5 *l*pm to 15.5 *l*pm.
Corrosion control of the lining is very important, and for this reason a cathodic protection system was installed in both shafts to control corrosion of the tube lining. At this time it was the first installation in Saskatchewan of such a corrosion production system.[3]

7.4 MINE DEVELOPMENT WORKINGS

In the development of every mine, including salt mines, a grid of main roadways is set out to delineate the extraction units. The extraction unit is developed by secondary roadways laid out to serve the working places. The development workings serve to connect working places or stopes to the outlet on the surface, which may be either a shaft or a tunnel. In salt mining, all development workings tend to be in the salt strata. In regard to morphological and structural characteristics of salt deposits three basic methods of ore body development are in effect. Each method of development directly relates to systems of mining.

7.4.1 *Panel development*

The panel development is considered for deposits of medium thickness (1-4 m) which could be mined out by one extraction lift. Also panel development is used only for the salt deposits which are flat or dipping up to 15°. The limiting angle of inclination of deposits occurs because continuous mining machines are effective only up to this angle. For example, the prairie evaporites of which the potash deposits are a member, are 1000 m below surface and could be considered as a blanket of approximately 21.5 m thick, extending for miles in every direction. The logical method of mine development is by panel development.

An example of the layout of panel development is given in Figure 7.4.1. The potash deposit is opened by two shafts in the central part of the deposits. One shaft is for production and the other shaft is for services. Triple main haulages are driven from the shafts to open the salt deposit laterally. A main belt conveyor is installed in one of the haulages. Potash is discharged on to the conveyor from the working panels. Quadruple roadways are driven perpendicular to the main haulages to delineate the extraction panels. One haulage is used for housing the panel belt conveyor. The panels are mined by one of the mining systems used for extracting flat and moderately thick beds.[1]

In the potash mines of the Canadian prairies a continuous miner extensible unit is used to drive haulage ways and roadways of 1300-1500 m long up to 20 m wide, and 2.5 m high, using a series of three horizontal passes or slices. Most mines use a Marietta continuous miner which has a cutting height of up to 2.5 m and a cutting width up to 2.5 m. The penetration rate is 350 mm per minute for a face area of 50m m². The continuous miner

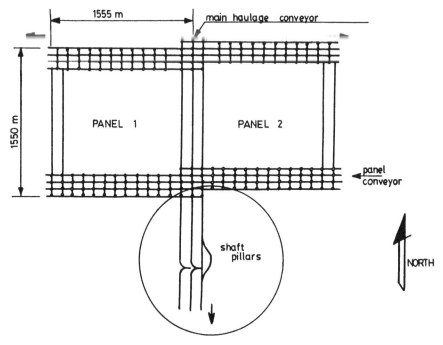

Figure 7.4.1. Panel development of potash deposits (Canadian prairie)

produces around 1000 tonnes per hour where production is limited by insufficient height of the openings. The conveyance system usually used in this operation is the extensible conveyor system of the 1050 mm width. The trailing tail of the conveyor is loaded with muck directly from the Marietta miner. The conveyance system consists of a belt bender which allows for varying conveyor bend angles. It is applicable in a two or three pass entry cycle which is used commonly in a potash mining cycle operation.[1]

The same technology is used in room excavation in long pillar mining, a common system of mining in the Canadian prairies.

7.4.2 *Block development*

Block development is mainly used for dome salt deposits where large room mining is applicable. The mineable blocks (rooms) have widths up to 35 m and heights up to 50 m. Block development is started at the top of a mineable block (trapeze room mining) or on the bottom of a mineable block (paralleloidal room mining).[10]

An example of the layout for development of mineable blocks (large rooms) is schematically given for the Klodawa salt mine in Poland (Figure 7.4.2). The exploitation block is opened from the shafts by two roadways, one connected to the haulage shaft and the other to the ventilation shaft. From these main roadways a circular roadway is driven around the ore body. The part of the roadways closest to the ore body is used for haulage and the part of the roadways furthest from the ore body is used for ventilation. Roadways throughout the ore body are used to block out extraction units. These roadways connect perpendicularly to the haulage road and ventilation road.[10]

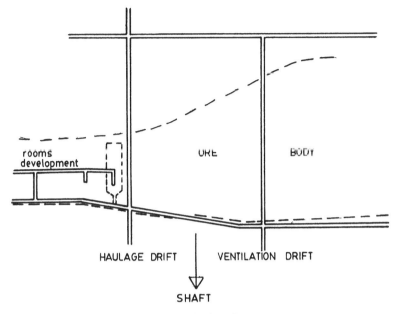

Figure 7.4.2. Block development of a halite deposit

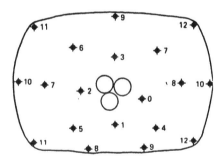

Figure 7.4.3. Drilling pattern of recovery with three bores of large diameter for ground openings and nineteen holes for ground blasting

The main and subsidiary development workings are excavated both by cyclic drilling and blasting and by continuous mining machines. In some salt mines in Germany the roadways excavation is advanced by drilling three holes of a large diameter (280 mm) and 7.5 m in length, and after that drilling of the blasting holes is commenced. In this case the centrally located three holes open the cut on the face and help ground breakage. An example of a drilling pattern of the roadway is given in Figure 7.4.3.

7.4.3 *Level development*

Pitching and steep pitching ore bodies similar to vein-type deposits are developed by levels. This is usually the case with folded potash beds located on the boundary or inner part of large halite bodies. Developing pitching potash bodies by levels is similar in many respects to hard rock mining.

In regard to the inner position of potash veins to the salt body boundary two concepts of development are implemented. The first concept considers the potash deposits in Romania which are located on the outer part of intensively folded salt bodies. Due to the larger lateral extension of ore beds, and their relatively shallow depth, their opening is by shafts which pass through the potash vein. The level development is from the shaft intersection with an ore body along its strike, as further discussed for the Tazlau salt mine in Romania. The potash deposit is opened by two shafts, at a distance of 25 m from each other. One is used as a production shaft and the other is a service shaft. The shafts are sunk to different depths (e.g. to the depth of each level). Horizontal drifts are driven from them along the strike of the potash beds.[2] The difference in elevation of the horizontal drifts is between 30 and 40 m. The two levels are connected by raises every 100 m by which extraction panels are delineated (Figure 7.4.4). The depth of the shafts is in the range of 100-150 m. It should be noted that the shafts are located at the top of an anticline structure, so that water flows away from the shaft.[2]

The second concept considers the development of potash veins located inside the domal deposits, as is the case in the Zechstein salt domes of Central and Eastern Europe. Kali & Salz A.G. is the only potash company in Germany with eight mines which are opened by shafts and developed by levels.[11] The layout of shafts and level drifts for one mine of this company is shown in Figure 7.4.5. As can be seen from the cross-section of the salt dome, the potash bodies are distorted a great deal by tectonic forces. The former flat deposits are folded almost vertically and consist of thick and thin mineable sectors, both laterally and vertically. Willi and Potthoff describe the surrounding sedimentary strata as mostly water-bearing. The cap rock covers the top of the salt dome which contains salt brines.[11] Consequently 150 m thick, safety pillars are left in place against the cap and surrounding rock. In flat potash deposits generally all phases of mining are carried out in the ore body. In salt domes, however, the classical methods of mining a pitched and irregular ore body have to be applied. At least two vertical shafts provide entries into the salt body, located as close as possible to the potash beds to reduce the amount of work through waste. From the shafts

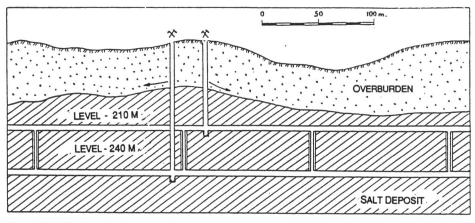

Figure 7.4.4. Horizon development of halite and potash deposits (Tazlau, Romania)

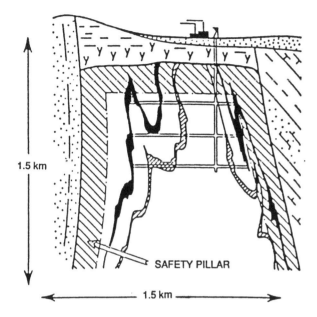

Figure 7.4.5. Typical access and development of the potash vein caused by the salt dome

the salt body is subdivided vertically by main levels at distances of 200 to 250 m. From them horizontal drifts are driven along the strike of the potash vein. Horizontal drifts are connected by raises every 100 to 300 m which block out the level panels.

Drill hole exploration is carried out at the lowest level and close to the limits of the salt body to locate mineable sections of the potash beds and to detect dangerous gases and brines. This is essential for the economic and safe development of the main levels and associated mine workings. The depth of holes reaches 1500 m.[11]

Excavation of the level drifts can be cyclic, either by drilling and blasting or by continuous mining machines. The width of the level drifts depends on the thickness of the ore body.[11]

7.4.4 *General consideration of development*

Comprehensive mine development should be carried out utilizing geological, geophysical and drilling exploration data as well as experience of operating conditions of existing mines in production, in order to mine out salt reserves safely and efficiently. Mine development must take into consideration the ground conditions which should be defined by closure measurements, stress and pressure monitoring and subsidence measurements to determine the method and concept of mining development and the optimum panel width or block height to establish a maximum safe extraction factor.

The development of potash deposits depends on the chosen mining method and the mining concept. For example, to achieve early production an advanced type of mining system should be applied in which only a limited amount of mainline development is required. Several panels would be pre-developed and completed working from the shaft pillar outwards. As Moore and Gauthier indicated, the problem with an advanced mining concept is that each time a panel is driven it reloads the main entry system and accelerates the closure rate and deterioration of ground conditions. This results in major problems in both haulage and service entries and their control is costly and sometimes difficult. The instabilities are related also the the multiple clay bands in the roof, in conjunction with high rock stresses created by great depth of excavation (1000 m). The unstable back conditions from time to time exhibit a large rock fall.[1]

To overcome the problem of adverse effects of advanced mining, most mines in Saskatchewan introduced a retreat mining concept. The mainline entry system is developed to a predetermined maximum travel distance or to the property boundary and then to the shaft pillar. This allows the reloaded main entry system to be abandoned as the conveyor is retreated and minimizes the amount of costly ground control, as is the case with advanced mining systems. Retreat mining is capital intensive, but when an adjacent mine block is developed, costs decrease. It is felt that retreat mining development is essential to enable an operation to recover the maximum volume of ore reserves around each production shaft. Driving the development system in advance of production mining also allows an operation to explore conditions and ore grade in advance of production mining of panels and assists in the long range planning process. In the mining planning of development block 2 and block 6 will be extended, and for other blocks development is in the inception stage. Medium range planning is based on a five-year program, where the mining units are laid out on a year by year basis. Short-term planning is based on a one-year program which includes exploitation parameters such as location and tonnage from each machine on a month by month basis. Also included is the planning of ventilation, capital projects, rock mechanics and major construction and maintenance work.[1]

REFERENCES

1. Moore, G.W. & R. Gauthier 1983. Long-range planning is Saskatchewan potash mine. In *Potash technology*, pp. 91-98. Toronto/New York: Pergamon Press.
2. Jeremic, M.L. 1964. Possibility of utilization of Romanian experience for exploration and development of salt deposits in former Yugoslavia. *Technol. Acta* (Tuzla Univ.), 3-4: 15-27 (in Yugoslav).
3. de Korampay, V. 1971. Shaft sinking. *Western Miner* (Vancouver, B.C.), 44 (11): 3-7.
4. Annett, S.R. 1969. The chemical and physical aspects of grouting potash mine shafts. *CIM Bulletin*, July 1969: 715-721.
5. York, L.A. 1964. Grouting the prairie sediments. *CIM Bulletin*, January 1964: 63-67.
6. Ostrowski, W.J.S. 1967. Design aspects of ground consolidation by the freezing method for shaft sinking in Saskatchewan. *CIM Bulletin*, October 1967: 1145-1153.
7. Roesner, E.K. & S.A. Poppen 1978. Shaft sinking and tunnelling in the oil sands of Alberta. In *Proc. of AOSTRA Conf. on U/G Exc.*, May 19, 1978, Edmonton, Canada.
8. Jeremic, M.L. 1987. *Ground mechanics in hard rock mining*, pp. 443-452, 503-524. Rotterdam: Balkema.
9. Domke, O. 1915. *Ice wall stress during shaft sinking using the freezing process* (in German). Thyson Mining Corporation, Germany.
10. Jeremic, M.L. 1965. Methods of development and exploitation of salt deposits in Poland. *Min. & Metall. Bulletin* (Belgrade), 32 (2): 264-274 (in Yugoslav).
11. Willi, A.H. & A.H. Potthoff 1983. Potash mining in steep deposits (salt domes). In *Potash '83. Potash technology*, pp. 79-84. Toronto/New York: Pergamon Press.

CHAPTER 8

Risk of water inflow

The presence of water in evaporite-bearing strata may cause adverse effects in salt mines due to the high solubility of salt. For this reason salt mining is singled out for the hazard of water inflow into the mines since interaction between salt and water cannot be avoided due to the easy dissolution of the salts. The groundwater inflow may be to such a degree that dry mining becomes impossible and only solution mining could be implemented. The hazard due to water penetration into salt mines is discussed in four separate topics

8.1 GROUND WATER EFFECTS

The ground water regime affects the water inflow in salt mines. Farther consideration is given to the origin of the groundwater and its relation to penetration to the salt deposits and mines as discussed below.[1]

8.1.1 *Meteoric water*

Meteoric water originates from the atmosphere. It is the product of run-off water at the ground surface due to rain and snowfall. The water from the surface percolates into the ground and supplies the aquifers with fresh water. 'Aquifer' is the term for a bed which yields appreciable quantities of water.

The meteoric water is present in two geological structures: surficial deposits and bedrock strata. The water horizons can be found in surficial deposits which are Quaternary in age. They are represented by unconsolidated and semi-consolidated sand and gravel. Meteoric water percolates from surface to aquifers below through permeability channels (Figure 8.1.1).[2] The surficial deposits are either unsaturated due to water outflow or are saturated in closed sand bodies. In the latter case, it could result in a water inflow into the shaft during sinking, as has happened in some cases in the Saskatchewan potash mines.

The water-bearing horizons in the underlying strata of bedrock consist of aquifers with fresh water. The aquifers might be intersected by fractures to ground surface to emerge as springs or seepages. They also might be intersected by a fractured zone and come in contact with connate water. The aquifers formed by saturated sand at high hydrostatic pressure (in mining circles called 'quick sand') are a primary consideration in shaft-sinking technology.

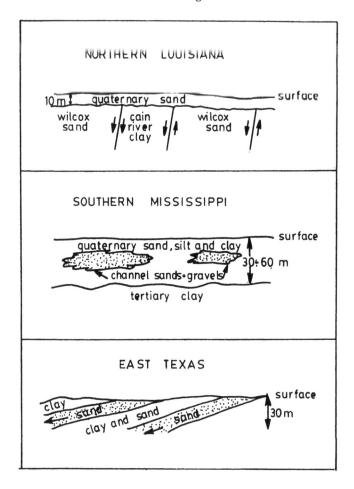

Figure 8.1.1. Quaternary geology and structures which facilitate water collection and flow in direction of gravity

8.1.2 *Connate water*

Connate water refers to water entrapped in the interstices of a sedimentary stratum at the time of its deposition. This water may be derived from either ocean water or land water.[3] The connate water represents a hydrostatic saline water horizon, the pressure of which depends on its depth. If these waters come in physical contact with the salt deposits and if salt solution is in effect, they will become brines. In the case of their intersection with the ground surface they emerge as saline springs and wells which can be used for hydrochemical prospecting and investigations (Figure 8.1.2). Saline aquifers might effect water inflow in the mines either through natural fracture zones or through man-made fractures due to mining excavations.[3] Connate water can be identified by its higher content of $MgCl_2$.

Water of connate origin can play an important role in the hydrology of overburden strata, as is known from present salt mining of domes in the USA. Connate water aquifers can form an artesian system. The water confined with impervious beds or aquicludes, which are inclined so that the catchment rises to the surface, forms free-flowing wells of fresh or saline water correspondingly.[4]

Figure 8.1.2. Well of connate water (Sibosnica basin, former Yugoslavia)

8.1.3 *Trapped water in salt bodies*

Three distinctive types of trapped water in salt bodies can be identified as follows.[5]

1. *Fluid inclusions* are exhibited by brine bubbles or vapour bubbles and they are formed both during sedimentation of salt deposits and during metamorphic processes of salt deposits. The bubbles are locked in salt minerals and they remain even after rock salt breakage. Fluid inclusions do not have any influence on water inflow into salt mines.[6]

2. *Primary trapped water* fills pores and voids of some pockets in the salt bodies. This is a fossil water which represents a remnant of marine water locked in during evaporation processes and formation of salt deposits. Trapped water is associated with gas pockets. The state of fluids is characterized by the co-existence of a liquid phase with an infinitesimal quantity of gas phase in equilibrium. At the moment when this equilibrium is not in existence, there exists a risk of gas outbursts. This phenomenon is of much greater danger to mine stability than possible brine seeps in the mine. From the case histories of minor brine seeps in salt domes, all tended to stop with time, indicating that fossil water is eventually drained from isolated brine pockets in the salt. The primary trapped water in the salt bodies can be identified by its Cl/Br coefficient.

3. *Secondary trapped water* fills isolated caverns within a salt body. This water usually originates during post-burial metamorphic processes. For example, large vertical structural features formed during the growth of domes can include zones with sediments, gas and water. This volume of secondary trapped water varies from very small, such as inclusions, to very large, such as caverns filled with water. In the latter instance the risk of water inflow into the mine can be appreciable. If the caverns are filled with water and gas, then the water inflow into the mine could be followed by a sudden gas discharge and explosion. The release of water and gas from cavities is facilitated by the propagation of fractures and is due to mining operations or the weakening of an existing fracture zone. Secondary trapped water can be identified by a high content of $NaCl$ or $CaCl_2$. From isotopic chemistry analyses of brines it has been established that they originated from formational waters trapped in the salt during intrusions.[7] The pattern of caverns in the salt body is shown for the Klodawa salt dome, Poland (Figure 8.1.3).[1]

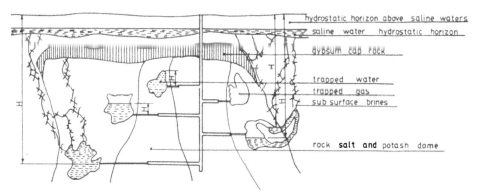

Figure 8.1.3. Categorization of water occurrences at Klodawa salt mine (Poland)

Figure 8.1.4. Salt crustification
on the walls and timber supports
of drift deposited from brines
leaks (Bochnia salt mine, Poland)

8.1.4 *Sub-surface brines*

Sub-surface brines can be derived from either types of water origin which has come in contact with the salt deposits and has undergone the process of salt solution. The sub-surface brines might be closed semi-circulating or circulating waters or percolating waters.[4] The mechanism of migration of the sub-surface brines is through permeability channels by bubble migration or by bubble transport through salt diapirism.[6] The chemical composition of the brines can be analyzed by the presence of major salt elements, trace elements, and isotopic composition.[8] The brine can identify the location of water leaks in underground mines; it is often recognizable by the distinct salt crustification on the walls of openings (Figure 8.1.4).

The brines circulating through permeability channels which are in physical contact with salt bodies, present a greater risk for water penetration into the mines. Salt mine flooding in the majority of cases is related to circulating brines. The solution of salt beds by circulating brines is progressive and can be appreciable, resulting in the formation of a great number of small or large caverns, usually filled with water the greatest risk of brine inflow and mine flooding is in the direction of the progressive solution of salt, as well as the area with caverns, since it offers no protection from water inflow into the mine.[5]

The circulating brines in some salt basins necessitate solution mining. Dry mining is prohibited in their vicinity, protected by barrier pillars. In some instances the adverse effects of circulating brines are such that dry mining must be converted to solution mining. The circulating brines in exploitation always make up by meteoric water the volume of pumped brines to the surface. This phenomenon indicates that circulating brines are in simultaneous contact with the salt deposits and with meteoric water aquifers.

8.2 NATURAL FACTORS

Natural factors of the water penetration in the mine have to be determined before mine design and layout takes place. These natural factors have been analyzed under four categories.

8.2.1 *Geological characteristics*

The geological characteristics with regard to water flow and its inflow into the mine are considered below.

1. *Erosion phenomena* are exhibited by the formation of narrow and steep-sided valleys with flowing streams, wide valleys with more gentle slopes, or very wide valleys with flat slopes and slow water flow (rivers). In the case of deep erosion, the overburden strata are shallow or completely eroded so that evaporite-bearing strata outcrop or are close to the surface and are covered by Quaternary sediments. The erosion process usually results in fissured strata below ground surface. Under these circumstances the evaporite might come in direct contact with surface waters as run-off water, creeks, rivers or sometimes lakes. Louisiana salt domes (USA) located below lakes experience water leaks known to have drained from above, since isotopic analysis found the water to be meteoric in origin. In any event, salt deposits close to the surface, either as a result of erosion of overlying strata or from salt body uplift to ground surface, are sensitive to water inflow. This particular geological feature is an important consideration in determining the risk of water penetration into the mines.

2. *Stratigraphic sequence of formation* has to be considered with regard to the relationship between aquifers and salt beds. It is particularly important to establish a distance between horizons of water-bearing strata and mineable salt beds. Close spacing between water-bearing horizons and mine workings creates a risk of flooding. A stratigraphic column of the Saskatchewan evaporite-bearing strata given in ascending order is: Ashern

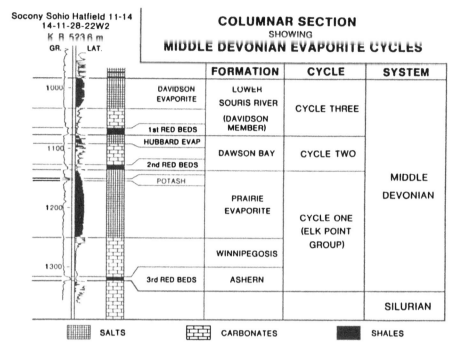

Figure 8.2.1. Stratigraphic column of prairie evaporite strata (Canada)

formation, Winnipegosis formation, prairie evaporite formation and Dawson Bay formation, represented by second red beds, Dawson Bay limestone, and first red beds (Figure 8.2.1). The mineable potash beds are located on the top of the prairie evaporite and below the water-bearing strata of the Dawson Bay formation. This particular sequence is not favourable to mining, because of the possibility of water penetration into the producing mine. Under these circumstances a protective mining method must be implemented. Careful layout and ground control must be established in order to keep water out of the mine workings.

3. *The lithology of the strata* must be considered from an engineering geology point of view rather than from the petrological aspect. There are various rock types with different mineral compositions which greatly affect the porosity and permeability of strata. Permeable rocks allow the passage of water under pressure through them. Their permeability is primarily dependent on the mineralogy, shape, size and degree of cementation of the grains. For example, clastic sediments such as sandstone, conglomerate and others are permeable because the pores are interconnected due to incomplete cementation of mineral grains. Rocks with pores which are not connected are impermeable, for example shale or claystone. The risk of water inflow into a mine is decreased when salt deposits are situated within impermeable strata. However, if these sediments are interbedded with permeable rocks then the risk is increased. In the Tuzla salt basin the water inflow is facilitated by tuffite which occurs interlayered between impermeable sediments. Under these circumstances the impermeability is disturbed and water penetration in the rock salt mine is possible. It is obvious that the permeability of individual lithological units within the stratigraphic column is an important factor for evaluation of risk of water penetration.

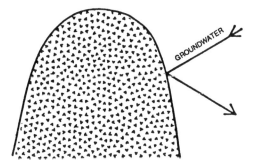

Figure 8.2.2. Hydrologic stability of salt domes (after Martinez)

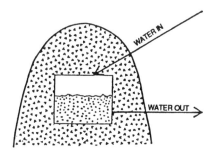

Figure 8.2.3. Hydrologic isolation of salt domes (after Martinez)

4. *Perviousness phenomena* of course are related to permeable rocks but also relate to impermeable rocks if they are fissured, jointed and faulted. In this case they could contain and permit the passage through them of great amounts of water. The degree of perviousness or permeability of the rocks is related to fluid transmission, and it is measured by the rate at which a fluid moves a given distance within a given interval of time. The salt domes are considered to be dense with a porosity almost zero and impervious,[2] because of the rehealing process of fissures and fractures. This physical phenomenon is particularly important when considering a mass of salt domes for waste disposal excavation. The investigations carried out by J. D. Martinez modelled the resistance of the salt stock by external water inflow and internal water outflow, namely:[6]

– hydrologic stability: a resistance of the salt stock to external brines (Figure 8.2.2);
– hydrologic isolation to block inflow into and outflow of water from the excavation opening (Figure 8.2.3).

The hydrological stability and isolation of the salt bodies are the significant geological factors with regard to the risk of water penetration.

8.2.2 *Structural features*

Structural features which control water flow and water inflow are related to tectonic events, as discussed in other chapters of this book, and to chemical processes of dissolution which are also discussed in various chapters of this book. Structural features are one of the most important factors which might control water inflow in mines. Several factors in this regard are discussed.

1. *Folding factors* exhibit various influences on the risk of water penetration as, for

example, squeezing salts toward ground surface. As a result of this deformation some parts of salt deposits may be shallow and their boundaries may be in contact with water-bearing horizons, which is the case with some domes. Of particular interest are overturning folds, which have rotated footwall water-bearing strata in lateral positions and overhang salt deposits. Under these circumstances the salt deposits are enveloped with permeable sediments saturated with water. Generally, overfolded beds contain artesian water under high pressure. In these cases water inflow into the salt mines is under pressure and the loss of mining operations is unavoidable. In addition, clastic strata due to extension deformation can be fractured during folding. The open fractures establish very permeable zones with a high rate of water flow. They may also facilitate water inflow into salt deposits, via fold structures.[5]

2. *Faulting factors* must be taken into consideration as water feeder channels which can bring water into the vicinity of salt deposits, and in some cases into the salt deposit itself. P.C. Kelsall and J.W. Nelson reported on the Avery Island mine which has a long history of mining water inflows, all of which have been successfully controlled, though not eliminated. The dome contains a major fault zone which includes a high proportion of clastic rock such as sandstone (Figure 8.2.4). The upper surface of the salt is highly irregular, with a deep trough along the fault zone and sediment filled crevices developed along banding parallel to the fault. Brine seepage into the mine appears to be associated with these crevices and related fractures that extend at least 150 m below the top of the salt. In addition, caverns in the salt near the top of the dome were identified in the exploratory well at the Bayou Choctaw dome, where a 4 m void was found 30 m below the top of the salt and a 0.6 m void was found 44 m below the top.[7]

3. *Fracturing factors* are related to shear and breccia zones, which are responsible for the fractural porosity of salt deposits. They are of great significance for water penetration and circulation in salt deposits. A particularly critical feature is developed when fractured zones occur in gypsum strata, as discussed in Subsection 8.2.3 regarding the cap rock of domes. The analysis of fractured zones is undertaken through engineering geology using parameters such as fracture indices and others. Water circulation through fractures within evaporites results in chemical alterations and cavitations as discussed in the next paragraph.

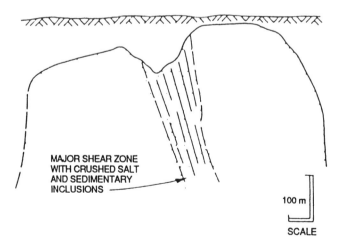

MAJOR SHEAR ZONE
WITH CRUSHED SALT
AND SEDIMENTARY
INCLUSIONS

100 m

SCALE

Figure 8.2.4. Fault zone of the Avery Island dome, USA (after Jacoby)

4. *Carstification factors* refer to the formation of caverns and voids within evaporite beds by dissolution of saline minerals. This process could occur equally in zones which have not been affected by tectonic events and zones where structural fracturing took place during mining activities. Under the latter circumstances, zones of loose evaporite beds may be formed where the continuity and strength are eroded by chemical action. In addition, the porosity and flaking of the salt may be variable. In this case, the original salt deposits, because of the very discontinuous nature, actually become a loose evaporite rock mass, which cannot be considered for underground mining extraction, but only for solution mining.

8.2.3 *Caprock water considerations*

The geohydrologic environment above and around the dome must be investigated in order that the risk of water penetration into the future excavation can be evaluated. The investigations are done by hydrological and geophysical downhole logging, to establish the lithology, porosity and permeability, the chemistry of pore water, water inflow and other factors. Of particular interest is the caprock water consideration. A case history of the Richton dome (Perry County) in the Mississippi salt basin, USA, has been described by M. L. Wouch and J. D. Martinez.[8]

The crystalline anhydrite caprock at the Richton dome grades into 1.5 to 2.5 m of unconsolidated anhydrite sand immediately above the salt. This sand proved to be a poor aquifer when tested and caused no problems with lost circulation of drilling fluid. Other wells which penetrated the salt also identified a porous contact with a caprock of 2.1 m of granular anhydrite above the salt encountered at 231 m (Figure 8.2.5).

Caprock waters were found to be slightly saline, being about 11 000 mg/l TDS, but the specific conductance decreased during testing, indicating a possible hydraulic connection to the less saline aquifer that overlies the caprock. The permeability of the caprock appeared to increase upward, and water movement in the upper portions of the caprock was through fractures, joints and solution channels.

Although the base of the fresh water was shallower in the vicinity of the Richton dome than in the surrounding areas, this may have been due more to the structural uplift of the salt water-bearing strata and the trapping of saline water above clay layers than to ongoing dissolution of the salt stock. The sub-surface flow is believed to have been southerly, but there was also an upward component to the flow which was weaker than the southerly component. The base of fresh water appeared to be higher in the down dip sections south of the dome than elsewhere in Perry County, which could have been due to the dissolution of the salt stock, but could also have been due to incomplete flushing of saline water originally in the aquifers, or to upwellings from deeper saline aquifers.

The upward component of flow may have been responsible for slightly saline waters found in water wells west and northwest of the town of Richton over the shallowest part of the dome. Chloride concentrations from wells in the town of Richton were more than 100 mg/l higher than the average for Miocene aquifers in the area. Also, wells south of the Richton dome near Beaumont and New Augusta contained chloride concentrations more than 60 mg/l above the average. Whether the cause of these anomalies was from the dissolution of the salt stock or upward movement of waters from deeper saline aquifers has not been determined.

The evidence supporting dissolution at the Richton dome is weak. The saline anomalies found in aquifers adjacent to the Richton dome cannot be conclusively linked to the

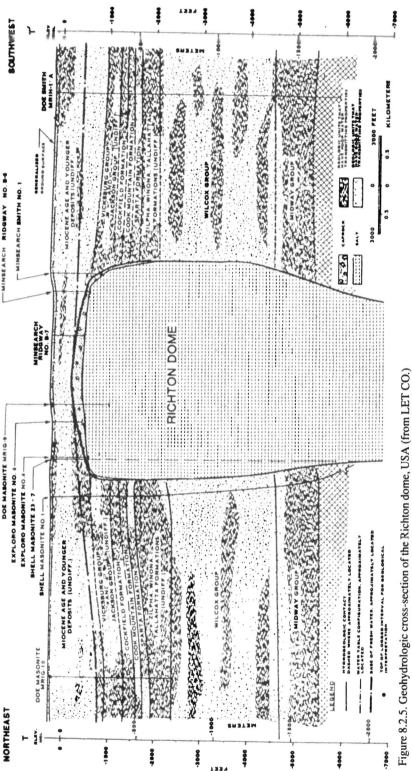

Figure 8.2.5. Geohydrologic cross-section of the Richton dome, USA (from LET CO.)

dissolution of the salt stock. The anomalous salinites may be due to the failure of the slow-moving groundwater in this area to flush out relict brines. Therefore, the presence of anhydrite sand at the salt-cap rock interface cannot be correlated with a plume as evidence of salt dissolution at the Richton dome.

In some instances the entire caprock is extremely variable in hydrologic characteristics. For example, the caprock water at the Tatum dome is relatively fresh and hard (calcium sulphate and sodium bicarbonate type). Saline (> 500 mg/l of salt) to fresh water (< 1000 mg/l of salt) is being produced from fractures and cavities in the calcite and anhydrite caprock, which is hydrologically connected to surrounding aquifers. The presence of anhydrite sand at the salt-cap rock interface and dissolution cavities at or near the interface suggests that the Tatum salt stock is being dissolved by circulating sub-surface waters. The caprock is more than 180 m thick at the centre of the dome and consists of anhydrite with occasional gypsum bands. Lost circulation of drilling fluid in 22 of 30 wells drilled through the caprock, cannot be related totally to the caprock, since some of them lost water above and below the anhydrite.[8]

8.2.4 *Salt deposit water consideration*

Geohydrologic consideration of salt deposits can also be understood through a case history as in the previous sub-section. The Tetima salt deposit at the Tuzla basin, former Yugoslavia, has had hydrologic investigations carried out by surface and sub-surface (drill holes) studies. The investigations delineated three regions of water-bearing zones, as discussed by J. Stojkovic et al.,[3] and are represented below.

Overburden consists of thick younger Miocene sediments mainly of clay, claystone and sandstsone. The surface morphology of the overburden is rather hilly with erosion valleys and outcropping sediments with intensive structural defects. The water is present in the claystone, sandstone and conglomerate, and it is controlled by the fracture porosity with limited circulation along fracture systems due to fracture filling. Water inflow in drill holes is at maximum 3.5 l/sec decreasing to 0.5 l/sec. Mineral content in the water is up to 1 g/l but the water could not be classified as potable because of the high content of H_2S. The water temperature is 12-13°C. The water horizon could be classified as closed or semi-closed with weak connections to the ground surface. However, in the region of Jala Creek the closed water horizon occurs along a fault zone, where water is mineralized up to 50 g/l with an increased gas content and a temperature of 22°-24°C. Maximum inflow in drill holes is 7 l/min, with a tendency to decrease. The hydraulic state of the water-bearing horizon is artesian with a gradual transition to sub-artesian. The water of this horizon is older than 37 000 years.

The salt deposit is of irregular elliptic shape with a long axis over 2 km and a short axis of about 1 km. The salt lens, with a strike NW-SE and dipping to the west, has a depth varying from 400 m (SE) to over 1000 m (NW). The maximum thickness of the salt bed is 150-180 m (Figure 2.1.3). Along the boundary of the salt deposit there is a water-bearing zone as well as along the contact between the overburden strata and anhydrite breccia. The water is sub-artesian and artesian in the area where the peisometric level is higher than the level of the ground surface. Water along the boundary of the salt body effects a solution of the salts. The water contains 40-50 g/l of minerals, primarily NaCl. Temperature of the water is 16-17°C. Pumping tests have determined a coefficient of transmissibility in the range of:[3]

$$T = 0.1 - 0.05 \ m^2/day$$

Rejuvenation of the water-bearing horizon is by ground water infiltration on the west, and surface water inflow on the north-east. Drainage of the water horizon occurs in the eastern part in the valley of Kovacica Creek and the southern part within the fault zone along Jala Creek.

Footwall strata, below 20 30 m of the salt deposit, contain a water-bearing zone with partially fractured banded claystone. The water comprises salt brine with 315-320 g/l NaCl. It is sub-artesian, originating during the sedimentary processes of the evaporite strata. Hydraulic connection with other water-bearing horizons is not established.

Hydrogeological characteristics of the Tetima salt deposits do not favour underground mining because there is a possibility of water penetration into the mine workings. However, for solution mining there is no danger of the water jeopardizing this type of technology. Generally, the depth of some portions of the deposits and some zones with waste partings favours solution mining.

The hydrogeological characterization of water-bearing horizons both within overburden strata and evaporite-bearing strata is equally important for underground mining as well as solution mining. Of particular concern are caverns in the salt deposits filled with a large accumulation of water which might be released during mining operations if fractures are propagated into them.

8.3 TECHNOLOGICAL FACTORS

The risk of water penetration into salt deposits due to natural factors mainly depends on the geologic-tectonic characteristics of the salt basin. The equilibrium of virgin evaporite strata could be disturbed or destroyed by technological activities such as drilling, shaft excavation, ore extraction. Some mine floodings are directly related to the insufficient adjustment of technological factors to natural factors. A good knowledge of the geohydrology of salt deposits could result in technological change which might prevent water inflow in a salt mine. The technological risk of water inflow also depends on the human factor, since avoidable errors are often the cause of water penetration in the mine.

8.3.1 *Exploration drilling*

Exploration drilling is an essential tool in obtaining data from the sub-surface with regard to various aspects such as the geologicl structure, grade and tonnage of the ore, geotechnical characterization of strata, geohydrology, mining, strata control etc. (Figure 8.3.1). Exploration drilling with casing plays a large role in hydrologic investigations such as determining the capacity of water inflow or outflow, chemical composition of water and isotropic analysis, hydrostatic pressure, water temperature and others. All these hydrologic data are used for mine design and mine development. Also these data are used to evaluate the possible risk of water penetration in the mines and to design protective measures to avoid water inflow.

Drilling through the water-bearing horizons of overburden strata and evaporite-bearing strata should be carried out with great care. If there is a possibility of distorting a natural equilibrium and damaging the salt deposits with drilling fluid, then drilling must be undertaken without water circulation as discussed in Chapter 6. Each hole, after completion, should be sealed off with cement, in order to avoid potential channels for water inflow in the mines.

Figure 8.3.1. Deep exploration drilling of the Zechstein's salt-bearing strata (Poland)

The drill hole itself could present a risk for water penetration in the mine. This could be indicated by the loss of circulation of drilling fluid or by the penetration of drill holes into the mine workings. For example, the Jefferson Island mine was flooded in November 1980 by waters draining from an overlying lake. This occurred immediately after a drilling rig operating in the lake had lost circulation, but the exact cause of the catastrophe was not identified. Prior to the flooding, several major leaks had been exhibited in the mine. Two leaks were associated with old exploration drill holes, and a third occurred when the mine approached to within 30 m of the edge of the dome. The latest inflow was proven through chemical and isotropic analysis to be meteoric water. A fourth leake occurred when an exploration drill hole on the 400 m level encountered a cavern which drained 200 m^3 of $CaCl_2$ brines (secondary trapped water in the salt body).[7].

From this example and others it is obvious that any penetration of a drill hole in the salt body creates a high risk of water inflow. To avoid this, the best solution is to stop drilling when nearing the top of salt deposits as discussed in Chapter 6.

8.3.2 *Mine access workings*

This subsection on mine access working with regard to the risk of water penetration is based on the facts represented in the previous chapter on the same topic. The risk of water inflow through the mine shaft is much greater than through tunnels, as discussed below.[5]

The great majority of salt mines are accessed by shafts, which usually results in a high risk of water inflow. The inflow might occur during shaft collaring (surface or close to surface water inflow) or after completion (percolating water inflow). Some European countries (Germany, Poland, Romania) developed criteria for shaft location based on mining experience over many years.

The alternatives are to locate the shaft in the central part of the salt body or out of the salt body at the safe distance to its contact with the rock strata. The contact zones of the salt body and surrounding strata are contaminated by so-called periphery waters, which could threaten mine openings (Figure 8.3.2). Also one of the criteria is to sink the shaft, if possible, through impermeable rocks such as claystone, marl and others. However, shaft-sinking through the gypsum-anhydrite strata should be avoided. In reality, the majority of shafts are sunk through claystone-gypsum strata. Unavoidable sinking of shafts through pervious sediments requires a safe distance between the shaft and exploration drill holes. To decrease the risk of water penetration it is necessary to have thorough knowledge of water-bearing horizons, particularly highly saturated ones containing unconsolidated sand.

A great majority of salt basins all over the world contain water-saturated horizons above the salt deposits. They have to be intersected by shafts in order to access the salt body. These horizons have to be stabilized by freezing before shaft-sinking, since ground stabilization through cementation is not possible due to the high perviousness of the sediments. The fundamental rule is that before a shaft is sunk, a pilot hole along its axis has to be drilled. This hole should produce data on hydrology, geotechnology etc. with regard to shaft excavation and ground control.[6] Percussion dry drilling is recommended in order to access the hydrology and control of the hole deviation. The limited deviation of percussion drill holes favours their drilling around the shaft for ground freezing at–35°C to–40°C. The risk of water penetration controlled by ground freezing increases when the ground is thawed. There is always a danger that water between the shaft support and the rock could percolate into the mine. This is controlled by shaft rings and other measures, as discussed in Chapter 7.[5]

A small number of salt deposits are accessed by tunnels and some by haulage ways from the shaft sited out of the salt body. The risk of water inflow into those workings is lower because they are more parallel to water-bearing horizons and not to intersect them. However, they could intersect percolating and circulating brines, which could represent a danger for water inflow in mines. There are two causes of water inflow in tunnels and haulage ways; firstly, the unintentional intersection with sub-surface brines; and secondly,

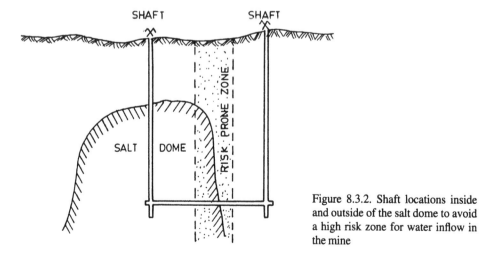

Figure 8.3.2. Shaft locations inside and outside of the salt dome to avoid a high risk zone for water inflow in the mine

the crossing contact between the rock strata and the salt body. The control of water inflow zones is effected by cementation and grouting. Of primary interest are contact zones, because they usually have a maximum leak rate when first exposed, which declines with time and stops after a few months. When leaks occur close to the edge of the dome their leak rate increases with time and ground cementation is required. The leaks are generally brines with $NaCl_2$ and $CaCl_2$ indicating their connate origination.

8.3.3 *Development and extraction*

To minimize and avoid the risk of water penetration in producing mines it is necessary to take a certain number of protective measures.

Designated mining areas of underground mines have to be protected by barrier pillars, which enclose the area. The thickness of the barrier pillars, depends on the salt density, the structure of the salt body, the hydrology of evaporite strata and others, but is generally in the range of between 100 to 200 m (Figure 8.3.3). In addition, if a mining area is intersected by a drill hole, then it should be protected from water inflow by a cylindrical barrier pillar around the drill hole with a diameter of between 25 and 30 m. The pillars around drill holes should not be disturbed by operational mining. For example when mine workings come in contact with drill holes, the water inflow in the mine is increased and this might have a grave consequence such as flooding of the mine.[9]

Mine development workings should be driven with caution, particularly in the initial phase when they are also used to explore the ore body. The ore body contours and layout of barrier pillars are usually based on extrapolation of a few drill holes and geophysical surveying. Under these circumstances the deviation of boundary lines could be appreciable, and development workings could come close to the contact between the salt deposit and the rock strata or even intersect such a contact, as has happened in the Klodawa mine in Poland, for example. The work to effect ground consolidation by cementation in order to eliminate a high risk of water inflow in underground mines could be difficult, expensive and sometimes with only temporary success. The elimination of adverse effects on mine development workings in zones prone to high risk of water inflow is much more difficult in flat and folded deposits than in domal deposits. When development takes place in virgin ground toward the potential zone of water inflow, then long holes should be drilled in advance of the roadway face. In this case drilling must be under strict control where the

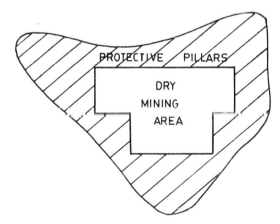

Figure 8.3.3. Protective pillars around designated dry mining area

drilling shuts down automatically if it suddenly intersects water and/or gas occurrences. A special technology of drill hole cementation has been developed, which can cope with any water or gas pressure.[10]

The room extraction of salt can be equally affected by adverse effects as the mine development workings if the natural hydrologic equilibrium is disturbed by mine excavation. Harmless ground water could create a dangerous situation if ore extraction ignores its existence. The great majority of salt deposits are mined in rooms of various shapes and sizes, which depend on the mining methods. To decrease the risk of water inflow in mining it is necessary to accomodate the room size and layout to the hydrologic characteristics of the deposit. There are rooms with a height between 20 and 40 m and rooms with a height ten times smaller (2 to 4 m). The large rooms are located in domal and flexural deposits where ground water should not represent a danger for penetration in the mine, due to optimal natural and technological factors such as leaving a large protective pillar, as in trapeze rooms in Romania. However, small rooms of flat beds with an inadequate thickness of protective pillar and close proximity to water-bearing horizons could experience a water leak, where control could become a difficult task. This is discussed in the last part of this chapter under the case of a water inflow in a potash mine in Saskatchewan.[11]

Inflow in underground salt mines can also occur due to changes in mining technology and mine layout. The Weeks Island mine had a history of minor brine leakages, all of which tended to stop with time, indicating drainage from isolated brines in the pockets and caverns within the salt dome. The separation between the mine and the upper surface of the salt had been maintained by a barrier pillar of a thickness greater than 100 m. Salt excavation started using a DOE miner in a new mine which was developed adjacent to the old mine and at a slightly higher elevation within the dome. One of the drifts of the new mine encountered a water inflow to the extent that the drift was abandoned and sealed off with a bulkhead. Substantial grouting from surface and underground has appreciably reduced water inflow but not completely eliminated it. Kelsall and Nelson concluded on the basis of drilling data and dye tests from the surface that the water entering the drift originated from the overburden aquifers above the dome at a point where the salt protective pillar is approximately 90 m thick.[7]

8.3.4 *Interaction between dry and solution mining*

The risk of water inflow in underground mines becomes a particular problem when in a single ore body there is simultaneous salt exploitation by dry mining and solution mining. The coincidental salt mining by two different technologies is known in Poland (Wieliczka-Baric mine), Romania (Muresh mine), former Yugoslavia (Tusanj-Hukala mine, Figure 8.3.4) and others.

The risk of water penetration depends on the type of solution mining used, i.e. controlled solution mining or uncontrolled solution mining.

Controlled solution mining is based on the technology of transmitting fresh water from the surface in the salt deposits, dissolving the salt into brines, and pumping it back to the surface. In this case the risk of water inflow into the underground mine is controlled by the drill hole and protection pillar layout.[9]

Uncontrolled solution mining is based simply on the pumping of sub-surface circulating brines, which are created by the meteoric water inflow from the surface. The circulating brines might occur around the periphery of a dry mining area, and consequently, represent a high risk of water penetration into the mine. The main risk of water inflow is the

Figure 8.3.4. Part of the salt body named Hukalo where solution mining is in effect (Tuzla, former Yugoslavia)

uncontrolled penetration of brine into protective pillars. Under these circumstances the circulating brines might advance through a protective pillar and form a cavernous salt zone as a channel for water inflow in the area of dry mining. The risk of water inflow is especially high when there is not knowledge of the flow direction of the circulating water, the velocity of circulation and the geometrical relation of their flow pattern to the morphology of the barrier pillars. The process of water circulation and salt solution is controlled by natural factors and is out of the control of technological factors. The final solution for the elimination of the risk of water inflow in a dry mine is to change the salt exploitation to solution mining technology.[10]

The technological risk could vary not only due to the chosen design and technology but also due to human factors, where errors might cause a high risk of water inflow in the mine.

8.4 CASE HISTORY OF BRINE INFLOW

A representation of a case history of brine inflow in a mine is outlined in the 'Interim report on controlling roof brine inflows through diamond drilling and grouting at Central Canada Potash', written by R. Tofani.[11]

8.4.1 *General considerations*

Central Canada Potash is located approximately 76 km south-east of Saskatoon, near the Yellowhead highway. There are two shafts in operation. Number 1 is equipped with skips and is used as a production shaft, while number 2 is used to transport personnel and equipment. Mining began in 1969 at a depth of 1021 m below the surface. The average ore grade is 27% K_2O and the production capacity is 1.3 million tonnes of nuriate of potash (finished product).

Till now there have been two back inflows in the mine. The first occurred in 1974 in Panel 109 Room 24W and the second in 1978 in Panel 209 Room W53N, as illustrated in Figure 8.4.1. The subject of consideration in this case is the second brine inflow, when potash excavation was to the north of it. Within three hours of discovering the inflow approximately 1.5 m of roof in the area had collapsed. The rate of inflow was estimated at 227 lpm. It was also noted that other ancillary leaks had developed in the vicinity of the initial brine occurrence. The subsidence of the overlying strata, associated with rock creep and pillar compaction caused the original inflow area in Wing N7 to dry up over a period of five months. As this took place a new leakage area developed in Wing N9 and in the present location of the inflow. Flow rates have varied considerably since December 1978, the highest rate being 341 lpm in October 1981. At present, leakage brine is retained in adjacent

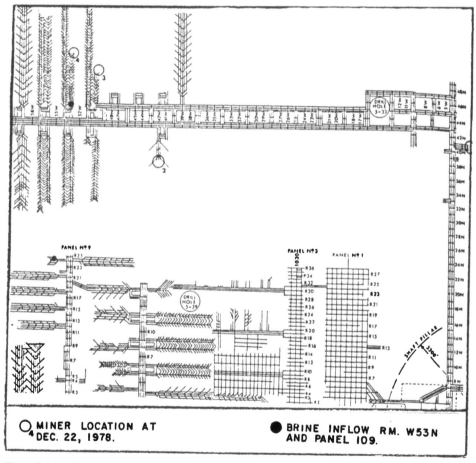

Figure 8.4.1. Mine layout of blocks 1 and 2

rooms, and runs into a sump which is proximal to the inflow. A weir was built, so that approximate flow measurements could be taken. From the sump, the brine is pumped to the shaft station and then to the surface. The first brine leakage in the roof of Panel 109 R24W (1974) decreased in flow rate from 86 to 38 *l*pm, when the brine inflow in W53N occurred. In order to confirm that the two inflows were associated, a laboratory analysis of the brines was done. The results indicated that both brines are a sodium chloride saturate, containing approximately 100 000 ppm sodium and 200 000 ppm chloride. Specific gravity, pH level, and temperature are very similar.

8.4.2 *Contributing factors*

The brine had been identified in the fracture of the Dawson Bay formation, encountered during diamond drilling in 1981. There is no accepted theory as to how the water inflow occurred in the mine. There are several factors which might have contributed to the brine entry.

Relative to the brine inflow, at the west end of the mine, seismic surface survey indicated a local up in the overlying Winnipegosis formation. Underground electromagnetic survey was conducted in the same general area. A high conductivity anomaly was identified in the back. Topographically the no. 3 zone, in the immediate area of the inflow, is on a ridge having a local NE-SW trend. This ridge may have had some flexural effect on the subsequent deposition of the Dawson Bay. Diamond drilling had indicated an abnormally thin salt back thickness in the immediate area of the W53N inflow. The extraction ratio in W53N was relatively high, with a two-wing spacing on 22 m centres. A similar mining pattern was used in the rooms surrounding W53N. Observations of the condition of the mine back in the area of both inflows showed anomalous clay concentrations. The clay beds in some instances are weak and loosely compacted.

None of the factors seemed significant alone, but it was considered that some combination could explain the rapid non-uniform closure, causing a zone of weakness. This zone may have permitted the migration of brine from the Dawson Bay limestone through the poorly stratified second red beds in the mine.

8.4.3 *Investigation of water inflow*

The investigation program essentially was performed by diamond drilling and dye testing. Diamond drilling investigated the formation thicknesses above the mine back, the location of the brine and fracturing systems. The location of diamond drill holes (shown in Figure 8.4.2) was based on the position of brine leakages into the mine and was intended to probe target areas indicated from electromagnetic surveys.

Core recovered from diamond drill holes indicated no fracture planes which were connected between drill holes. A second and more important result indicated that the average salt back thickness from the mining zone to the second red beds was abnormally thin, the average thickness being 1.8 m less than normal. Drill holes confirmed that the brine was located in the Dawson Bay. Initially, eleven drill holes were completed over a period of thirty months. All penetrated the Dawson Bay, nine of which tapped the brine zone. During drilling, precautions were taken as safeguards; 2.4 and 5.0 m packers were used for pressure testing the holes, followed by grouting in instances of pressure test failures. Then 9.0 m of NW casing was installed and pressure tested so that the hole at this depth could be cased through the salt back securing it from possible incursion. The second

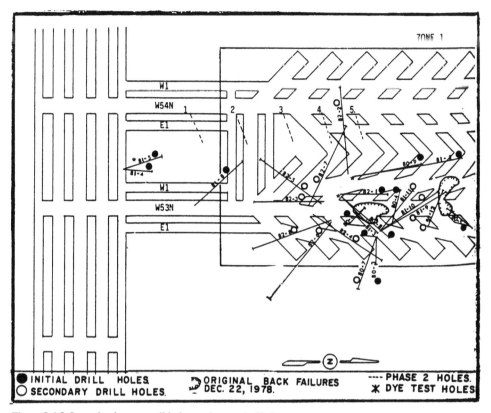

Figure 8.4.2. Investigation, consolidation and control of inflow area

red beds lower 3.0 m of the Dawson Bay were secured in a similar manner, using BW casing. The packers and casings were tested at pressures ranging from 3450 to 12 400 kPa. Engineering supervision was maintained throughout the drilling operation to ensure that these high standards were kept.

Dye testing was implemented in order to determine the flow direction and possible coupling between drill holes. Green sodium fluorescent dye and an ultra-violet light to detect low concentration of dye were used for selected initial diamond drill holes, as indicated in Figure 8.4.2. With all other holes opened slightly, checks were made to see if the dye migrated to them. The results of the investigation indicated a north-easterly flow direction with good connection between most drill holes and the mine back flow. With this information, a carefully designed grouting program was used to seal the brine flow fractures. The dye testing program took 72 hours to complete.

8.4.4 *Ground consolidation*

The ground consolidation was performed by grouting using the initial diamond drill holes. As a result of batch testing it was decided that the prime grouting medium should be a cement-zeogel mix. Testing indicated that CSA type 30 cement had the optimum setting time. The zeogel was added to provide plasticity. Its proportion was 10% by weight cement.

The roof brine was used to mix the grout since setting times were not significantly effected. Brine pressure, with relief through the mine back, was measured at 1380 kPa. Pumping pressures would have to exceed this pressure and were therefore set at a minimum of 2070 kPa to minimize the fluid pressure build-up in the mine back. In an effort to obtain the most effective seal in the brine-bearing fractures, drill holes were grouted individually. The brine inflow in drill holes was pumped at approximately 11 lpm. A thin mixture was used to ensure a good flow of cement along the entire length of the brine-bearing structures. After grouting the drill holes were closed off and the grout in the hole allowed to set. After a period of between 24 and 48 hours, cement in the casing was drilled out in order that chemical grout could be injected into the formation at a later date to provide the best possible seal. After grouting all the intitial drill holes, the inflow decreased from 341 lpm to 121 lmp. The decision was made to drill additional holes which would penetrate the brine-bearing zone. In all, twelve secondary diamond drill holes were drilled and grouted (Figure 8.4.2). Testing was temporarily terminated. The inflow was still active but it was felt that continued drilling and grouting would not further reduce the inflow volume. This had been indicated by the last three secondary holes grouted. The grouting of water-bearing ground resulted in a 76% reduction of the brine inflow into the mine and a brine pressure relief through the mine back up to 150 kPa. Most drill holes have been equipped with the pressure gauges.

8.4.5 *Mine back support*

The initial ground control started with timber cribs support. The cribbing supported the mine back around the inflow area in W53N to arrest the rate of roof closure and related deterioration and to maintain safe access to working faces. The timber cribbing was also installed north of the leak, to prevent the advance of roof closure and possible roof falls. More effective ground support and control was undertaken using mine fill. With progressive subsidence throuthout the Dawson Bay formation and beyond, it is generally accepted that further movement of the limestone strata might result in continuous leakage into the mine. By placing fill in as much of the area as possible, the closure would be controlled and further fracturing of the Dawson Bay arrested. In the long term, closure may eventually seal off the inflow within the clay strata.

The mine fill in conjunction with grouting should permanently seal off the W53N room in zone no. 1 (Figure 8.4.2). Diamond drill holes 81-4 and 81-5 can be used to relieve pressure in the back and sustain a constant pressure of 138 kPa or less.

The combined effects of subsidence and mine fill should stop bed separation and lamination in the mine back due to closure deformation and stress relation. The brine inflow in the affected area resulted in difficult access through the mine and some suggestions have been made to abandon this block. However, the concept of ground reinforcement and support as well as careful monitoring of mine back movement and brine inflow into the mine was accepted. Development of an active brine inflow into the mine and its potential can be predicted in advance from monitoring.

REFERENCES

1. Jeremic, M.L. 1964. Salt deposits of evaporite bearing strata in Poland. *Min. & Metall. Bulletin* (Belgrade), 3: 60-66 (in Yugoslav).

2. White, R.M. & C.A. Speirs 1983. Characterization of salt domes for storage and waste disposal. In *Proc. of 6th Int. Symp. on Salt*, The Salt Institute, Virginia, Vol. 1, pp. 511-518.

3. Stojkovic, J. et al. 1986. Basic geological characteristics of Tetima rock salt deposits. Manuscript, Research Centre for Mining Investigation, Tuzla (in Yugoslav).

4. Stojkovic, J. et al. 1987. Salt deposits of Tuzla basin and prospect of future geological exploration. *J. Min. & Geol.* (Faculty of Mining and Geology, Tuzla Univ.) 15\16: 9-18 (in Yugoslav).

5. Jeremic, M.L. 1966. Risk of water penetration in the salt deposits. *Min. & Metall. Bulletin* (Belgrade), 8: 175-182 (in Yugoslav).

6. Martinez, J.D. 1983. Energy programs – A contribution to salt dome knowledge. In *Proc. of 6th Int. Symp. on Salt*, The Salt Institute, Virginia, Vol. 2, pp. 235-245.

7. Kelsall, P.C. & J.W. Nelson 1983. Geologic and engineering characteristics of Gulf region salt domes applied to underground storage and mining. In *Proc. of 6th Int. Symp. on Salt*, The Salt Institute, Virginia, Vol. 1, pp. 519-542.

8. Wouch, M.L. & J.D. Martinez 1983. Salt dome plumes and dissolution features: are they related. In *Proc. of 6th Int. Symp. on Salt*, The Salt Institute, Virginia, Vol. 2, pp. 159-175.

9. Jeremic, M.L. 1964. Dry mining and solution mining of rock salt deposits in Romania. *Min. & Metall. Bulletin* (Belgrade), 12: 264-274 (in Yugoslav).

10. Jeremic, M.L. 1964. Methods of development and exploitation of salt deposits in Poland. *Min. & Metall. Bulletin* (Belgrade) 2: 32-42 (in Yugoslav).

11. Tofani, R. 1983. Interim report on controlling roof brine inflows through diamond drilling and grouting at Central Canada Potash. In *Potash '83. Potash technology*, pp. 105-111. Toronto: Pergamon Press.

CHAPTER 9

Elasto-plastic deformation and failure

The conventional mechanical properties and behaviour are complex and difficult to correctly define because of the plastic nature of salt. The salt exhibits elastic and inelastic behaviour in the function of stress magnitude, loading conditions and time scale of loading. The salt failures due to incremental loading have a specific feature which could be explained by cataclastic deformation. The shear displacement of salt along preferential structures has been studied from a point of view of plasticity rather than elasticity, but its mechanics should be considered from an engineering point of view. The four basic topics of mechanical behaviour and failure are discussed as follows.

9.1 ELASTIC STRENGTH PROPERTIES OF SALT

The parameters of strength properties are considered for an unconfined compressive strength, triaxial compressive strength, tensile strength and Young's modulus. The testing procedures are carried out at room temperature with a short duration of loading known as quasi-static. The strength properties and elasticity parameters should have a careful consideration in regard to mine design and ground control.

9.1.1 Unconfined compressive strength

The unconfined compressive strength test is most common for determination of salt strength. The test is exposed to the uniaxial loading of a cylinder or prism of salt to failure. The salt specimens usually fail in shear, as indicated by the characteristic cones formed on the conjuncted planes of failure. The compressive strength so obtained depends to some degree on a certain number of variable factors such as the sample shape, size, porosity, moisture, mineral composition and others. Further consideration of unconfined compressive strength in regard to three factors is discussed.

   1. *The relationship between size effects and strength* has been studied since the end of the last century by some investigators. Their investigations suggest a number of seemingly different predictive equations.

   The size effect and strength relationship from various salt deposits of Europe indicated a differential strength increase with the length of the cube edge increase.[1] The function of strength increase is similar, as illustrated in Figure 9.1.1. The diagrams indicated that rock salts exhibit a certain point of sample size after which its strength becomes constant. The limiting point of the size effect on the strength increase is between 10 and 25 cm of the

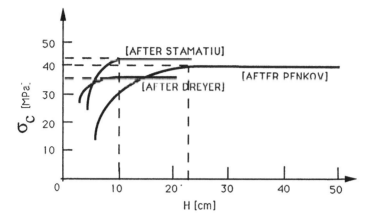

Figure 9.1.1. Cubical compressive strength as a function of the edge length of the rock salts

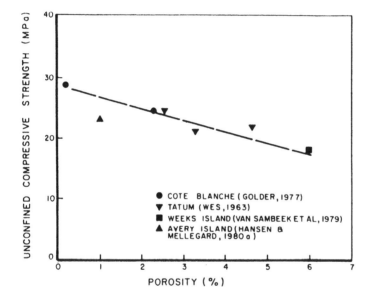

Figure 9.1.2. Unconfined compressive strength of dome salts as a function of porosity (USA)

length of the cube edge, which is much lower than in the case of other mineral compositions (35-50 cm).

The variations of the unconfined compressive strength of the salts exist not only between various salt deposits due to different type and structure of the mineral components, but also in the same deposit as in the case of Zechstein's dome for older rock salt and younger rock salt.

2. *The relationship between porosity and strength* is illustrated in Figure 9.1.2 for the salt of the Gulf region dome (USA).[2] The unconfined compressive strength appears to decrease almost linearly with the porosity increase. It is noted that high porosity may be the cause for some salts to be noticeably friable in the mines (Weeks Island). Alternatively, the friability may be the result of different crystal fabric occurring in recrystallized salt.[2] The discussion in regard to porosity itself has been briefly represented in Chapter 4, and further comments in this regard at this place are not required.

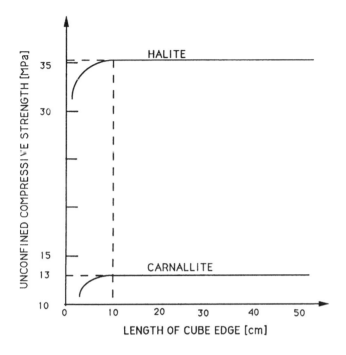

Figure 9.1.3. Cubical compressive strength of halite and carnallite in regard to the edge length (Germany)

3. *The strength variation of salt minerals* is in the wide spectrum due to their physical properties and chemical composition. For example, Fairhurst et al. tested salt samples of the various mineralogical compositions of the Sergipe salt deposit (Brazil) and reported the following unconfined compressive strengths:[3]

tachyhydrite 3.0 MPa;

sylvinite 42.3 MPa;

halite 37.3 MPa;

carnallite 19.7.

In regard to potash ore it is important that the main mineral component is sylvinite because the presence of carnallite deteriorated its strength. For examples in the potash deposits of Saskatchewan, the presence of 5 to 9% of carnallite in the ore alters its unconfined compressive strength to less than 50% of the mono-mineral salt strength. In this regard the average uniaxial compressive strength of potash ore is in the range between 16 and 19 MPa, where the average carnallite content ranges from 2 to 4%.

The comparative analysis between cubical compressive strength of halite and carnallite has been done as a function of the edge length of German salt mines. This analysis indicated that carnallite has a three times lower strength than halite, but it has the same functional relationship between size effect and strength as a rock salt, as discussed in the previous paragraph (Figure 9.1.3).

9.1.2 *Triaxial compressive strength*

The consideration of triaxial compressive strength is represented on the basis of the publication 'Elasticity and strength of ten natural rock salts by F. D. Hansen et al.[4] The test results are obtained from quasi-static triaxial compression experiments and petrographic studies. The only load path used in these experiments was application of axial stress differences at a constant confining pressure. The experiment has been carried out on the rock salt samples from ten different locations in the United States.

Since experiments on all ten of the salts produced qualitatively similar stress-strain response, Jefferson Island data will be used to illustrate deformational characteristics of natural rock salt. A set of typical triaxial compression experiments conducted on Jefferson Island salt produced the stress-strain curves shown in Figure 9.1.4. This figure reveals that both ultimate strength and ductility increase with increasing confining pressure. However, this is not a proportional effect. For example, a confining pressure of 1.7 MPa increased the ultimate strength with 65% over the unconfined case. A confining presure of 20.7 MPa only increases the ultimate strength with an additional 53%. Ductility also is increased with confining pressure. At a pressure of 20.7 MPa, the salt underwent 25% shortening before it failed to hold increasing loads.

Another characteristic of rock salt is that its yield strength is essentially zero, as shown in Figure 9.1.5. Deformation is not linearly elastic at the outset; even at very low strains and streses, time-dependent deformation occurs. The initial concave upward nature of the curves probably is due to crack and interface closure. At strains of 0.1%, the deformation is nearly linear, about half elastic and half time-dependent. However, the only real measure of elastic properties can be made on the unload/reload excursions. The unload/reload cycles were time-dependent; the specimens neither recover nor creep when held at a constant stress of less than about 90% of the previous maximum.

The data in Figure 9.1.6 were obtained at room temperature. At higher temperatures plastic flow is promoted by decrease of strength and increase of ductility. Elevated

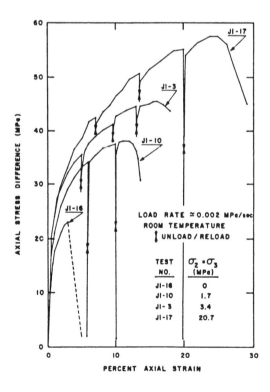

Figure 9.1.4. Axial stress difference as a function of percent of axial strain for Jefferson Island salt (after Hansen)

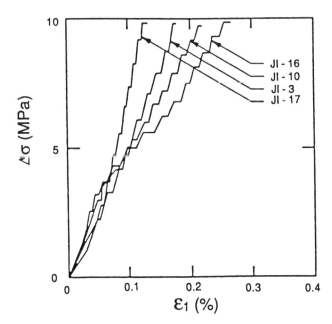

Figure 9.1.5. Detail of low axial stress difference and low axial strain for Jefferson Island salt (after Hansen)

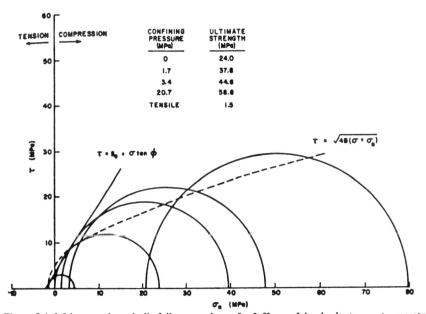

Figure 9.1.6. Linear and parabolic failure envelopes for Jefferson Island salt at room temperature (after Hansen)

temperature data are not available from all ten sites, so discussion of mechanical properties will centre around only room temperature data.

Ultimate strength values were obtained from stress-strain curves as in Figure 9.1.4. for each of the ten sites. Sufficient data were available to construct the Mohr's circle diagrams for stresses at failure. Typically, ultimate shear strengths of all ten salts increased to

Table 9.1.1. Failure envelope parameters at room temperature for ten rock salts (after Hansen)

Mines	Experimental data		Linear		Parabolic	
	σ_r	σ_c	S_o	ϕ	σ_o	B
SE New Mexico (1900' level)	1.26	16.9	2.31	59.5	1.26	2.46
SE New Mexico (2700' level)	1.63	25.7	3.24	61.7	1.63	3.90
Lyons	1.56	25.2	3.14	62.1	1.56	3.85
Permian	1.72	22.1	3.08	58.8	1.72	3.18
Paradox	2.61	33.6	4.68	58.9	2.61	4.84
Jefferson Isl.	1.54	24.0	3.04	61.6	1.54	3.63
Weeks Island	1.24	13.9	2.08	56.7	1.24	1.93
Côte Blanche	1.93	25.2	3.49	59.1	1.93	3.65
Avery Island	1.17	23.1	2.60	64.6	1.17	3.70
Richton	1.32	13.3	2.10	55.0	1.32	1.79
Vacherie	1.12	15.3	2.07	59.7	1.12	2.24

approximately 30 MPa at the highest confining pressures. The values determined for the unconfined compressive strength σ_c, and the indirect tensile strength, T_o, were used to construct two failure criteria:

1. The linear Mohr-Coulomb failure envelope, $\tau = S_o + \sigma \tan \phi$;
2. The parabolic Mohr failure envelope, $\tau^2 = 4B(\sigma + \sigma_o)$.

Fitting parameters for the two failure envelopes are presented in Table 9.1.1. The characteristic linear and parabolic failure envelopes fitted to Jefferson Island data are presented in Figure 9.1.6. While neither of the two failure criteria may be appropriate for all situations, they do provide an expedient method for comparing the ten salts.

On a percentage basis, the strength values in Table 9.1.1 vary considerably more than do the mean values of the elastic parameters given in Subsection 9.1.4. No conclusive correlations could be made between the salt characterization based on petrofabric studies and the variability of strengths. However, on the average, σ_c is about 20% higher for bedded salt than for dome salt. Similarily, σ_t is about 10% higher for bedded salt than for dome salt.

In conclusion, it could be stated that the ultimate shear strength values were found to increase with increases in confining pressure. Values of about 30 MPa were common at the highest confining pressures. The unconfined compressive strength, σ_c, and indirect tensile strength, σ_t, were used to fit both a linear and parabolic failure envelope for each site. Additionally, there were sufficient data to construct Mohr's circles of stress at failure for each site.

9.1.3 *Tensile strength*

Tensile strength of the salts is usually obtained by the Brazilian test, which determines indirect tensile strength. This is an indirect testing method in which a diametrical line load is applied over the length of a rock cylinder ($L/D < 1/1$). The loading rate of the salt disc is in a range between 0.005 and 0.007 MPa/sec. At this rate crack initiation and propagation could be easily observed. The tensile strength of salts determined by the pull test differs to some degree from tensile strength determined by the Brazilian test, due to the testing procedure and the nature of salt itself.

The tensile strength of salt of the Gulf region ranges from 0.2 MPa to more than 3 MPa, with 1.0 MPa being a typical value. The unconfined compressive strength shows a narrower range of variation, with 23 MPa being a typical value. It may be noted that in salt, the ratio of compressive strength to tensile strength is typically more than 20, whereas a value of 10 is typical for brittle rocks.[2] However, as can be seen from Table 9.1.1, the ratio of compressive strength to tensile strength is in the range between 10 and 20, where six samples of a total of eleven have a ratio below 15.

The general conclusion that salts have a lower tensile strength than brittle rocks is quite logical because mechanical behaviour of salt differs from a great majority of petrological and ore units.

9.1.4 *Elastic parameters*

The representation of elastic parameters is also based on the publication 'Elasticity and strength of ten natural rock salts' by F.D. Hansen et al. and other authors listed in Hansen's paper.[4] The results of the experiment indicated that the elastic moduli do not vary substantially among sites.

The elastic parameters are obtained as an integral part of quasi-static triaxial compression experiments, as discussed in the previous subsection. Axial stress differences were applied to the specimen in a load-hold procedure whereby the stress was manually applied in even increments.

Young's modulus and Poisson's ratio were determined from the load/unload cycles. An increase in confining pressure tended to increase Young's modulus and reduce Poisson's ratio, but that tendency was not well defined. Therefore, any confining pressure effect was neglected and average values of Young's modulus and Poisson's ratio were calculated. Values determined for each site are presented in Table 9.1.2 along with the standard deviation and range for value. Eliminating the Côte Blanche value which was based on a single experiment, the elastic modulus means ranged from 26.6 GPa to 31.5 GPa and averaged 30 GPa. Similarly, Poisson's ratio means ranged from 0.29 to 0.38 and averaged 0.35. The elastic parameters appear to be insensitive to the petrographic variations.

In conclusion it could be stated that the elastic parameters did not vary appreciably

Table 9.1.2. Elastic parameters at room temperature for ten natural rock salts (after Hansen)

Mines	Young's modulus GPa			Poisson's ratio v		
	Mean	δ	Range	Mean	δ	Range
SE New Mexico	n.a.	n.a.	29.6 – 36.5	n.a.	.17 – .26	
Lyons	n.a.	n.a.	n.a.	n.a.	n.a.	n.a.
Permian	26.6	3.7	19.0 – 33.4	.33	.05	.24 – .41
Paradox	31.0	3.4	25.2 – 36.3	.36	.10	.09 – .50
Jefferson Island	29.5	3.5	25.0 – 34.4	.29	.07	.17 – .39
Weeks Island	30.5	7.1	21.5 – 42.3	n.a.	n.a.	n.a.
Côte Blanche	24.1	Only one test		.41	Only one test	
Avery Island	30.6	5.8	21.0 – 38.2	.38	.06	.31 – .47
Richton	31.5	3.0	26.7 – 36.4	.36	.09	.21 – .55
Vacherie	31.1	3.5	26.7 – 37.6	.34	.03	.29 – .39

n.a. = not available; δ = standard deviation

among the ten sites. The strength value showed large variations, especially at low confining pressure. From the experiments it becomes evident that, rather than the petrographic composition, the internal structure of the salt influences its strength.[4]

9.2 ELEMENTARY BEHAVIOUR OF SALT

The salt behaviour depends on several factors such as salt composition, mechanism of salt loading, type of acting stress, stress differences, temperature and others. Elementary behaviour of salt is discussed in the following four topics.

9.2.1 *Characterization of salt behaviour*

The mechanical characteristic of the salt behaviour is that, at low stress magnitude, it exhibits elasticity. There is a general opinion that the majority of salts exhibit elastic behaviour for the stress field up to 7.0 MPa. The limit of elastic behaviour of the salt is influenced by its mineralogical composition as well as a rate of incremental loading. The elastic failure takes place when the boundary contact between granular crystals is weakened. Most salts under uniaxial compression, up to elastic stress limit, exhibit a markedly brittle behaviour.

However, when loading overcomes the elastic limit, the plastic behaviour is exhibited by flow deformation. The intensity of flow deformation depends on the level of shear stress present. The plastic behaviour is represented by ductility and interionic sliding inside the crystals. When plasticity is introduced, then both types of deformation takes place simultaneously and salt exhibits elasto-plastic behaviour. The deformation which predominates, depends primarily on the magnitude of the loading stress. Also, there is a general opinion that the majority of salts exhibit elasto-plastic behaviour for the stress field up to 14 MPa, and become virtually plastic (ductile) over 14 MPa. Due to increase of loading magnitude with the mine depth, some deeper portions of the salt deposits are in the state of ductile or plastic deformations.

It is very well known that elastic deformations are recoverable and plastic deformations are irrecoverable. The degree of deformation recovery when imposed load is removed, will depend on the ratio between the elastic and the plastic component present. Some deformation recoveries due to shear stress relief are present for a moderate magnitue of loading. Large unrecoverable deformations are present only for large magnitude of loading (high shear stress concentration).

Loading and reloading of rock salt samples during testing procedures will influence their particular behaviour during deformation and failure. For example, if the uniaxial loading stress is decreased below 7 MPa, the rock salt become increasingly friable and tends to fail more readily by brittle failure under relatively small stress. Under these circumstances, the salt fails and deforms almost entirely by brittle fracture. Although this indicates that the strength of salt material is severely decreased in the brittle region, the brittle fracture strain rate is increased. The diagrams of axial stress-strain relationship for loading and reloading salt samples are present in Figure 9.2.1. Further remarks on the salt behaviour are given for each elementary type of deformation described as follows.

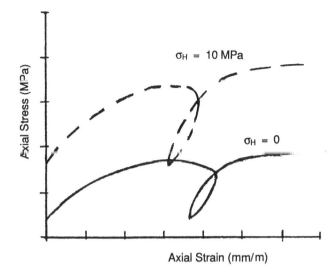

Figure 9.2.1. Stress-strain relation for loading and reloading of salt material (uniaxial and triaxial compression)

9.2.2 *Elastic behaviour*

The salts from evaporite sequences and domes throughout North America and Europe have been experimentally investigated in regard to their elastic behaviour. The principal model of elastic behaviour is obtained from stress-strain diagrams. In regard to elastic behaviour two principal types of salt could be distinguished.

1. The ratio of elastic to total strain of salt is between 86% and 100% at room temperature. This salt is considered as linear elastic with brittle failure (Figure 9.2.2).This type of salt occurs in domal deposits which had undergone intensive metamorphic changes. The elastic salt is a source of salt and gas outbursts, as it is a very well known case in German salt mines.

2. The ratio of elastic to total strain of salt is between 10% and 35% at room temperature. This salt is considered as linear elastic only for a low magnitude of loading, for example up to 3.5 MPa, for the majority of rock salts.

The degree of elasticity of the salt, of course, will depend mainly on the content of elastic strain. It would be fair to say that the majority of salt deposits all over the world exhibit a low degree of the elastic strain component. For example, laboratory investigations of the salt from the Canadian prairie indicate that the representative value of the ratio of elastic to total strain is 10%.[5]

The linear elasticity of salt due to a short term of loading was discussed by Hansen and co-workers in regard to mine design and other conventional applications. Design and analysis of a salt structure based on linear elasticity and a falure envelope are valid only for the salts with very high elastic components or for limited loading conditions (below the plastic region). The stresses from the elastic analyses around openings are substituted into an expression for failure to assess the potential of salt fracturing. If failure is likely to occur, then the salt structure has to be redesigned or the loading of the salt structure should be reduced.

For many cases, a simple analysis based on linear elasticity and failure criteria is adequate if stresses and temperatures are low enough to say that non-linear deformations, developed with the time, are insignificant. A good example is the Avery Island mine

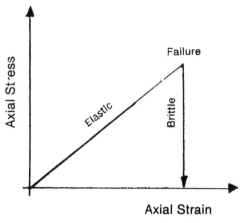

Figure 9.2.2. Model of elasto-brittle behaviour

(Louisiana salt dome, USA), with a shallow depth and a low extraction ratio, where the 150 m level was opened more than 75 years ago and the roof to floor closure is negligible. The elastic parameters for mine design are discussed and described in Subsection 9.1.[4]

9.2.3 *Elastic and plastic behaviour*

The elastic and plastic behaviour could be described by a simple model as illustrated in Figure 9.2.3. The confined rock salt sample at the beginning of incremental loading shows linear elastic deformation but with further load increases plastic behaviour is induced, which continues until yield failure.

Let alone that elastic deformation and plastic deformation are considered as separate modes of deformability in the great majority of cases, the salt material simultaneously exhibits both elastic strain and plastic strain, as discussed. The principal difference between elastic behaviour and plastic behaviour, as stated in the previous section, is that elastic deformation is temporary (recoverable) and plastic deformation is permanent (irrecoverable). The degree of permanent deformation depends on the ratio of plastic strain to total strain. For example, the potash from the Canadian prairies is known by its low strength and plasticity, where a plastic strain is between 65% and 95% of the total strain.[5]

With an increase of load at room temperature, the salt will come from elastic state to plastic yield, which is an irrecoverable deformation and is independent of any viscous or creep strain.[6] It has been suggested by Dusseault and Mraz[7] that plastic yield is an irreversible deformation which does not arise through brittle failure or diffusion creep, as illustrated in Figure 9.2.4.

The elastic and plastic deformations could be observed also by short-term loading, but at higher load magnitude. For example, some salts from domes of the USA indicated elastic microscopic fractures at a load of 3.5 MPa, but at a load of 13.5 MPa indicated plastic flow dominated by slip on (110) plane. Plastic mechanisms are underlain by dislocation motion which suggests that evolution of the free dislocation density is also important.[8]

At present the techniques of the elasto-plastic analysis are commonly used in solving soil mechanics problems. These analyses could be extended for applications in rock mechanics of salt mining and may be used as a tool for ground control and stability analyses.

For example, the finite element program (BOPACE) was utilized for analysis of cavern

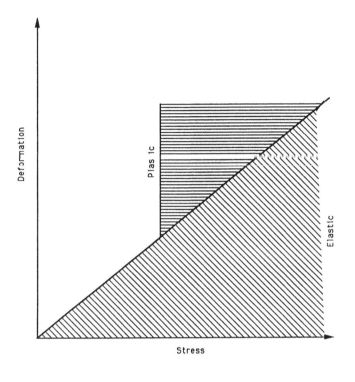

Figure 9.2.3. Model of elastic and plastic deformations

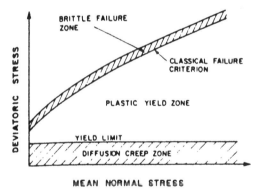

Figure 9.2.4. Salt behavioural zones (after Dusseault and Mraz)

openings assuming that the salt is an elastic-plastic material, and as such at low stress behaves elastically and above a critical stress (yield stress) behaves as a time-independent plastic material. In this case, because salt material was considered to be time-independent, the rate of cavern closure could not be evaluated. Further definition of elastic and plastic behaviour is elaborated in Subsection 9.3.2.

9.2.4 *Plastic behaviour*

The concept of an ideal plastic substance or a general plastic substance arises from the definitions of Prandte, which has been expressed by Nadai as below.[6]

$$\sigma_0 = K + \frac{\sigma_1 - \sigma_3}{2} = \text{constant; ideal plastic substance}$$

$$\sigma_0 = \frac{\sigma_1 - \sigma_2}{2} = f\left(\frac{\sigma_1 - \sigma_2}{2}\right); \text{ general plastic substance}$$

The ideal plastic material will not deform if applied stress is less than σ_0 and it will deform permanently without limit if it is equal to σ_0 (Figure 9.2.5). Also the ideal plastic material will not support a stress greater than σ_0.

As clearly mentioned, the plastic or ductile behaviour is induced by increase of incremental compressive stress or increase of temperature or both. Most rocks including salts have ductile behaviour in the plastic region similar to that of metals. A great deal of attention has been paid to ductile behaviour of salts because of geological and mining importance. During the last decade or so particular attention to salt ductility has been paid in regard to nuclear waste disposal in salt domes. One of the reasons for it is the high plasticity of the salt, which should prevent crack formation and possible escape of radio-nuclides from the repository to the biosphere. However, cracks are found in virgin salt of domes and the question arises how they are induced. Plastic or ductile deformation of salt observed at mine openings could be related to two phenomena.

1. Firstly, due to mining depth. With the increase of depth of salt below ground surface two processes are exhibited: increase of gravitational stress and increase of temperature due to the geothermal gradient, as illustrated in Figure 9.2.6. In this regard there is the opinion that underground mining is limited to a depth of 1200-1300 m due to natural ductility of salt below this depth, which can sustain permanent deformation without resisting to load.

2. Secondly, due to increase of temperature. The appreciable increase of temperature should be expected due to radiation processes of nuclear waste deposited in the salt structure. The increase of temperature will result in the transition of brittle to ductile salt, or to magnification of plasticity of already ductile salt.

Ductile behaviour could be expressed by yield criteria. A yield criterion is a relationship between the principal stresses such that, if it is satisfied, the material becomes ductile. Two general requirements for such a criterion appear at once.[9]

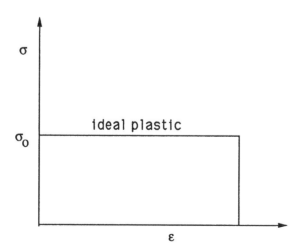

Figure 9.2.5. Model of ideal plastic deformation

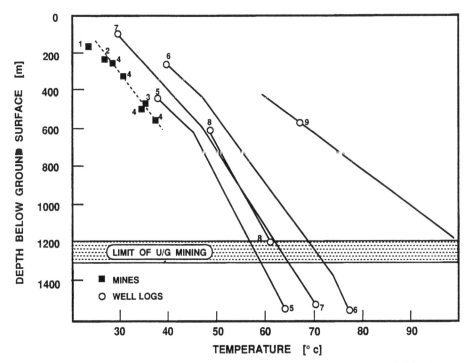

Figure 9.2.6. Salt domes temperatures in function of depth (modified after Kelsall and Nelson)

1. Since the process of yield must be independent of the choice of axes in an isotropic material, any yield criterion must be independent of choice of axes, and thus should be expressed in terms of the in variants of stress deviation ($J_2, J_3, ...$).

2. It has been found experimentally that hydrostatic stress $\sigma_1 = \sigma_2 = \sigma_3$ does not cause yield in salt and crystalline rocks. This suggests that only stress deviation should enter into a criterion for yield.

Various simple criteria for yield were proposed, as some commented briefly.[9]

The 'maximum shear stress of Tresca's criterion' is that yield occurs when the maximum shear stress attains a value of $1/2\,\sigma_0$, characteristic of the material. This may be written as:

$$\sigma_1 - \sigma_3 = \sigma_0$$

The 'von Mises criterion' is obtained by taking J_2 to be a constant, given by:

$$J_2 = \frac{\sigma_0^2}{3}$$

In soil mechanics, the criteria:

$$\sigma_1 - \sigma_3 = \alpha \left(\frac{\sigma_1 + \sigma_2 + \sigma_3}{3} \right)$$

and

$$(\sigma_1 - \sigma_2)^2 + (\sigma_2 - \sigma_3)^2 + (\sigma_3 - \sigma_1)^2 = \alpha^2 \frac{(\sigma_1 + \sigma_2 + \sigma_3)^2}{9}$$

are called 'extended Tresca and extended von Mises criteria' (Bishop).[10] These criteria can be used for mine design and structure stability when the salt is in the ductile state.

9.3 TRANSITIONAL DEFORMATIONS OF SALT

In this section, the transitional behaviour of salt due to short-term loading (incremental load) will be discussed. Long-term loading analyses will be discussed in Chapter 10. Transitional mechanics of salt is controlled by strain rate changes which are induced due to increase in temperature and load. Transitional behaviour of salt is of great interest for the mine design and ground control, because its adverse effects could be utilized for mine stability and salt recovery. Four topics of transitional salt behaviour are further discussed.

9.3.1 *Brittle-ductile deformations*

At a certain strain rate which depends on the corresponding temperature, the halite deformed by tensile stress exhibits a sharp transition from brittle to ductile behaviour, as Skrotzki described in his publication.[11] This behaviour is of particular interest in the case of storage of gas and oil and for disposal of radioactive waste, because the salt at high plasticity prevents crack formation and a possible escape of radio nuclides. On the basis of Skrotzki's work,[11] a summary of brittle-ductile behaviour is presented as follows.

Deformation of the salt at a certain temperature, marked as T_{bd}, exhibits transitional strata of brittle to ductile behaviour, which could be defined as below:

1. Brittle salt has an ability to resist load decreases with an increase of deformation.
2. Ductile salt sustains permanent deformation without losing resistance to a critical load.

Below temperature T_{bd}, the strains at fracture are too small to produce slip steps, but above this temperature the strains are large enough to produce a wavy slip. With the increase of ductility, the density of river patterns observed on fracture surfaces rises markedly. When grain separation occurs, the fracture surface is no longer planar.[11]

The changes from planar to wavy surface takes place in the proximity of temperature T_{bd} when screw dislocations are able to cross slip, as illustrated in Figure 9.3.1. Wavy slip is the result of cross slip of screw dislocation. The stress concentration which was needed for the intergranular crack initiation was relieved and suppressed when crack propagation took place.[11]

Strain rates during diapirism of salt domes are estimated to lie outside the brittle field above a room temperature (Figure 9.3.2).Thus, if the existing cracks formed due to slip mechanism, the tectonic movement would last a short period of time. The crack formations were limited to weak zones such as the layers with a high content of anhydrite, where the brittle field may be much wider.[11]

Strain rates caused by thermal expansion and contraction of salt domes around the disposal of radioactive wastes are expected to be in the ductile field. Disposal conditions could be estimated more accurately than conditions which prevailed during diapirism. The crack formation due to diapiric mechanism should be localized in certain zones. In this regard, the layout of the disposal must be out of the cracked zones, e.g. the area of homogeneous salt. Further, it should be pointed out that the disposal of radioactive waste must be on a certain distance from the cracked tectonic zone to obtain a barrier pillar, which will not permit the escape of radio-nuclides from a repository.

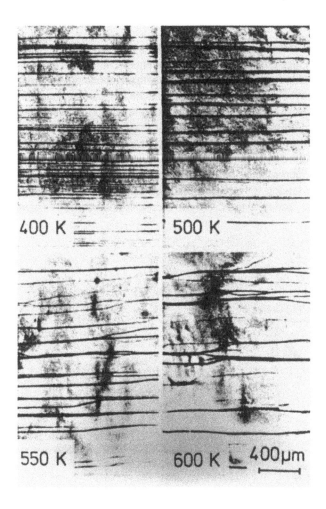

Figure 9.3.1. Influence of temperature on the slip lines of screw dislocations on (110) planes

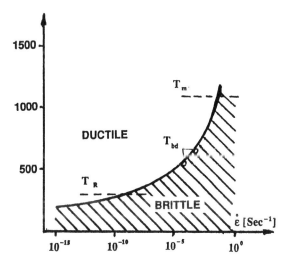

Figure 9.3.2. Strain dependence of the brittle to ductile transition due to temperature T_{bd} (after Skrotzki). The transitional zone, where slip bonds of screw dislocation gliding on (110) planes become wavy, is given between room temperature (T_R) and melting temperature (T_m) (after Skrotzki)

9.3.2 *Elasto-plastic deformations*

The elasto-plastic behaviour integrates elastic strain and plastic strain and both of them make a total strain. As discussed in the previous section the percent of either strain can vary in such a wide range as 1% to 99%. In function of the particular strain content, the salt could behave like an elastic or plastic body. The model of elasto-plastic behaviour is presented in Figure 9.3.3, which clearly indicates the recoverable strain of elastic deformation and the non-recoverable strain of plastic deformation.

A simple example of elasto-plastic behaviour is given for the typical Saskatchewan potash sample ($D = 4.7$ cm). The average crystal size was approximately 8 mm ranging up to 15 mm. The rate of loading was approximately $0.2\ \sigma_c$/sec (σ_c is the uniaxial compressive strength). Figure 9.3.4 shows the relationship between stress and strain which indicates the elasto-plastic behaviour of potash, with great similarity to the model illustrated in Figure

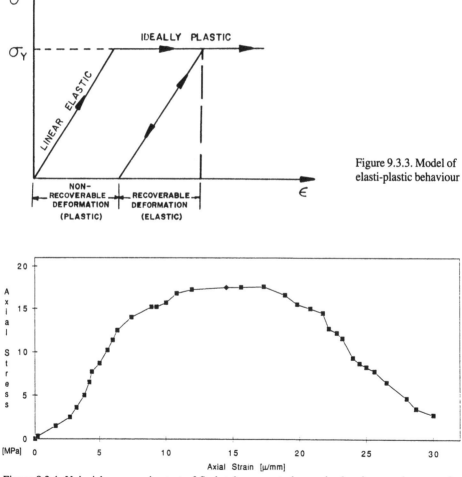

Figure 9.3.3. Model of elasti-plastic behaviour

Figure 9.3.4. Uniaxial compressive test of Saskatchewan potash sample showing reaction at strains beyond the maximum strength (after Coates et al.)

9.3.3. The uniaxial compressive strength of the potash sample is indicated to be 17 MPa at a strain of 1.4% with the sample reaction at strains beyond this strength. At failure, the reaction of the sample decreased considerably but the material remained intact. In a series of steps, the load was then increased again until peak resistance, or reducing yield strength, was obtained, which is represented by the point beyond the maximum on the curve (Figure 9.3.4). The test was stopped when about 20% of the outer surface of the sample had flaked off and the strain was approximately 3%.[5]

Similar tests were carried out on the rock salt samples at the Tuzla basin (former Yugoslavia).[12] During incremental compressive loading the rock salt samples showed deformations and fractures which could be distinguished in four regions:

– Region I: elastic deformations, not visible to the naked eye.

– Region II: plastic deformations, visible by lateral extensions at the central part of the samples.

– Region III: reaction at strains beyond the critical strength, due to shear resistance along fissuring plane with some flaking.

– Region IV: shear failure of salt samples and formation (in the majority of cases) of the hour glass structure (Figure 9.3.5).

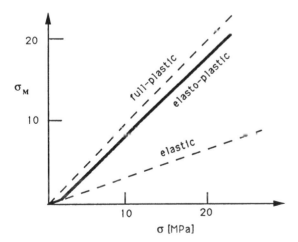

Figure 9.3.5. Deformation and failure pattern of rock salt sample due to uniaxial compressive loading (Tuzla basin, former Yugoslavia)

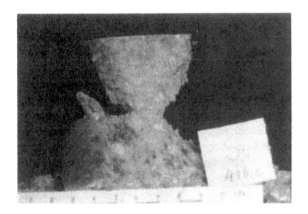

Figure 9.3.6. Confining pressure as a function of the axial stress – samples of the older rock (after Dreyer)

The testing of rock salt samples of Tuzla basin indicated that the ratio of plastic to total strain varied from 33 to 68%.

A tectonic history of orogenic loading indicated the increase of elasticity which has been observed by experimental testing of salt samples. The degree of elasticity of older salts, which often have undergone more extensive orogenic changes, is higher than the elasticity of younger salts.

Elasto-plastic deformations of rock salts under confining pressure have been investigated by W. Dreyer.[1] He showed that above approximately 3 MPa, the confining pressure characteristics are already sufficiently linear, so that rock salt reacts plastically when lateral expansion is prevented. Figure 9.3.6 illustrates diagrams of rock salt behaviour in relation to the elasto-plastic curve defined by laboratory testing. The curve very clearly indicates that elasto-plastic behaviour differs from both elastic and plastic behaviour. However, it is more close to plastic than to elastic behaviour.

9.3.3 *Strain hardening deformation*

A number of hypotheses have been developed to explain the increase of yield strength. In the early days it was assumed that during incremental loading the crystals gradually transformed into an amorphous mass which was responsible for strain hardening.[1] A. Seeger characterised the deformation due to incremental loading of crystalline solids as 'plastic' and that of the amorphous mass as 'viscous'. Integrated plastic and viscous deformation of rock salts can be modelled as a strength hardening during yielding (Figure 9.3.7).

The main element in strain hardening mechanics is the distribution factor of micro-structural defects of the crystals, which originated due to irregular crystal growth and loading deformations. The micro-structural defects usually form planar dislocation arrays and thus they provide the prerequisite for development of internal stresses. These internal stresses could explain the hardening effect. According to some dislocation theories, strain hardening is designed as an impediment of dislocation motions and takes place with increasing deformation.[1]

W. Dreyer considers that for polycrystalline rock it is very difficult to predict the amount of strain hardening from a knowledge of the internal stresses generated during plastic

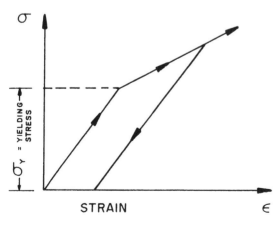

Figure 9.3.7. Model of strength hardening

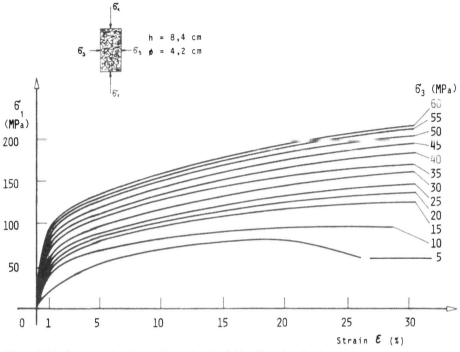

Figure 9.3.8. Stress-strain relationship under triaxial loading of rock salt samples

deformations. The grain boundaries themselves consist of planar dislocations, and the influence of deformation on strain hardening is affected by the presence of impurities.[1]

The cylindrical rock salt samples from the Tusanj salt mine, Tuzla, former Yugoslavia, have been tested under a short-term compressive uniaxial and triaxial loading.[13, 14] The stress-strain curves indicated three regions of deformation. The first region showed a steep and linear curve, which corresponds to the elastic field of deformations. The second region exhibits deformation with a small radius of curvature at limited distance, which corresponds to the transitional e.g. elasto-plastic field. The third region after yield point shows further but gradual strain increase, which corresponds to a strain hardening region (Figure 9.3.8).

The curves of the stress-strain relationship of tested samples exhibit an increase of the strain hardening according to increments of axial strain and the principal stresses. It should be noted that the maximum strain hardening effect of confined cylindrical rock salt samples is achieved when they are contracted for 30% of the original height. In the case of uniaxial compressive loading of cylindrical rock salt samples (H/W = 2) the strain hardening effect is almost nil.

For the strain hardening characterization, it is necessary to determine the elastic modulus E (steep and linear section of the curves) and the elasto-plastic modulus E_T (gently dipping section of the curves), as illustrated in Figure 9.3.9. The determination of the modulus is based on the changes of magnitude of minor principal stress (σ_3) and an increase of axial stress (σ_1), as presented in Figure 9.3.10.

The strain hardening characterization is given by calculation of the strain hardening

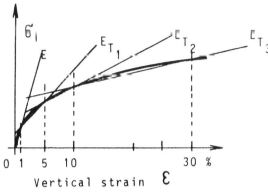

Figure 9.3.9. Graphic determination of elastic modulus and elasto-plastic modulus

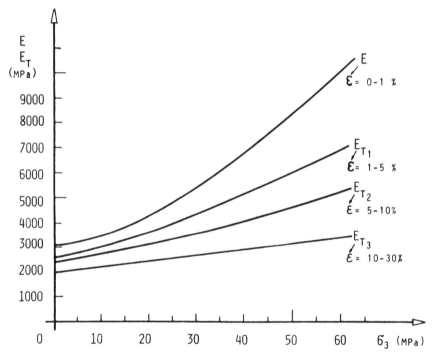

Figure 9.3.10. Functional relationship between minor principal stress and axial deformations of rock salt

parameters as written below:

$$\sigma = H\,(\varepsilon_p);\; \frac{d\sigma}{d\varepsilon_p} = H'(\varepsilon_p) = \frac{d\sigma}{d\varepsilon - d\varepsilon_e} = \frac{1}{\dfrac{d\varepsilon}{d\sigma} - \dfrac{d\varepsilon_e}{d\sigma}}$$

$$H' = \frac{E_T}{1 - \dfrac{E_T}{E}}$$

These equations show that characterization of strain hardening behaviour depends on the relationship between axial deformation and elastic constants of proportionality (H').

The stress path curves indicate different relationships of the stress-strain curves, as well as the failure envelopes or yield surfaces. In the subsequent analysis, it has been assumed that the intermediate principal stress does not occur in the yield criterion. In this case the yield conditions can be expressed by a general form, as written below:

$$f(p, q) = 0$$

where:

$$p = \frac{(\sigma_1 + \sigma_3)}{2};$$

$$q = \frac{(\sigma_1 - \sigma_3)}{2}$$

The construction of the curves for the yield criterion is given by the relationship between p and q for several cases of axial deformations, namely 1%, 5%, 10% and 30%, as illustrated in Figure 9.3.11. This example indicates that it is possible to define linear strain hardening on the basis of the stress-strain relationship for the case of triaxial loading. Also, as for the elastic and elasto-plastic response it is possible to construct a failure envelope for different strain conditions.

Finally it could be concluded that the rock salt under triaxial conditions of loading becomes the strain hardening material when 1% of initial deformation is in effect.

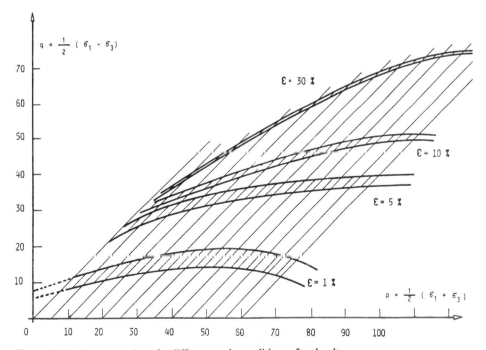

Figure 9.3.11. Failure envelope for different strain conditions of rock salt

Rock mechanics analysis in underground mining should consider the theory of strain hardening an important factor for better prediction of the mechanical behaviour and bearing capacity of rock salt structures. The strain hardening phenomena should be utilized for mine design and mine layout. For example, the strain hardening of potash is quite remarkable. Its strength could be increased three to four times which could be utilized for the increase of the bearing capacity of mine pillars.[15]

9.3.4 *Tempering of salt (work softening deformations)*

The tempering properties of salt have been studied by Dreyer in both natural salt crystal and salt crystalline mass. A single crystal of the salt under uniaxial compression behaves brittle at room temperature. Crystals of rock salts lose their brittleness after extension tempering at approximately 600°C and exhibit a critical shear stress up to 1 MPa. During testing, glide and translation were observed along crystal dislocations which originated in undeformed crystals due to irregular crystal growth.

The thermally treated monocrystalline samples of rock salt showed appreciable decrease in strength. This phenomenon is described as the crystal's loss of memory of the in situ stress history. This effect of work softening is mostly due to reduction of dislocation density in crystal lattice because of higher temperature.

The relationship between stress and strain illustrated in Figure 9.3.12 represents the behaviour of untempered and of tempered rock salts. The samples of rock salt have been loaded at rates of 0.1 MPa per increment up to a maximum of 20 MPa. Tempering leads to the complete recrystallization of rock salt which assumes mechanical properties much like those of an ideal crystal, because all the effects of the strain hardening due to geological loading are eliminated.[1]

Influence of the temperature on the strength deterioration of the salt has been studied by Vouille et al. They tested eight samples of Tersanne rock salt, which have been submitted to two cycles of loading and unloading deformation, one at room temperature and the other at 80°C. The increments of confining pressure were at an average rate of 0.1 MPa/min from 0 to 30 MPa.[1] The yield criterion is defined by a zero angle of internal friction regardless of the temperature. It appears that the temperature has an influence on the cohesion, because

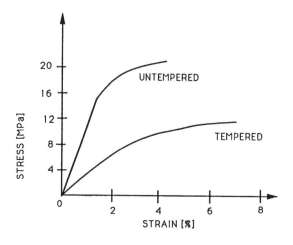

Figure 9.3.12. Stress-strain relationship of rock salts before and after tempering

with an increase of temperature from 30°C to 80°C the average cohesion decreases from 2.8 MPa to 1.9 MPa. The load relaxation test indicated that the viscosity of rock salt decreases when the temperature increases. This means that a state of equilibrium is reached more rapidly when the temperature is high than when it is low.[1]

The work softening of the salt due to increase in temperature has significant influence on the properties and behaviour of this material. This phenomenon of deterioration of strength of rock salt is of particular interest for large underground excavations within rock salt stocks where it is proposed to store material which might rise the temperature and cause heating of surrounding rocks. Under these circumstances, deterioration of the strength of rock salt and changes of its properties could result in a failure of rock salt structure.

Finally it should be noted that the melting point of rock salt is 800°C and the tempering temperature is 600°C in order to avoid the adverse effects of thermal stress. At this temperature a tempering time should be eight hours and will eliminate the effects of previous geological loading during orogenic events.

9.4 TIME-INDEPENDENT FAILURE

There is a general opinion that rock salt and potassium salts always behave only plastically with a permanent deformation without fracturing and without volume increase. The underground mine observations clearly indicated that this is not the case, because failures involving various tensile stress mechanisms have been observed, particularly in greater mining depths. At the present time, it is established that rock salts have semi-brittle properties to which the cataclastic deformations have to be related.

9.4.1 *Micro-cataclasis structures*

Micro-cataclasis mechanisms around mine openings are generated by mining stresses, restricted to the very small induced 'minute micro stress fields'. Formation of micro-cataclasis around mine openings could be indicated by volume increase which causes corresponding increase of the convergence. Development of micro-cataclasis probably belongs to a mechanism of quasi-plastic flow, accompanied by porosity and permeability.

The mechanics of formation of the cracks could be related to indirect tension similar to that in the case of axial cleavage generated in compression where stress concentration will occur within the induced tensile stress field. These induced tensile stress fields may occur at a notch in the boundary of a 'Griffith flow', which will form a rim crack in the oblique position.

According to the mechanics of crack formation they should occur systematically; however, obvious lack of the lineation of the micro-cataclastic system within the rock salt openings disagrees with this postulation. It is most likely that the irregularities of the system occur due to the presence of previous cracks in rock salts formed by endogenous and exogenous processes. However, the main trend of the micro-cataclastic system formed by induced mining stress still could be noticed (Figure 9.4.1). Also, the data of the velocity analysis showed a faint element of the expected linearity.[16]

In addition to micro-cataclasis by mining operation, the cracks could originate also by temperature changes as would be the case in a nuclear waste repository. In this regard, an outer part of the mine structure is heated and exerts pressure on the inner part, which is at a lower temperature (Figure 9.4.2).[16] As a result of this mechanism, concentric brittle cracks

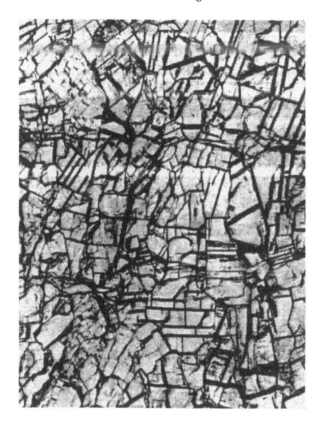

Figure 9.4.1. Slide of rock salt
from the roadway with micro-
cataclasis of closed cracks
(Asse salt mine, Germany)

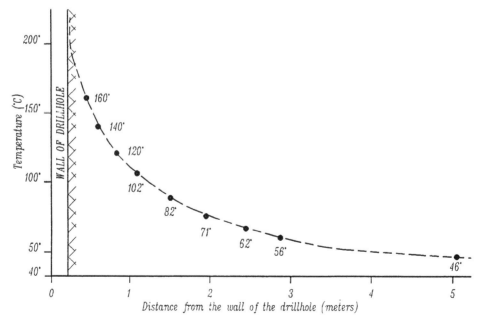

Figure 9.4.2. Temperature decrease at distance from the drill hole wall.

are formed due to the induced tensile stress by thermal expansion which corresponds to the axial cleavage theory.[17, 18]

It should be noted that the cracks induced by mining operations or thermal effects as well as the cracks of geological origin are mostly closed and a significant proportion of them reheeled. The closed cracks cannot be indicated by the changes of acoustic velocity and accordingly they are not detectable. The microclastic salt structure of the closed cracks exhibits a cataclastic-plastic-elastic equilibrium which permits a stable behaviour of the rock salt structure. Under these circumstances, satisfactory maintaining of the mine structure is by self-supporting capabilities.

The length of the cracks varies from several microns to several millimeters and it is expected that they are closed. However, due to stress changes, they could be reopened and that will result in an increase of the salt porosity and relevant permeability. Under these circumstances, rock salt structures will have a deteriorated bearing capacity, particularly when moisture is present.

The conditions for the formation of a stress level for opening of cracks are not known. It is known only that at a low stress level, in spite of abundance of cracks, they are tight, particularly if they are not in mutual contact. It could be assumed that at a low stress level the pure plastic properties of rock salt play a role in keeping the micro-cataclastic mass compact. Finally, as other rock structures, the salt mine structure will change its properties with stress increase and failure formation will be in effect, as further discussion.

9.4.2 *Mechanics of fracturing*

Fracture of salt structures corresponds mainly to split failure mechanics. These split failures are generated by indirect tensile stress which have been induced by uniaxial compression.[18] The salt fails by semi-brittle tensile fracture under the following circumstances.

1. The rock salt structure is in the clastic-elastic state due to micro-cataclasis with open cracks, which could be indicated by acoustic velocities.[19]

2. The acting lateral stress named as minor principal stress is equal to or exceeds the tensile strength of the microclastic rock salt structure ($\sigma_3 \geq \sigma_T$), as illustrated in Figure 9.4.3.

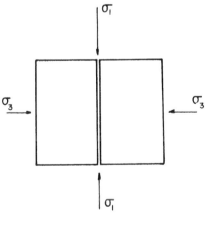

$\sigma_3 = \sigma_T$ Figure 9.4.3. Model of axial split failure

Figure 9.4.4. Axial fracturing of rock salt pillar at greater depth (Poland)

3. The rock salt splits parallel to the direction of the major principal stress, and perpendicular to the direction of the maximum tensile strain.[17, 18]

4. The micro-cataclastic rotation of the grains oblique to the gravitational force is altered by further grain rotation in a direction parallel to axial failure.

The axial macro-cataclasis could be observed in some deep rock salt mines, particularly in the case of underdesigned rock salt pillars (Figure 9.4.4). This type of fracturing, however, is not frequent, because the salt pillars are mainly designed properly.

The split failures are very close to the skin of the rock salt pillars (outer zone) and this could be quickly visualized, because deformation progresses within a short period of time. The mechanism of axial splitting is easily recognizable from the fracture surfaces and fracture shapes, which occur in a more or less systematic pattern forming a characteristic structural element of macro-cataclasis.

Until the present time, the mechanics of axial cleavage in the salt structures is not very well understood, particularly for the circumstances which contribute to the generation of the cracks, their opening, and propagations.

9.4.3 *Progressive wall failure*

The mechanics of macro-cataclasis of slab or slough failure in the walls of underground excavations in rock salt mining could be better understood by using an analogy with the same type of failure, for example in coal mining.

The sloughing of salt mine structures is similar to soft coal yielding deformations, which could be analyzed by elasto-plastic stress analysis.

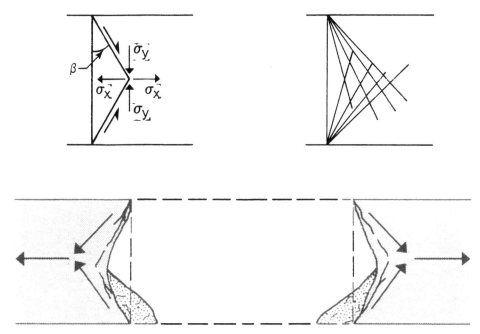

Figure 9.4.5. Model of slabbing failure of rock salt structure

The slabbing failure of rock salt structures is also generated by indirect tensile stress, which is induced by vertical compressive stress, as further discussed.

1. The plastic-elastic state is located in the immediate inner zone of the rock salt structure, where initiation and propagation of the failure will be in effect. The failure position corresponds to the action of the maximum vertical stress on the rock salt structure $\sigma_{V(\max)}$).

2. The induced tensile stress is perpendicular to the transverse profile. It has a variable magnitude which reaches a maximum at the centre ($\sigma_{T(\max)}$) and vanishes at contacts between the roof and floor of the mine structure ($\sigma_T = 0$), as illustrated in Figure 9.4.5.

3. The yielding deformation of the micro-cataclastic zone is by quasi-plastic flow, which slowly progresses with time. The progressive slabbing of the rock salt pillars usually results in the formation of hour glass structure.

The macro-cataclasis of slabbing is mainly exhibited in the drift walls of deeper rock salt mines (Figure 9.4.6). For example the location and configuration of the slabbing failure to the drift have been determined in the Asse salt mine (Germany). The categorization of the sloughing has been indicated by the increase in the amplitude of the acoustic velocity and by failure observations by endoscope.[19]

The cross sectional profile of the drift wall obtained by acoustic measurement indicated a large failure with a fracture opening of 4-8 mm. The configuration of the identified failure corresponds to the model of slabbing failure, with a typical fracturing characteristic (Figure 9.4.7). The centre of the profile of failure is between 0.5 and 1.0 m from the drift's free face.

The longitudinal sectional profile along the drift wall obtained by an endoscope observation indicated that the large failure is continuous and locally it is followed by the

Figure 9.4.6. Progressive salt slabbing along elongated fissured planes

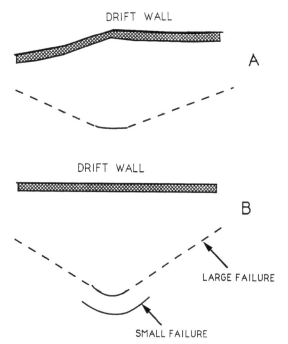

Figure 9.4.7. Cross-section of drift wall with the indication of slabbing failure by acoustic monitoring (Asse salt mine, 490 m level)

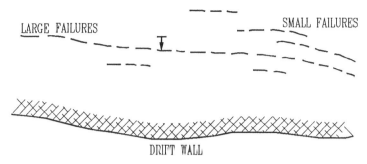

Figure 9.4.8. Longitudinal section of drift wall with failures indication by endoscope (Asse salt mine, 490 m level)

small failure with a fracture opening of 1-2 mm. The small failure had small extension laterally as well as vertically. The large and small failures indicate a microclastic zone in the state of the plastic-elastic deformations. The width of this zone is up to 0.4 m (Figure 9.4.8).

It could be concluded that sloughing of rock salt structures is a continuous process, which is clearly indicated by the initiation and propagation of small failures behind a main slab failure. Most likely, initiation of the main failure started during transformation of rock salt from the brittle-elastic state to the plastic-elastic state, e.g. during micro-cataclastic processes. The higher density of the micro-cataclasis is rather inward than outward of the salt structure.

To avoid continuous slabbing and related drift widening or pillar narrowing even a small confinement of the free faces of mine structures could help because the lateral tensile stress would be offset to a certain degree.

9.4.4 *Rock falls failure*

The rock falls from the roof of the mine openings, particularly from the stope back, are also present in salt mining operations.

The mechanics of macro-cataclasis related to the roof falls is based on the action of pure tensile stress in the semi-brittle material.

The mechanism of rock salt breakage and its fall could be defined in two stages, as commented below.[18]

1. Firstly, rock salt separation is exhibited along structure continuities (sedimentation surface) or structural discontinuities (fissure elongation plane). The separation failure is generated by the vertical pull force which equals to a gravity of induced stress, e.g. dead load of rock salt block. In this case, a physically delineated rock salt beam or plate is erected which will further deform (Figure 9.4.9).

2. Secondly, bending of a delineated salt structure is exhibited under its own weight. In this case, the failure is generated by the lateral pull force, followed by progressive opening and cracking until the final rock salt falls. The mechanism of stope back convergence and fracture could be related to the failure criteria of a beam with clamped edges. The duration of rock salt beam deflection before it falls could be related to beam thickness and length as well as its strength properties.

The majority of observed rock falls in underground mine rooms indicated rather local macro-cataclasis of roof falls than large collapse of the stope back (Figure 9.4.10). This phenomenon suggests that only locally the cataclastic-plastic equilibrium of the salt back is disturbed, probably due to maximum tensile stress as well as presence of a denser micro-cataclasis structure. Under these circumstances, a self-supporting ability of the stope

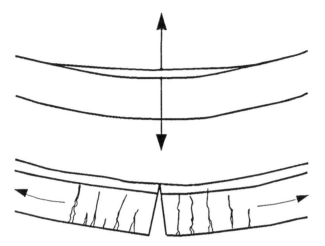

Figure 9.4.9. Model of 'rock salt beam' separation, bending and break

Figure 9.4.10. The area of rock falls is delineated by hanging rock bolts from the roof (gypsum mine, Southern Ontario)

back is affected only locally with thinner slab separation rather than thick rock beam.

It should be pointed out that maintaining self-supporting abilities of the stope back is of great importance because its support is not integrated into the mining extraction technology: for example, mining of thick and very thick salt deposits.

Generally speaking, the majority of cataclastic deformations of salt mine structures are in the axial direction rather than in the lateral direction. This phenomenon could be related to the predisposition of the micro-cataclasis within salt deposits, the character of fissuring due to mining activities, and possibly to particular properties and behaviour of the salt mass itself.

Finally, cataclastic deformations of the mine openings in the salt deposits are a very important factor, not only with respect to mine stability and ground control but also relating to safe nuclear waste repository so that radio nuclides could not migrate into the biosphere, and others.

REFERENCES

1. Dreyer, W. 1973. (1) Relationship between cubical compressive strength and the edge length, and (2) Approaches to a quantitative formulation of hardening salt. In *The science of rock mechanics*, Part 1, pp. 61-63, 105-126. Clausthal: Trans. Techn. Publ.
2. Kelsall, P.C. & J.W. Nelson 1983. Geologica and engineering charcteristics of Gulf region salt domes applied to underground storage and mining. In *Proc. of 6th Int. Symp. on Salt*, The Salt Institute, Virginia, Vol. 1, pp. 519-542.
3. Fairhurst, C. et al. 1979. Rock mechanics studies of proposed underground mining of potash in Sergpipe, Brazil. In *Proc. of 4th Int. Congr. on Rock Mech.*, Montreux, Switzerland, Vol. 1, pp. 131-136.
4. Hansen, F.D. et al. 1984. Elasticity and strength of ten natural rock salts, In *Proc. of 1st Conf. on Mechanical Behaviour of Salt*, pp. 71-83. Clausthal: Trans. Techn. Publ.
5. Coates, D.F. et al. 1986. *Potash creep and pillar loading*. CANMET internal report MR 68/4-LD, Ottawa, pp. 1-18.
6. Nadai, A. 1950. *Theory of flow and fracture of solids*, Vol. 2, pp. 458-465. New York: McGraw-Hill.
7. Dusseault, M.B. & D.Mraz 1983. Salt constitutive behaviour. In *Potash'83. Potash technology*, pp. 311-319. Toronto/New York: Pergamon Press.
8. Hansen, F.D. & P.E.Senseny 1983. Petrofabrics of deformed salt. In *Potash'83. Potash technology*. pp. 347-354. Toronto/New York: Pergamon Press.
9. Jaeger, J.C. & N.G.W.Cook 1969. Behaviour of ductile materials. In *Fundamentals of rock mechanics*, pp. 215-227. London: Nethuen.
10. Bishop, A.W. 1966. Strength of soils as engineering materials, *Geotechnique* 16: 91-130.
11. Skrotzki, W. 1984. An estimate of the brittle to ductile transition of salt. In *Proc. of 1st Conf. on Mechanical Behaviour of Salt*, pp. 381-387. Clausthal: Trans. Techn. Publ.
12. Jeremic, M.L. 1969. Angle of break and internal friction of rock salt. *Min. & Metall. Bulletin* 6: 119-124 (in Yugoslav).
13. Mandzic, E. & M.L.Jeremic 1989. Utilization of strain hardening phenomena to alleviate ground control problems in dry salt mining. *J. Min. & Geol.* (Faculty of Mining and Geology, Tuzla), 28 (1) (in Yugoslav).
14. Jeremic, M.L. & E.Mandzic 1989. Strain hardening characteristics of rock salt. In *Proc. of Jubilee Symp. on Strata Mechanics*, pp. 89-92. Academy of Science, USSR, Novosibirsk, June 5-9, 1989. Rotterdam: Balkema.

15. Serata, S . 1976. Stress control technique – An alternative to roof bolting. *Min. Engng.*, May 1976: 87-91.
16. Gramberg, J. & J.P.A. Roest 1982. Laboratory and in-situ measurements of cataclastic effects in rock salts. Manuscript of final report of first phase, pp. 1-97. Delft Univ. of Technology, Mining Department.
17. Stacey, T.R. & C.L.Johngh 1977. Stress fracturing around a deep-level cored tunnel. *J of S. Afr. Inst. of Min. & Met.*, December 1977: 124-134.
18. Jeremic, M.L. 1987. Failure of rocks. *Ground mechanics in hard rock mining*, pp. 45-67. Rotterdam: Balkema.
19. Roest, J.P.A. 1986. Acoustic P-wave velocity measurements of cataclastic effects in rock salt. Manuscript of report WAS-335-83-700NL, pp. 1-46. Delft Univ. of Technology, Mining Department.

CHAPTER 10

Time-dependent deformations and failures

The deformations and ruptures are related to the time-dependent behaviour of rocks subjected to constant stress and/or elevated temperature. The majority of rock types, and specifically evaporites, show time-dependent mechanical properties which could cause serious problems associated with the stability of underground openings. The flow deformations of the salts could be characterised by several aspects, as discussed below.

10.1 RHEOLOGICAL CHARACTERIZATION

Rheology follows the definition of Bingham (1928), who concernes the world rheology not only as the science of the actual flow of matter (viscous, plastic flow) but also its deformability in general. Thus rheology is a branch of mechanics of deformable bodies with an attempt to come closer to the real deformations of material in function of time and temperature.[1,2]

10.1.1 *Correspondence principle*

C. D. da Gamma, in his publication,[3] stated that a great number of stress analysis problems in linear viscoelasticity can be solved by using the so-called correspondence principle. This principle is based on the utilization of elastic solutions of similar problems, after applying Laplace transformation analysis. As it is well known, the stress-strain relations for isotropic elastic materials (Hooke's law in three dimensions) may be written as follows:

$$\sigma = 3\,B\,\varepsilon$$

$$\sigma_{ij} = 2\,G\,\varepsilon_{ij}\ (i,j\text{ - }1,2,3)$$

where σ and ε are the mean stress and strain, respectively, σ_{ij} and ε_{ij} the components of stress and strain deviations, and B and G the bulk and shear moduli of the material. The first equation is valid in dilatation or hydrostatic compression, and the second one is applicable to distortion or shear, so that both equations fully define the stress-strain relationships.

On the other hand, in most viscoelastic materials the stress-strain relations may be written in the form of differential equations of the type:

$$f_1\left(\frac{\partial}{\partial t}\right)\sigma = 3f_2\left(\frac{\partial}{\partial t}\right)\varepsilon \qquad \text{(partial derivatives)}$$

$$f_3 \left(\frac{\partial}{\partial t} \right) \sigma_{ij} = 2 f_4 \left(\frac{\partial}{\partial t} \right) \varepsilon_{ij} \quad \text{(ordinary derivatives)}$$

where the functions f_1, f_2, f_3 and f_4 are polynomials in

$$\left(\frac{\partial}{\partial t} \right) \quad \text{(differentiation with respect to time)}$$

This system of ordinary linear differential equations can be solved in conjunction with the stress equations of equilibrium and the boundary conditions.

C. D. da Gama also showed[3] that the correspondence principle may be expressed by stating that a class of stress analysis problems in viscoelastic materials can be solved by taking the inverse Laplace transformation of the elastic solution of the same problem, after the applied stresses or displacements are replaced by their Laplace transformations, and the moduli B and G by

$$\frac{f_2(p)}{f_1(p)} \quad \text{and} \quad \frac{f_4(p)}{f_3(p)}$$

respectively.

The validity of this principle requires that the material departs from undisturbed conditions at time zero.

It is normally accepted that rocks behaving viscoelastically remain elastic in hydrostatic compression, so that in most cases we have:

$$f_1 \left(\frac{\partial}{\partial t} \right) = 1$$

and

$$f_2 \left(\frac{\partial}{\partial t} \right) = B$$

as well as Poisson's ratio is assumed to be approximately constant.

For representing their mechanical behaviour in shear, appropriate rheologic models have to be assumed, and one of the best adjusted to most rock creep curves is the generalized Kelvin substance.[3]

10.1.2 *Generalized Kelvin model*

The rheological behaviour of the salt material has been characterized as that of a viscoelastic substance. In this respect, the salts are best explained by the so-called generalized Kelvin rheological model, that consists of a Kelvin element: spring K_1 in parallel with dashpot η_1, approximating Hookean and Newtonian substances respectively, and spring K'_1 in series with it, as illustrated in Figure 10.1.1.

Upon application of a constant stress σ_0 at time $t = 0$ and when the strain is also zero, the instantaneous elastic strain is:

$$\varepsilon_0 = \frac{\sigma_0}{K_1}$$

The value of the viscous strain develops subsequently, and is given by:

$$\varepsilon = \sigma_0 \left[\frac{1}{K'_1} + \frac{1}{K_1} (1 - e^{-t/t_1}) \right]$$

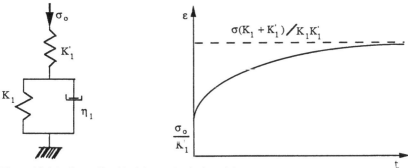

Figure 10.1.1. Generalized Kelvin rheological model

where:

$$t_1 = \frac{\eta_1}{K_1}$$

After a long time the viscous strain approaches asymptotically the value:

$$\sigma_o \left[\frac{K_1 + K_1'}{K_1 K_1'} \right]$$

This model of an instantaneous deformation followed by the time-dependent curve simulates the primary or transient and the secondary or steady state creep. However, the tertiary or accelerating creep is not well represented in any of the classical rheological models.

If a generalized Kelvin behaviour is accepted, it is not difficult to deduce the creep curve of a heterogeneous body formed by several viscoelastic materials connected in series.[1]

10.1.3 *Heterogeneous body model*

The most simple case of viscoelastic behaviour of a heterogeneous salt is given by two viscoelastic models in series. In this case, each substance possesses a spring constant K' linked in series to a spring of constant K and a dashpot of constant η (Figure 10.1.2). The corresponding rheological equation of the model subjected to a constant stress σ_0 is given:

$$\sigma_0 = \frac{\eta_1 d\varepsilon_1}{dt} + K_1 \varepsilon_1 = K_1' \varepsilon_1' = K_2' \varepsilon_2' = \eta_2 \frac{d\varepsilon_2}{dt} + K_2 \varepsilon_2$$

where ε_1 and ε_2 are the strains developed in the springs in parallel, respectively. The total strain is therefore:

$$\varepsilon = \varepsilon_1' + \varepsilon_1 + \varepsilon_2' + \varepsilon_2$$

$$\varepsilon = \frac{\sigma_0}{K_1'} + \frac{\sigma_0}{K_1} \left(1 - e^{-\frac{K_1 t}{\eta_1}} \right) + \frac{\sigma_0}{K_2'} + \frac{\sigma_0}{K_2} \left(1 - e^{-\frac{K_2 t}{\eta_2}} \right)$$

In general, for n elements the total strain is:

$$\varepsilon = \sum_{l=1}^{n} \frac{\sigma_0}{K_1'} + \sum_{l=1}^{n} \frac{\sigma_0}{K_1} \left(1 - e^{-\frac{K_1 t}{\eta_1}} \right)$$

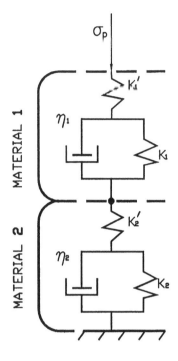

Figure 10.1.2. Generalized Kelvin rheological model of heterogeneous salt body

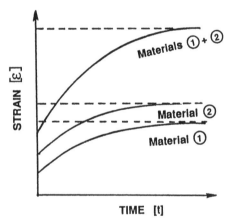

Figure 10.1.3. Strain variations in function of time for each material and for the complex body

Figure 10.1.3 illustrates the variations of strains with time for each material and for the series combination. This is considered valid for uniaxial deformation. The two curves representing the behaviour of the individual viscoelastic materials are different, where one is more superior than the other. It should be noted that for the three-dimensional case, this approach is not valid and so the correspondence principle has to be applied.

The rheological characterization of rock salt and potassium ore is important for the proper design of mine structures. For underground salt mining, where the viscoelastic rock constitutes pillars and stope backs, their rheological behaviour should be determined experimentally in the laboratory or in situ, as well, as predicted through analytical or computer models, as will be discussed later.

10.2 CREEP PROPERTIES

The greatest complexity and variability of salt creep properties is observed by the time-dependent deformations. These properties have great significance for mine design and ground control. The exact mechanism and quantitative descriptions of creep are a subject of considerable debate. Further discussion of this topic is by discussion of several generally recognized factors of creep occurrence.

10.2.1 Concept of creep

The creep is generalized as a time-dependent deformation under constant load. Creep deformation occurs in three different phases, as shown in Figure 10.2.1, which represents a model of rock properties undergoing creep deformation due to sustained constant load. Upon application of a constant force on the material, an instantaneous elastic strain ε_e is induced. The elastic strain is followed by a primary or transient strain, shown as a Region I in Figure 10.2.1. Region II, characterized by an almost constant slope in the diagram, corresponds to secondary or steady state creep. Tertiary or accelerating creep leading to rather sudden failure is shown in Region III in the diagram. Laboratory investigations showed that removal of applied load in Region I at point P will cause the strain to fall rapidly to the Q level and then asymptotically back to zero at R. The distance PQ is equal to the elastic strain ε_e. No permanent set is caused. If the removal of stress takes place at point T in the secondary creep region, a permanent set will be observed: VO. From the point of view of stability, salt structure deformations after constant load removal have only academic significance, since the stresses imposed underground due to mining operations are irreversible.

Creep properties of individual salts will be greatly affected by the content of soft or hard accessory minerals. The different creep behaviour of similar types of rock salts in salt deposits could be explained in a similar way. It is well known that fracture and flow occur in the saddle around the crest, while such phenomena are not observed in the vicinity of flat rock salt deposits. During geological times, these differences are manifested by a preferred swelling of the saddles, as discussed by K. Stocke and H. Borchert. This swelling could go

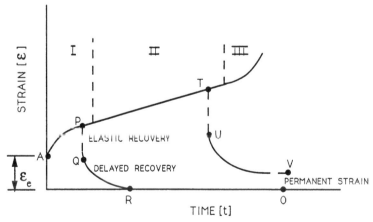

Figure 10.2.1. Model of the creep deformation in function of time

so far that the anhydrite and younger rock salt may be pierced and fractured.[4]

W. Dryer carried out a creep testing under constant axial load of a variety of rock salt types. For example, older salt rock has a greater strength, because of strain hardening and a higher content of anhydrite. A lower content of hard accessory minerals and coarsely crystalline structure results in higher ductility, as shown for sample no. 5 in Figure 10.2.2. This sample will probably reach the asymptotic strain value at a constant load of 20 MPa after several years. Creep almost ceased after 7.5 months in the two hard rock salt samples no. 1 and no. 2 as well as in the sylvinite-halite specimen no. 1, which undergoes the smallest strain.

The behaviour of the salts with time-dependent deformation under constant load is characterised as a viscoelastic phenomenon. Under these conditions the strain criteria are superior to the strength criteria for design purposes, because failure of most salt pillars occurs during accelerated or tertiary phase of creep, due to the almost constant applied load. The dimensions of a pillar in viscoelastic rock should be established on the basis of a prediction of its long-term strain, to guard against adequate safety factor accelerating creep.

10.2.2 Analytic expression of creep

Hansen and Carter make a contribution to better understanding the creep behaviour of salt by a valuable discussion of various analytic expressions.[5] Many different expressions have been applied in the past as described by Kelsall and Nelson, frequently indicating very different types and magnitudes of creep strain.[6] It appears that a creep equation of the following form is gaining acceptance as a reasonable accurate expression to describe both primary and secondary creep:

$$\varepsilon = e_t \left[1 - \exp\left(-rt \right) \right] + \dot{\varepsilon}_s t$$

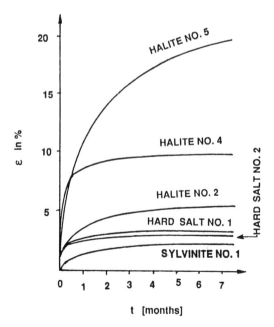

Figure 10.2.2. Creep curves of various rock salts under constant load (after Dryer)

where:
  ε = total creep strain;
  e_t and r = experimental fitting parameters;
  t = time;
  $\dot{\varepsilon}_s$ = secondary creep rate.

The first term, which predicts a decrease in creep rate with time, represents primary creep. The second term, which predicts a constant creep rate (under constant stress and temperature) represents secondary creep. The secondary creep rate is described by:

$$\dot{\varepsilon}_s = A\sigma^n \exp\left[-Q/RT\right]$$

where:
  A, Q and n = experimental fitting parameters;
  σ = measure of shear stress;
  T = absolute temperature.

Kelsall and Nelson further described that under equal conditions of stress and temperature, the variation in the rate of strain among different salt dome deposits in the USA is up to two orders in magnitude. Figure 10.2.3 shows data points from a test performed on various Gulf region salt domes and New Mexico bedded salt deposits. It appears that although the average creep rates vary greatly among sites, collectively the data lie roughly within the same very wide band. This raises the question of whether all sites exhibit a large variation in

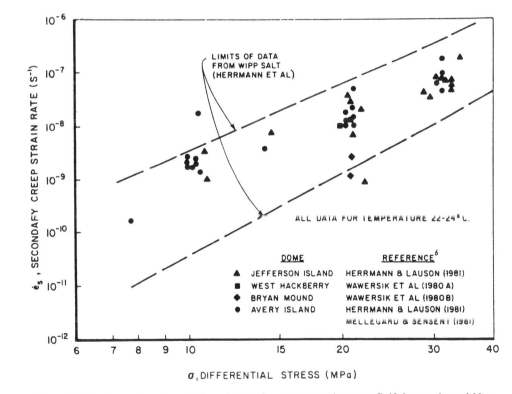

Figure 10.2.3. Comparison in variation of secondary creep rate between Gulf dome salts and New Mexico salt beds (after Kelsall and Nelson)

creep rate, which would be revealed only by additional testing. The reasons for the variation in creep rate among sites and within a single site are not apparent. Probably the variation does reflect inherent differences among the samples (although variation in test technique may have some influence), but no correlation has been established with factors such as fabric or the nature or degree of impurities.

The large variability in the creep data and the laws which are derived from the data is particularly significant when creep must be extrapolated over long time periods. Creep tests which have been run for only days or weeks are sometimes used to predict rates of deformation over periods of tens of hundreds of years. These is clearly a need to run creep tests for much longer durations than have currently been performed in order to validate long-term predictions. Furthermore, it is important to determine whether the variability observed in the laboratory is reflected by in situ behaviour. To date, there have been few comparisons between predicted closures of underground openings based on laboratory testing and analysis and actual measurements.

The comparison analysis between laboratory testing results and in situ results would be extremely useful in regard to prediction of salt structure behaviour in situ.

10.2.3 *Creep in function of temperature*

Hansen and Senseny[7] reported that hundreds of creep experiments have been conducted, simulating conditions which could be expected in the salt surrounding a nuclear waste repository, such as:
- temperature between 25° and 200°C (T);
- vertical stress between 5 MPa and 50 MPa (σ_v);
- lateral stresses between 0 MPa and 20 MPa ($\sigma_2 = \sigma_3$).

The observed processes by which salt deforms are a function primarily of stress and

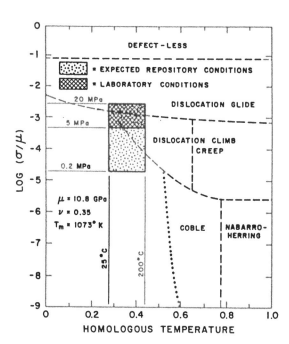

Figure 10.2.4. Deformation mechanism map for salt

temperature. The regime in which a particular mechanism governs is conveniently displayed in a deformation mechanism map rectangle. Typical laboratory experiments shown in Figure 10.2.4, encompasses the stresses and temperatures given above. Many of the experiments were well deformed into steady state based on a constant flow rate under constant stress and temperature conditions. At high stresses, dislocation glide is likely to govern deformation and at high temperatures, the dominant mechanism is a diffusion-controlled dislocation climb. Boundaries between mechanisms are where two processes contribute equally to deformation. It is clear that deformation mechanisms in much of the stress-temperature regime of interest are unidentified.

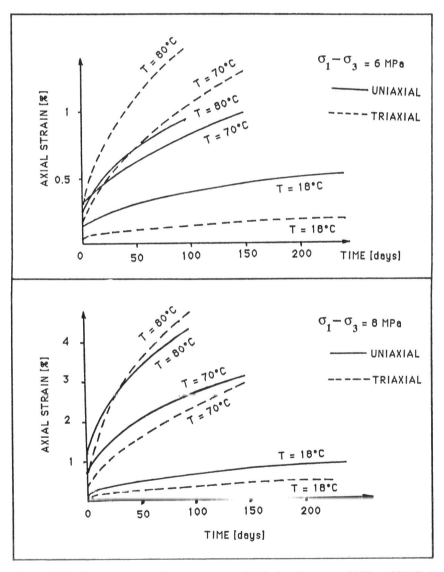

Figure 10.2.5. Creep curves at various temperatures for deviatoric stress at 6 MPa and 8 MPa (modified after Vouille et al.)

Vouille et al.[8] performed a test on a Tersanne rock salt sample in regard to influence of temperature, confining pressure, and deviatoric stress on the creep deformation. Twenty-four creep tests were performed under varying conditions:

temperature at 18°C, 70°C and 80°C (T);
- confining pressure between 0 and 20 MPa (σ_3):
- deviatoric stress at 6, 8, 10 and 12 MPa ($\sigma_1 - \sigma_3$).

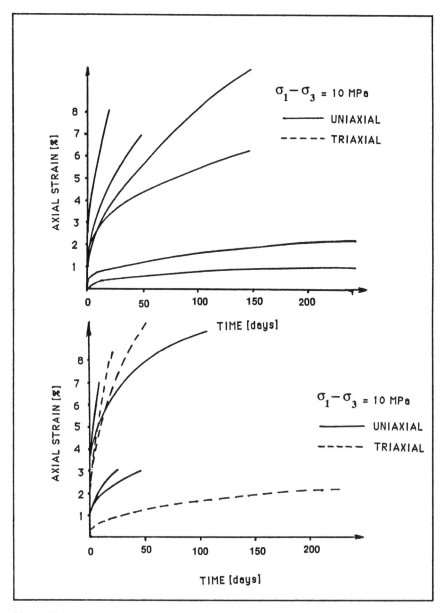

Figure 10.2.6. Creep curves at various temperatures for deviatoric stress at 10 MPa and 12 MPa (modified after Vouille et al.)

The result of the test was represented by plotting a strain versus time, as shown in Figure 10.2.5 and Figure 10.2.6. From the diagrams in these figures Vouille et al. concluded that:

1. the creep rate decreases with time and its initial value is practically infinite;

2. for a given temperature and confining pressure, creep strains and creep rates increase with the deviatoric stress;

3. for a given state of stress (either uniaxial or triaxial), creep strains and creep rates increase with the temperature; and

4. the influence of the confining pressure is more complex: at 18°C, for a given deviatoric stress, creep strains and creep rates for the uniaxial tests are greater than for the triaxial tests; at 70°C the contrary is observed; at 80°C it depends on the deviatoric stress.

The creep testing of tempered and untempered rock salt samples, conducted by Dreyer, consisted of square prisms (4 × 4 × 10 cm) exposed to a constant load of 10 MPa. The samples were coarsely grained and consisted of up to 8% of anhydrite and up to 1.5% of kieserite and clay. Some individual rock salt crystals were loosely interlocked and certain cleavage and glide planes could be seen. The curves indicated that deformation of the untempered sample ceased after twelve hours (strain approaches the asymptotic value of 0.6%). However, the deformation of tempered samples continued even after eighteen hours (Figure 10.2.7). The test indicated that deformation ceased at 3.7%, which is approximately five or six times the limiting creep strain of the untempered halitic rock sample.[4]

The large creep strain of the tempered sample was associated with the so-called recrystallization during tempering, where diffusion processes lead to growth of new crystals until all defective crystals are melted. The elimination of slips and dislocations is a very important factor in strain hardening mechanics and resulted in work softening. The dislocation climb by diffusion results in newly grown completely healed crystals. Under these circumstances, in situ strain hardening is cancelled because the memory of the in situ history is annihilated.

The influence of temperature on the creep properties and behaviour is significant and cannot be ignored in considerations of mine layouts and ground control, specially in the case of deep rock salt excavation, which will be used as the repository of chemical and nuclear waste.

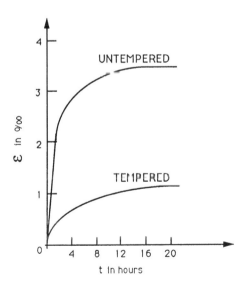

Figure 10.2.7. Creep curves of annealed and unannealed halite samples, Germany (modified after Dreyer)

10.2.4 *Creep around circular openings*

At the Tušanj mine, Tuzla, time dependent deformations of rock salt around circular openings were studied. The circular openings were excavated without blasting, and a smooth surface of opening walls was obtained. The creep deformations of the rock salt mass were monitored from eight boreholes drilled in a radial pattern. The length of each borehole was 3 m. The displacement of each measuring point was recorded on a five-day basis.

The circular openings are located at approximately 500 m depth in massive rock salt. It is assumed that the stress in the rock salt mass is hydrostatic and has a magnitude of 10 MPa.[9]

If one considers the general creep curve of each measured point, it could be linearized using different time segments, as illustrated in Figure 10.2.8. For these segments, the deformation velocity of each point could be calculated. By interpolation, contours of equal velocity known as isopaches were constructed. Velocity contours were constructed for three characteristic linear deformation segments: from 0 to 60 days, from 60 to 180 days and from 180 to 1100 days.

The monitoring time-dependent deformational characteristics of rock salt around circular openings indicated a rather complicated mechanics of the ground displacement in the beginning, and with time this progressed to a true creep deformation (Figure 10.2.9), as discussed below.[9]

1. The first phase of deformation (0-60 days) exhibited displacements which are rather the product of the geological-structural features of rock mass than of creep itself. For example, the maximum deformation took place at the floor of the mine opening. This could be related to the heterogeneity of rock mass and the stress concentration of the redistributed primitive stress. This indicates that in the first phase, after excavation of mine openings, the creep is integrated with relaxation deformations governed by structural features of rock salt mass.

2. The second phase of deformation (60-180 days) exhibited to a certain degree creep represented by asymmetrical contours of isopaches. The asymmetry of the isopaches is governed by remnants of structural stresses which acted very strongly in the first phase of

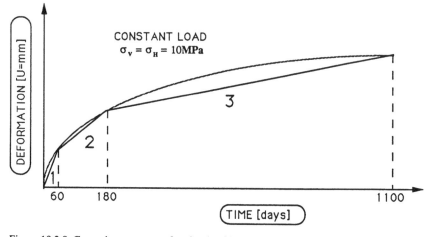

Figure 10.2.8. General creep curve of rock salt with linear segmentation

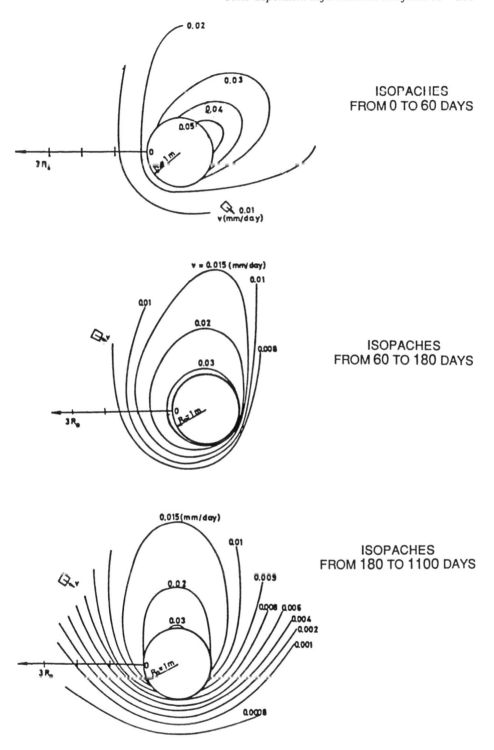

Figure 10.2.9. Isopaches of time-dependent deformations of rock salt mass around circular openings (Tusanj salt mine, Tuzla)

deformation. In addition, the isopach pattern suggests that the final relaxation deformations have been in effect around the mine openings.

3. The third phase of deformation (180-1100 days), which has a long time duration, indicated that displacement due to original structural stresses and rock salt mass heterogeneity has been assimilated by creep. This phenomenon could be observed by the symmetrical distribution of deformations. The isopach pattern is symmetric with respect to the vertical axis. The maximum creep deformation is in the roof of the mine opening and the minimum deformation is in the floor of the mine opening.

The deformations of the salt mass around mine openings are an important factor for mine stability. The investigations in Tusanj rock salt mine indicated that in folded evaporite deposits, heterogeneity, complexity of primitive stresses, and structural features are very important factors for short-term stability of mine openings (up to six months), and that creep properties of rock salt come into effect in the case of longer-term stability of mine openings.

The analysis of salt deformations around mine openings by isopaches could be an important tool in predicting the stability of underground openings and their deformation with time, particularly at the beginning of creep when structural heritage of rock mass is present.[9]

10.3 CREEP MECHANISM

In this section creep mechanism will be discussed on the basis of Lang's paper 'The rheological behaviour of rock salt'.[1] Rock salt is an ideal material for consideration because the creep phenomenon and plasticity are well produced, as well as the crystal and grain structure are relatively simple. The deformations mechanics of salt are governed by the dislocations (more rapid creep) within crystal and grain boundaries, and it could be represented by dislocation glide, dislocation climb, diffusion creep (slow-creep) and grain boundaries glide, as further discussed.

10.3.1 *Dislocation glide*

Dislocation glide dominates at high stress and relatively low temperature. The deformation is the result of the dislocations which migrate in the crystal grains blocking each other so that hardening and strength increase as well as a dislocation density. The decrease in deformability corresponds to the primary creep (transition creep). The deformation does not stop completely during recovery. It could be distinguished two types of recovery:

1. Static recovery depends upon the internal stress effected by the number of dislocations and the temperature.

2. Dynamic recovery depends upon the external stress and the temperature, which is more important.

The dislocation glide may be represented by the following deformation law, which has been developed by several authors:[1]

$$\dot{\varepsilon}_s = C \times \exp\left(-Q_{eff}/R \times T\right) \sin h\,(B\sigma)$$

where:

$\dot{\varepsilon}_s$ = stationary strain rate;
C = constant;

Q_{eff} = effective activated energy;
R = universal gas constant;
T = temperature;
B = constant;
σ = stress.

The experimental investigations of several authors confirmed the validity of the equation for a temperature between 23°C and 400°C.

10.3.2 *Dislocation climb*

Dislocation climb or polygonization is illustrated in Figure 10.3.1, which represents a most important deformation mechanism. When the dislocation glide area has a constant temperature and stress is reduced below 16 MPa, then polygonization will dominate. Now the blocked dislocations have enough time to climb, thermically activated, out of their glide planes and further migrate. The step dislocations align themselves on the basis of their interaction (attraction of dipoles) and form small angle grain boundaries. In this manner, subgrains with sizes from several 10-100µ will arise. The subgrains arising during secondary creep show very little or no tendency to deform or change in size, regardless of magnitude of deformation. Thus, there is a dynamic equilibrium between formation and

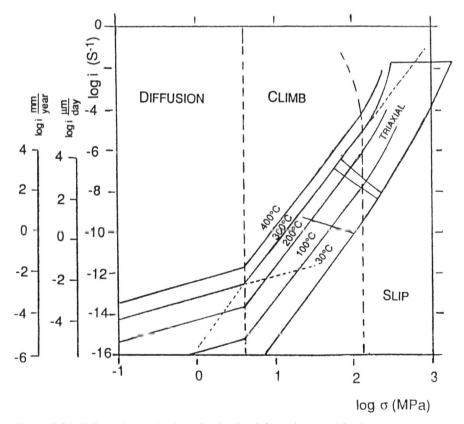

Figure 10.3.1. Deformation mechanism of rock salt – deformation map (after Langer)

destruction of the subgrain boundaries. The gradual adjustment of the stationary subgrain density or the stationary average dislocation density corresponds to the primary range of the creep curve.

The mechanism of the polygonization is very much dependent on the temperature. However, there is the opinion that a temperature below 100°C does not affect this mechanism.

$$\dot{\varepsilon}_s = A \times \exp\left(Q_{eff}/R \times T\right)\sigma^n$$

where:

$\dot{\varepsilon}_s$ = stationary strain rate;
A = constant;
Q_{eff} = effective activated energy;
R = universal gas constant;
T = temperature;
σ = stress;
n = constant.

From a certain number of creep experiments of rock salt samples, it is concluded that the value for the constant n is between 4 and 7, and if $n = 5$ then it is independent of temperature.

10.3.3 *Diffusion creep*

With a decrease in the deformation rate or temperature the diffusion creep dominates. According to some investigations at high temperature a diffusion creep does not occur. On the basis of experimental data the diffusion creep in function of temperature could occur for stress values between 0.1 and 0.5 MPa.

Nabarro-Hering creep regards a diffusion creep which is represented by the following law:

$$\dot{\varepsilon} = \left(D_s \, \Omega \, \sigma / k \, T\right) L_k^{-2}$$

where:

D_s = self-diffusion coefficient;
Ω = atomic volume;
k = Boltzmann's constant;
L_k = grain diameter.

The characteristics of the diffusional creep are:

$\dot{\varepsilon}$ is strongly dependent upon the grain diameter (L_k);

$\dot{\varepsilon} - \sigma$ relation gives a viscous behaviour;

$\dot{\varepsilon} - \dfrac{D}{T}$ relation strongly depends on temperature;

the transient creep does not occur.

Nabarro-Hering creep considers transport and deposition of atoms in the free space of the single crystals or crystal boundaries.

10.3.4 *Grain boundaries glide*

The grain boundaries have no influence on the deformation during dislocation creep and polygonization, when the average grain diameter is considerably larger than the average distance travelled by the dislocations. During polygonization the conditions are satisfied when the grain diameter is larger than the subgrain diameter of up to 0.3 mm. In natural rock salt, the grain diameter is usually about 1 cm so that this requirement is fulfilled. During dislocation creep the average travelling dislocation distance is still considerably shorter, so that also the grain boundaries have no influence.

With small enough grains and small stresses, the grain boundaries glide does play a role. At high enough temperatures the deformation law is probably very similar to that of diffusion creep:

$$\dot{\varepsilon} = D_v \cdot \frac{\sigma}{L_k}$$

Also, the diffusion coefficient D_v plays an important role, because volumetric changes cannot be avoided due to compatibility reasons during the grains glide.

The grain structure of the undeformed and deformed salt samples was examined by the X-ray diffractometer, in order to obtain knowledge of the structure changes and the deformation mechanism (Figure 10.3.2). The examination showed a clear arrangement of the (110) planes which are in close relationship and symmetrical to symmetry of the planes

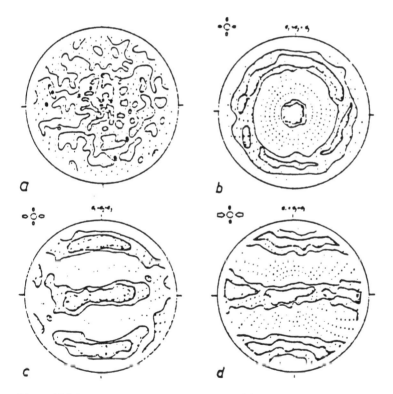

Figure 10.3.2. Petrographic diagram by X-ray diffraction of undeformed and deformed rock salt samples. The plane of projection is perpendicular to the direction of principal stress. Temperature 100°C. The (110) planes are represented (after Langer).
a. undeformed initial sample; b. deformed under axial pressure; c. deformed under true triaxial (rhombic) load; d. deformed under axial strain

of external stress. Furthermore, it has been seen that the plastic deformations in the range between 20°C and 200°C mainly occur by means of translative glide for 110 planes and minimally for 100 planes.

Consideration of the creep mechanism is given with the academic point of view which should help to better understand mechanical behaviour of salt which is strongly time-dependent, and its constitutive model, as discussed in the next section.

10.4 CREEP RUPTURE

The long-term load-bearing behaviour of the salt structure is of particular importance, because salt possesses material properties that are strongly time-dependent. Long-term stability of the salt structure is dictated by particular mining conditions or time-extended use of caverns for storage. In this regard, in the last couple of decades considerable efforts have been undertaken to obtain an insight into the viscous material behaviour of salt under mechanical and thermal loads. In addition, to quantitatively determine the long-term behaviour of the salt structure and the possibility of rupture. The further discussion is continued by four topics, as follows.

10.4.1 *Creep deformation and rupture*

Langer summarised a model of creep deformation and rupture on the basis of the laboratory experimentation and phenomenological observations.

1. *Creep deformation*: an irreversible change of shape is modelled as listed.[1]

– Variations in stress lead initially to transitional creep (primary creep), i.e. changes in shape with a rate of distortion (creep rate) decreasing by time (Figure 10.4.1).

– With constant deviator stress and constant temperature a stationary state of deformation velocity ('secondary creep') is reached at a sufficiently large stress difference ('yield point'); the creep rate is an over proportional function of deviator stress and temperature (Figure 10.4.2).

– At constant temperature, a reversible relation between the stress and distortion variables is obtained. A certain constant deviator stress condition leads to a stationary rate of distortion and, conversely, a constant rate of distortion condition will bring about the corresponding deviator stress level.

– The magnitude of the rate of distortion of the change of shape depends generally upon the material or structural parameter and upon the magnitude of the deviator stress and the temperature and their variation in time.

– Isotropic stress conditions exert no influence on creep deformations.

– With every change in stress and temperature reversible (elastic) deformations occur, which appear to be independent of the particular stress and temperature level.

2. *Rupture deformation:* irreversible changes of shape, independent of the velocity, with discretely distributed planes of fracture and dilatation.[1]

– Under monotonous stress, the rupture occurs when the variables of state stress, creep rate and temperature reach a certain configuration regardless of the type of stress (imposed forces or deformations) or a particular stress sequence (Figure 10.4.3).

– The resistance to fracture increases with rising isotropic pressure; it reaches higher values even with an increasing rate of stress.

– The strength decreases at higher temperatures (Figure 10.4.4).

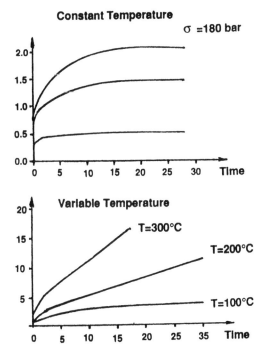

Constant Temperature

σ =180 bar

Variable Temperature

T=300°C

T=200°C

T=100°C

Figure 10.4.1. Creep curves of rock salt for various constant load conditions (after Langer)

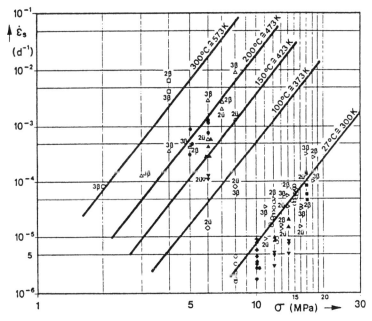

Figure 10.4.2. Creep data of rock salt (according to tests by the 'Bundesanstalt für Geowissenschaften und Rohstoffe' (BGR), Hanover

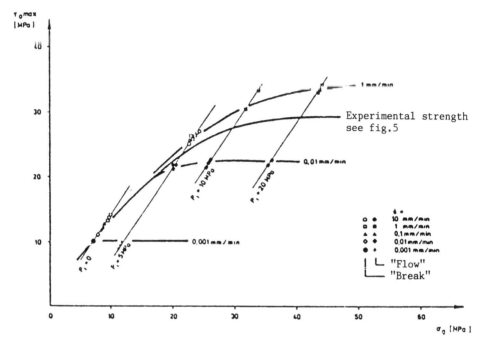

Figure 10.4.3. Rupture curves of rock salt as a function of velocity deformation (after Langer)

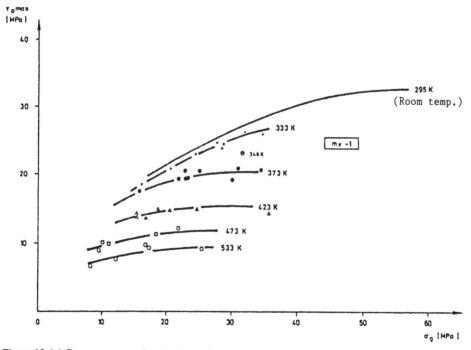

Figure 10.4.4. Rupture curves of rock salt as a function of temperature (after Langer)

– Tensile and compressive stresses lead to different strength values.

– Under uniaxial load, rock salt is distinctly anisotropic to strength. With increasing isotropic pressure the anisotropy to strength decreases greatly whereas the anisotropy to deformation on rupture is retained, although it is also reduced.

– Under constant isotropic pressure and temperature, the resistance to rupture depends upon the stress field (mean principal normal stress).

– The influence of the stress path and the isotropic stress disappears at higher temperatures.

On the basis of these phenomenological observations the following constitutive model is proposed, based on the splitting of the total deformation into three parts:

$$\varepsilon = \varepsilon_e + \varepsilon_{cr} + \varepsilon_f$$

where:

ε = total deformation;

ε_e = reversible (elastic) deformations;

ε_{cr} = irreversible (viscoplastic) creep deformations;

ε_f = irreversible (plastic) rupture deformations.

These criteria of the creep deformation and rupture deformation are affected by a number of natural and technological factors. They should be considered in combination with geological structural factors, rock mechanics factors and mining factors.[1]

10.4.2 *Multiaxial creep rupture*

Multiaxial creep rupture has been first investigated and evaluated by Nair and Singh. They assumed that the maximum main stress difference is the controlling stress and that the onset of creep rupture is mainly related to the occurrence of tensile strain. It should be noted that such strain conditions are not necessarily related to tensile stress, but in accordance with the in situ conditions can also occur under triaxial compressive loads.[10]

Rupture creep tests (Figure 10.4.5) clearly show the linear relationship between the stress difference ($\sigma_1 - \sigma_3$) and the time to failure t_B. The figure covers a time to failure of approximately 7 hours to over 1000 hours, i.e., the curve can also be used for very short tests, where most likely no tertiary creep takes place. The effective stress lies between 36.5 MPa and 66.9 MPa. Nair and Singh consider the effects of the average stress to be insignificant.[10] However, Rokahr and Standtmeister concluded on the basis of their own test results that the time to creep rupture may be different for the same main stress difference but with differing isotropic stress components. This phenomenon is particularly significant in the case of rock-mechanics considerations, because in the vicinity of the cavity, three dimensional stress conditions can occur with differing isotropic and deviatoric components, and thus if the effects of the isotropic stress component on the creep behaviour are ignored, this can lead to underestimation of the time to failure.[11]

Within the framework of comprehensive theoretical and experimental investigations, Menzel and Schreiner have formulated a creep rupture criterion that is based on consider ations of the energy involved.[12] It is principally based on the assumption that the total energy input required in a short-term test up to the point in time when the sample fails, is the same as the plastic deformation energy required in a creep test up to the time point of the start of creep failure. The stress/strain relationship for the two cases and the estimated plastic deformation energy with respect to unit volume are shown schematically in Figure 10.4.6.[11]

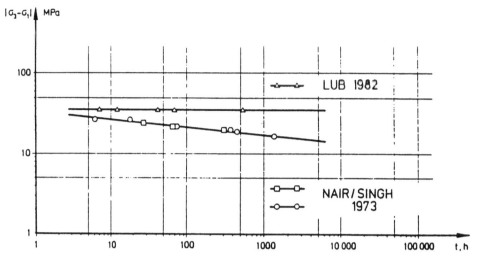

Figure 10.4.5. Creep rupture diagrams, plotted in double logarithmic form (after Nair & Singh)

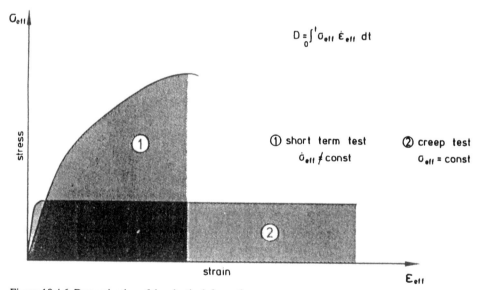

Figure 10.4.6. Determination of the plastic deformation energy

Whereas in short-term tests with constant load increase, the stress σ_{eff} increases constantly until the failure point has been reached, in creep tests a particular constant effective stress is applied for the entire test duration.[11]

The formulated creep rupture criterion by Menzel and Schreiner is expressed by the material law equation:

$$\varepsilon_v = K \cdot n \cdot \sigma_{eff} \cdot t^{m-1}$$

where:

ε_v = viscous strain;

$K = 2,31 \cdot 10^{-6}\ \text{MPa}^{-25}\ d^{0.258}$

$n = 2, 5;$

$m = 0.258.$

The viscous strain and strain rates measured during the test and the results obtained from the material law (equation above) could be represented by curve fitting.[11]

10.4.3 *Triaxial compressive load rupture*

The 'Bundesanstalt für Geowissenschaften und Rohstoffe' (BGR) in Hanover has obtained interesting results from creep rupture tests with triaxial compressive loading by Wallner.[13] The results of the tests, which were regulated in accordance to strain, are schematically represented in Figure 10.4.7. The influence of various strain rates $\dot{\varepsilon}_{eff}$ and the average controlling stress σ_o were investigated. The latter was kept constant for each test. For each average controlling stress σ_o, the results allow the boundary to be determined between stress conditions that lead to failure and stationary stress conditions that can be indefinitely accommodated without failure.[12]

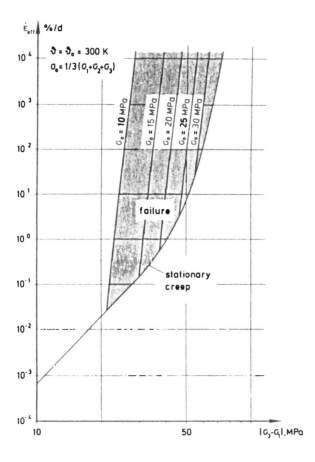

Figure 10.4.7. Creep rupture diagrams in triaxial compressive loading field (after Wallner)

10.4.4 *Creep rupture*

In this section, creep rupture test for Asse and Erslev salt deposits are discussed, on the basis of representation of Rokahr and Standtmeister.[11] The tests were carried out by triaxial extension conditions with a test duration from a few hours to approximately 500 hours. The results of these tests are shown in Figure 10.4.8 and Figure 10.4.9. Furthermore, in addition to the sample type, both figures illustrate the stress magnitudes (σ_1 and σ_3) for each test.[11]

The very large failure strains are noticeable. Even under the conditions of indirect tensile stress, they approached values of more than 20% of the natural logarithmic strain. The considerable influence of the isotropic stress components on the time to failure (t_B), and the corresponding maximum strain are also visible.

Rokahr and Standtmeister showed the evaluation of the few results of creep ruptures represented in a double logarithm scale, as illustrated in Figures 10.4.10 and 10.4.11. The

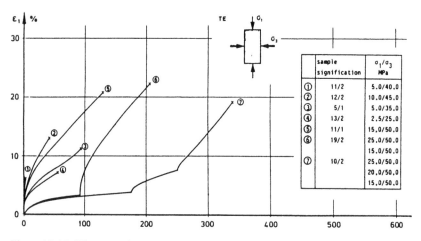

Figure 10.4.8. Diagrams of creep rupture test, Erslev salt deposit (after Rokahr and Standtmeister)

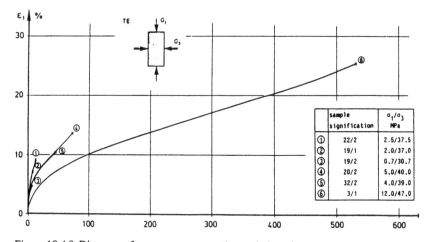

Figure 10.4.9. Diagrams of creep rupture tests, Asse salt deposit (after Rokahr and Standtmeister)

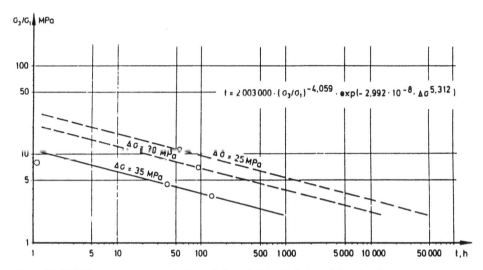

Figure 10.4.10. Creep rupture tests, Erslev salt deposit (after Rokahr and Standtmeister)

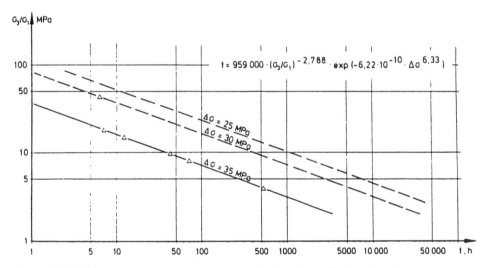

Figure 10.4.11. Creep rupture tests, Asse salt deposit (after Rokahr and Standtmeister)

diagrams indicated the relationship between the time to failure (t_B) and the ratio σ_3/σ_1 and the stress differences $\Delta\sigma = |\sigma_1 - \sigma_3|$.

Both figures also show corresponding mathematical formulations. However, it should be noted that these figures can only be used as an indication of the phenomenological description of the creep rupture behaviour of rock salt. Further tests are necessary, particularly with respect to the dependency of the time to failure on the differential stress $|\sigma_1 - \sigma_3|$ for smaller values of $\Delta\sigma$ too. Furthermore, it is necessary to investigate the validity limits of the diagrams. For practical purposes, it appears possible to satisfactorily describe the creep rupture behaviour of rock salt at a particular location with nine to twelve creep rupture tests at various stress differences and ratios σ_3/σ_1.

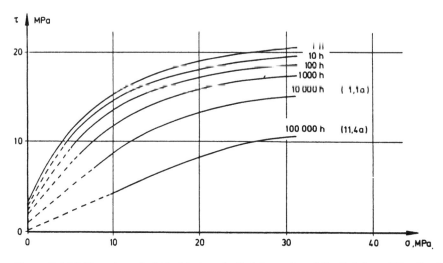

Figure 10.4.12. Time-dependent triaxial strength, Mohr's envelope (after Rokahr and Standtmeister)

These creep rupture diagrams then allow the construction of Mohr's envelopes, which describe the strength of the material with respect to time. Mohr's envelopes for rock salt samples from the Asse deposit are illustrated in Figure 10.4.12. The relationship between the time to failure and the maximum differential stress before failure is also clearly visible. If, for example, the time to reach a failure of 100 000 hours is required, then Figure 10.4.12 shows that the stress levels effective over this time period do not exceed approximately 40-50% of the short-term strength.

REFERENCES

1. Langer, M. 1984. The rheological behaviour of rock salt, pp. 201-230. In *Proc. of 1st Conf. on Mech. Behaviour of Salt*. Pennsylvania State Univ. Clausthal: Trans. Techn. Publ.
2. Kocar, F. 1973. *Analysis of rheological behaviour of rock salts from Tusanj deposit, Tuzla*. Ph.D. thesis, Faculty of Mining and Geology, Univ. of Tuzla (in Yugoslav).
3. daGama, C.D. 1979. Rheological behaviour of heterogeneous salt rocks. In *Proc. of 4th Int. Congr. on Rock Mech.*, Montreux, Switzerland, Vol. 1, pp. 107-110.
4. Dreyer, W. 1973. Flow at a constant stress. In *The science of rock mechanics*, Part 1, pp. 126-139. Clausthal: Trans. Techn. Publ.
5. Hansen, F.D. & N.L. Carter 1980. Creep of Avery Island rock salt. In *Proc. of 1st Conf. on Mech. Behaviour of Salt*, Pennsylvania State Univ. Clausthal: Trans. Techn. Publ.
6. Kelsall, P.C. & J.W. Nelson. Geologic and engineering characteristics of Gulf region salt domes applied to underground storage and mining. In *Proc. of 6th Int. Symp. on Salt*, The Salt Institute, Virginia, Vol. 1, pp. 519-542.
7. Hansen, F.D. & P.E. Senseny 1983. Petrofabrics of deformed salt. In *Potash '83. Potash technology*, pp. 347-354. Toronto/New York: Pergamon Press.
8. Vouille, G. et al. 1984. Experimental determination of the rheological behaviour of Tersanne rock salt, pp. 407-420. In *Proc. of 1st Conf. on Mech. Behaviour of Salt*, Pennsylvania State Univ. Clausthal: Trans. Techn. Publ.
9. Mandzic, E. 1981. Rock salt mass creep around circular openings, pp. 789-794. In *Proc. of Int. Symp. on Weak Rock*, Tokyo, 21-24 September, 1981.

10. Nair, K. & R. D. Singh 1973. Creep rupture criteria for salt. In *Proc. of 4th Symp. on Salt*, Houston (Texas), Vol. 2, pp. 41-50.
11. Rokahr, R. B. & Standtmeister 1983. Creep rupture criteria for rock salt. In *Proc. of 6th Int. Symp. on Salt*. The Salt Institute, Virginia, Vol. 1, pp. 455-462.
12. Menzel, W. & W. Schreiner 1977. Zum geotechnischen Verhalten von Steinsalz verschiedener Lagerstatten der DDR, Part 2, *Neue Bergbautechnik* 7.
13. Wallner, M. 1984. *Analysis of thermomechanical problems related to the storage of heat producing radioactive wastes in rock salt*, Pennsylvania State Univ. Clausthal: Trans. Techn. Publ.

CHAPTER 11

Outburst of violent fracturing

Outburst fracturing is related to the sudden and violent expulsion of salt or rock with gas. As a consequence of outburst, the cavities on the mine face or roof strata are formed. To describe this phenomenon, several topics are analyzed with the aspect of theoretical analysis and practical investigations.

11.1 CHARACTERIZATION OF GAS

The presence of gas in the salt-bearing strata creates the hazard of fire, gas explosion and gas outburst. These hazards could be manifested during the opening of the virgin ground as well as during mining operations, both in dry mining and solution mining. The gas might be present in the salt bodies and in the rocks of surrounding strata. Further consideration is given to the nature of the gas occurrences in the salt-bearing strata.

11.1.1 *Sorbed gas*

The term 'sorbed gas' refers to the mechanism of adsorption and absorption where molecules of gas may also be dispersed among the salt molecules. The amount of sorbed gas depends on the pressure, temperature, internal surface and the nature of evaporite beds. Highly porous evaporites of small radii require a pressure greater than 500 atmospheres for gas saturation, but at a lower porosity, a saturation is at 200 atmospheres. High porosity facilitates gas outburst, but lowers gas emission.[1] A sorbed gas occurs as intergranular bubbles. This type of gas probably occurs through salt in small quantities in many domes. Local concentration of sorbed gas could be a source of violent or gradual gas liberation. The gas concentration pockets are usually formed within anomalous structural-geological zones.

Desorption of gas is a reversible phenomenon. Any changes in the equilibrium of sorbed gas concentration (pressure, temperature, structure of pores due to cracking) will tend to reverse this state. The state of desorption of gas is proportional to the degree of the stability changes. Desorption at a slow rate results in gas emission (blowouts) in mine workings, what might cause a fire (methane) or explosion (air-methane mixture), but at a rapid rate might cause gas outbursts. Changes in desorption rate usually result in both: first a gas outburst and then gas emission and possible explosion.[2]

At the Belle Isle mine two zones of gas outbursts might be distinguished. the first zone is

226

composed by the black salt, which is a mineral mixture of halite and anhydrite (Figure 11.1.1).[3] This zone has not experienced severe gas outbursts, only small or large blowouts. Most likely, the sorbed gas of black salt was a source of both types of gas emissions triggered by the room and pillar mining operation in this zone. The second zone is represented by inclusions of clay-lime sediments with brines, which associate with the central shear zone. The gas concentration in the pockets most likely belongs to free gas, which is described and discussed in the next subsection.

11.1.2 *Free gas*

The occurrences of free gas in the evaporite-bearing strata could be physically classified in two groups, as briefly commented.[4]

The gas concentration within rock strata which contain free gas, is a well known phenomenon in salt mining industry. The gas fills cracks, fissures and fractures of the rock strata, usually the sediments above salt deposits. In those sediments, gas pockets could be formed, which could release gas violently at higher pressure, as for example in the Goderich salt mine (Ontario). However, if the gas is liberated through fractures, then its release is gradual at lower pressure.

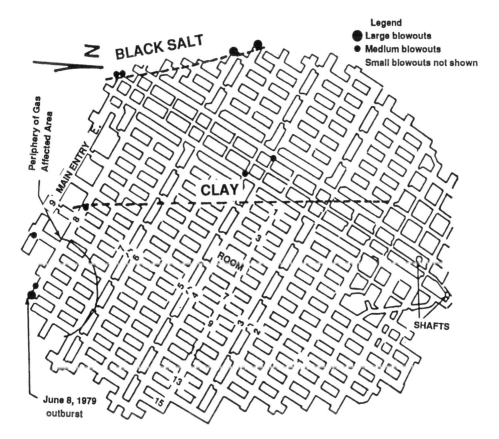

Figure 11.1.1. Plan layout of mining level at Belle Isle mine (USA), showing the locations of gas blowouts and gas outbursts (after Plimpton) with relevant anomalous zones (after Kupfer)

The gas concentration within salt bodies is restricted to the zones of fissure, cracks and fractures which are located in the proximity of tectonic disturbance. The practical example of the free gas concentration in salt bodies is represented for the Belle Isle mine, as discussed in the previous subsection.[3] The gas pockets within clay lime inclusions and fissure areas are associated with the central shear zone. They have been the source of gas outblows and gas outbursts. A massive gas outburst occurred on the face of no. 8 Main Entry East (Figure 11.1.1) where the face round was blased on June 8, 1979. The outburst expelled about 10 000 m^3 (16 000 tonnes) of broken salt and released gas (mostly CH$_4$) which encompassed a volume of about 110 000 m^3 within 150 m of the outburst cavity. An explosion occurred about ten minutes after the gas outburst, when an air-methane explosion mixture was ignited by electric arcing. The explosion killed five miners and widespread damage in the mine (Mahtab, 1981).[2]

11.1.3 *Gas in caverns*

The gas concentration in the salt caverns could be by itself or with brines. The case of caverns with entrapped brines and gas above them under high pressure is most common.[5] Due to mining excavations which considerably weaken the natural surrounding dam or shell by fissuring and fracturing, the gas liberation from the caverns could be rapid by outbursts or gradual by blowouts. However, the weakening of the protective rock around the cavern could be also due to shallow solution mining and surface ground subsidence. For example, the Hukalo solution salt mine (Tuzla basin) experienced gas outbursts in the solution cavern gallery, and the formation of a deep cavity on ground surface (Figure 11.1.2). When the explosive air-methane mixture in the cavity was ignited from the wire of fallen electric poll, then an explosion occurred.

The gas outbursts in the salt mines in great majority relate to free gas, which is not a very common case in coal mining, where the majority of outbursts are related to sorbed gas.[1]

Figure 11.1.2. Deep pot hole erected by gas outburst at solution salt cavern below (Hukalo solution salt mine, Tuzla, former Yugoslavia)

11.1.4 *Composition of the gas*

The composition of the gases of the evaporite-bearing strata has some influence on the gas desorption and liberation. It could be classified with two gaseous compositions, as briefly discussed.

1. Complex gas is composed by hydrocarbons such as methane, ethane, ethylene, n-butane, n-propane, pentanom and others. For example, the Tetima salt deposits, Tuzla basin, contain gaseous salt where gases such as CO_2, CO, H_2, SO_2, H_2S and NH_3 also occur beside those listed above. The nature of gas occurrence is a sorbed gas. The gaseous salt contains an amount of gas between 9.6 and 36.0 cm^3/kg. The average content of the gas in the salt deposit is over 20 ml/kg.[4]

2. Methane gas mainly occurs as a free gas in the gas pockets or in the caverns of the rock or salt bodies.[5] The gas liberation by blowouts could be followed by methane ignition and fire. The explosive mixture of the air and methane could be controlled by an increase of fresh air intake. For example, methane emissions from a drift in the zone of dirty salt at the Belle Isle mine ranged from 1 to 3 m^3 per ton of salt mined. This mine appears to be an example of a dome which is significantly more gassy than the other domes with developed mines.[6]

The majority of salt domes of the Gulf region (USA) are not all rich in methane. Kelsall and Nelson reported that the gas recovered from a well at Bryan Mound, for example, contained 92% CH_4 with 5% CO_2 and traces of nitrogen, ethane and propane. In other cases, the gas is richer in carbon dioxide; at Winnfield, a sample contained 47% CO_2 with 17% H_2O, 18% N_2, and small amounts of CO, O_2, SO_2, H_2, CH_4, A and C_2H_2. Gas encountered at Avery Island contains H_2S and N_2 but no CO_2. Gas from the Lake Hermitage dome was found to be 97% N_2 with 3% CO_2. Often the gas occurs only in small amounts and is barely noticeable during drilling or mining. In some domes gas occurs in large quantities and is a hazard to mining unless proper precautions are taken.

11.2 DYNAMIC PROPERTIES OF SALT STRUCTURES

Dynamic properties of the salts and their seismicity are important factors for consideration of mine stability and safe structure, particularly with the aspect of salt and gas outbursts. The dynamic properties of the salt are analyzed in a wider sense by several topics, as further described.

11.2.1 *Dynamic elastic modulus*

The dynamic elastic modulus of the material can be defined by measuring propagation velocities in the material. For an isotropic solid, there are two types of body or free-medium waves: a longitudinal wave which travels with a velocity V_p and a transverse wave which travels with a velocity V_s. These velocities are related to the elastic constants as expressed below:[7]

$$V_p = \left[\frac{Eg\,(1-v)}{\gamma\,(1+v)\,(1-2v)} \right]^{1/2}$$

$$V_s = \left[\frac{Gg}{\gamma} \right]^{1/2}$$

where:

V_p = longitudinal or compressional velocity;
V_s = transverse or shear velocity;
E – dynamic modulus of elasticity;
G = dynamic modulus of rigidity;
v = Poisson's ratio;
g = gravity acceleration;
γ = unit weight of rock.

The relationship of the moduli as given:

$$G = \frac{E}{2(1+v)}$$

provides the necessary equations for calculation of both the dynamic modulus of elasticity and Poisson's ratio. Thus:

$$E = \frac{V_s^2 \gamma}{g} \left[\frac{3(V_p/V_s)^2 - 4}{(V_p/V_s)^2 - 1} \right]$$

$$v = \frac{1}{2} \left[\frac{(V_p/V_s)^2 - 2}{(V_p/V_s)^2 - 1} \right]$$

These equations are valid for the isotropic and homogeneous and linear elastic rocks. However, some degree of anisotropy can be determined by measuring the propagation velocities in mutual perpendicular directions, as is commented in the next subsection.

In the laboroatory, the longitudinal velocities and shear velocities of salt samples can be determined by two methods:

1. *Seismic pulse*. By measuring the time required for the compressive or shear pulse to travel the length of the salt sample, the corresponding longitudinal or shear propagation velocity could be determined by the following relationship:

$$V = \frac{L}{\Delta t_p}$$

$$V_s = \frac{L}{\Delta t_s}$$

where:

L = length of the sample;
Δt_p = travel time for compressional pulse;
Δt_s = travel time for shear pulse.

The travel time can be measured by an oscilloscope having a calibrated sweep rate, by initiating the sweep with the driving pulse and recording the arrival of the pulse from the pick up (Figure 11.2.1).[7]

2. *Resonant frequency*. By measuring the resonant frequency of a vibrating bar or rod, the dynamic elastic constant can be determined. Bars or rods have three possible modes of vibrations, such as longitudinal, transverse (bending) and torsional. In this case of resonant frequency, the longitudinal bar velocity is related to the principal longitudinal frequency f_b:

$$V_b = 2 f_b L$$

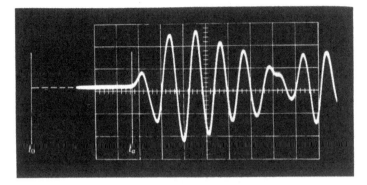

Figure 11.2.1. Oscilloscope recording of seismic pulse

where L is the length of the sample.

The dynamic modulus of elasticity is written as follows:

$$E = V_b^2 \frac{\gamma}{g} = 4 f_b^2 L^2 \frac{\gamma}{g}$$

Due to some heterogeneity of salt, the method is less satisfactory, because strain of vibrating bar reaches a maximum at the mid point and a minimum at the end of the bar.

As suggested by Obert and Duval,[7] both the pulse method and resonant frequency should be used to determine the propagation velocities V_p, V_s and V_b on the same sample of rock. If both methods are used, two independent measurements are obtained.

The dynamic constants of the rock salt from the Tuzla salt basin (former Yugoslavia) have been investigated in the laboratory and in situ. The laboratory testing of rock salt samples has been done by both the pulse method and the resonant frequency method. The modulus of elasticity has been determined in a range between 25 and 39 GPa.[8] It is interesting to note that dynamic elastic constants of rock salt determined in laboratory are close to static elastic constants of large rock salt samples tested also in laboratory, as illustrated in Figure 11.2.2. The elastic constant of the rock salt, however, determined by the micro-seismic method in situ at the Tusanj salt mine differs from data obtained in the laboratory, as discussed in the next subsection.[9]

11.2.2 *Seismic wave front instrumentation*

In situ instrumentation has an aim to detect and record a sub-audible noise, which represents seismic pulses of short duration and extremely small amplitude. The instrumentation is carried out by geophones and amplifiers as well as an oscillograph recorder to record seismic wavefront. In addition, a cassette type recorder is used to record micro-seismic effects and an electro-lab portable micro-seismic monitor for counting small local events.

Underground micro-seismic surveying is done by two methods. Firstly, the probing method is used, where a geophone is moved from point to point and the micro-seismic rate is either observed aurally or recorded. The intensity of a seismic pulse decreases with the distance from the point of origin, so that the micro-seismic rate will decrease with the distance from the centre of the stress area. Secondly, the monitoring is done by an array of

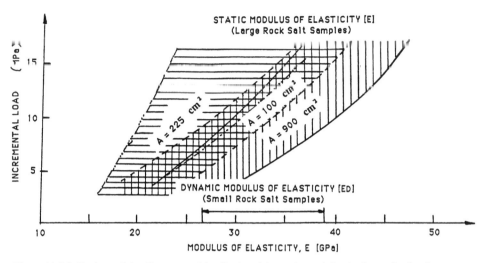

Figure 11.2.2. Static modulus (incremental loading) and dynamic modulus (pulse method and resonant frequency) of elasticity for rock samples

fixed geophones, which are placed around the stressed area, and they simultaneously record the micro-seismic noises. Comparing simultaneously recorded amplitudes from a given pulse, the geophone closest to the generating point could be identified.[7]

The micro-seismic pulses could be generated naturally and artificially. The natural micro-seismic pulses are self-generated with random intensities and at random times. The natural micro-seismicity is a product of particular geological-structural events. The artificial micro-seismic pulses are man-made as for example the pulse generated by the hammer impact on the rock that is used for a probing of seismic surveying. They are also a product of mining activities, either by blasting which produces ground vibrations or by dynamic stress impact induced by progressive salt excavation.

The micro-seismic surveying is further discussed with regard to two aspects, which have been studied at the Tusanj salt mine, Tuzla.

1. *Dynamic elastic modulus.* The micro-seismic surveying in this salt mine was by Terra Scout Model 150 (Sviltest, USA), as illustrated in Figure 11.2.3. Only longitudinal or compressional velocity (V_p) was measured, due to difficulties to read a transverse or shear velocity (V_s). The layout of a micro-seismic station is given in Figure 11.2.4. The longitudinal velocity was recorded along micro-seismic profiles for each geophone station. The magnitude of the longitudinal velocities varies significantly from station to station, because of the high unhomogeneity of the salt mass. The longitudinal velocity was used for the calculation of the dynamic modulus of elasticity. The higher magnitude of velocity corresponds to the higher magnitude of the dynamic modulus of elasticity and vice versa. On the basis of in situ monitored data, the dynamic modulus of elasticity had been calculated for each micro-seismic profile, as given in Table 11.2.1.

The measured longitudinal velocity of the pure white salt at station G_5 has been 3800 m/sec, which is in agreement with velocities of the majority of salt deposits of Europe.

It is necessary to point out that the variation of longitudinal velocities of the rock salt strata at the Tusanj salt mine has been influenced not only by petrological composition but also by the anisotropy of the beds, as discussed in the next paragraph.

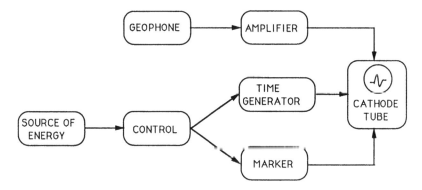

Figure 11.2.3. Scheme of the instrument – Terra Scout Model 150

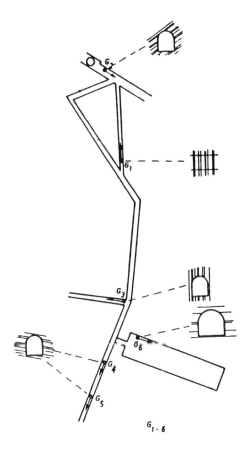

Figure 11.2.4. Mine layout with the geophones position and direction of microseismic profiles (Tusanj salt mine)

2. *Dynamic elastic anisotropy.* The elastic properties of many rocks vary with the orientation to their planes of structural continuities (bedding or foliation). In Chapter 4 of this volume , comments were made on the characteristics of anisotropy of salt deposits. The comments were given on the basis of laboratory testing of the salt samples under incremental compressive loading (static elastic anisotropy).

Table 11.2.1. Dynamic modulus determined in situ (Tusanj salt mine). Density of salt determined on the basis of 145 samples, $\rho = 2.15$; Poisson's ratio of salt determined by laboratory testing, $v = 0.27$

Station	Longitudinal velocity v_ρ m/sec	Modulus of elasticity E_D GPa
G_2	1700	4.9
G_4	1900	6.1
G_1	2400	9.8
G_3	2500	10.6
G_5	3800	24.6
G_6	5300	47.8

At this place, consideration is given to dynamic elastic anisotropy identified as an extension of micro-seismic investigations in Tusanj underground mine.[8, 9] The degree of dynamic anisotropy was determined by measuring the propagation velocity in three directions, namely: perpendicularly to bedding planes, at angle to bedding planes and parallel to bedding planes. The propagation velocity was recorded in the salt pillars for the 50 m profile, as illustrated in Figure 11.2.5. The deviation of the longitudinal velocities along profiles depends on the changes of the petrological composition of the beds as well as the variation of their thicknesses. The magnitude of the dynamic elastic modulus in regard to the strata anisotropy is in agreement with the static elastic modulus determined in the laboratory by incremental compressive loading.

The seismic surveying in the Plains region of Western Canada identified dynamic elastic constants of individual layers of evaporite strata. This investigation indicated that the dynamic elastic anisotropy ratio (E_1/E_2) increases with an increase of clay content and decreases with an increase of silt content (Figure 11.2.6).

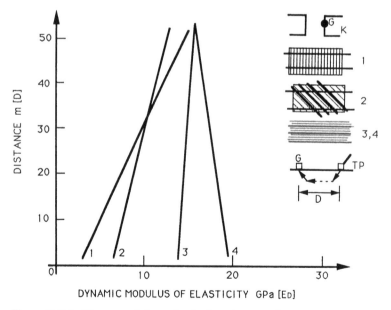

Figure 11.2.5. Diagrams of dynamic elastic anisotropy determined by orientation of longitudinal velocity to sedimentary planes (Tusanj salt mine)

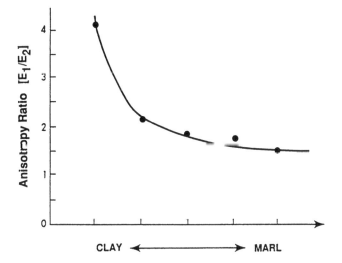

Figure 11.2.6. Diagram of dynamic elastic anisotropy ratio as a function of clay and silt content in the rock (Plains region, W. Canada)

From the rock mechanics point of view the geometrical relationship between the dynamic elastic anisotropy and layout of mine structure is an important factor, because it may control the strata stability and the salt pillar deformability.[1]

11.2.3 *Seismic characteristics of salt structures*

The seismic investigations of salt deposits and salt structure have been carried out in a certain number of salt basins all over the world. The seismic analyses have been centered to study both naturally induced seismicity (earthquakes) and mining induced seismicity (outbursts). The outburst produces an instantaneous failure of salt causing expulsion of material at the surface of mine excavation and a seismic disturbance to ground surface or underground mine. The seismic characterisation in the salt mine is of great importance in regard to mine stability. At this place the characterization has been given on the basis of strain energy phenomena, which may lead to the outbursts.

1. *Active seismicity* is considered on a rather large scale, as for example the seismicity of the salt regions and basins. Some salt basins do not experience any seismicity, but some of them do, as in the case of the interior salt basin of the Gulf coast region (USA). As far as mining is concerned, the regional potential strain energy is of lesser importance than local strain energy related to the faults within the parameter of an underground mine. The strain energy of the faults corresponds to the remnants of orogenic energy concentrated within the shear fault zone. These faults are called the active faults. The relief of strain energy is followed by the bonding stress reduce and by slip adjustment along fault planes. This could be observed in the mines by two phenomena: firstly, in virgin ground by some tremors and distant noise; secondly, in an extraction area by violent and sudden energy relief. The seismic energy liberated by outburst is in the form of elastic waves which is equal to the part of energy of the fault zone that has been liberated. The fault zones could be sources of repeated outbursts due to natural strain energy built up similar to the earthquake mechanics or artificial strain energy built up due to transfer of seismic energy erected by mining working operations.[10]

2. *Excess seismicity* is related to the anomalous zone in virgin strata such as a gas pocket

(Figure 11.2.7). The accumulated strain energy is locked in the strata and it is in a state of equilibrium. The strain energy of the anomalous circumscribed area can be released only when mining workings penetrate this area at the moment when the thickness of the dome or shell around the pocket is sufficiently decreased by the mine excavation, the perfect confinement of the gas pocket vanishes and the gas will be liberated. The gas liberation is by two mechanisms, namely: the gas blowout (gas emission) or gas outburst (violent fracturing of the salt). The seismicity has particular characteristics in the case of the gas outburst, because it takes the form of a chain reaction known for the gassy rocks. The excess strain accumulations could be detected with a micro-seismic survey, indicating the stress differences between the pocket and the surrounding strata.

3. *Induced seismicity* due to mining activities results from the change of the strain energy of the mine structure. The change of strain energy during mining operatings, surveyed by monitoring of seismicity, could be generated by two phenomena:

a) Blasting produces the seismic vibrations and seismicity. The most common monitoring of seismicity is by the peak particle velocity. Figure 11.2.8 shows the model of recorded seismicity around a room after blasting. The seismsicity was surveyed by an array of fixed geophones. Immediately after blasting a micro-seismic activity was heard over the monitor on the geophones located in the blast vicinity.

b) Impact stress known as a travelling abutment stress progressively induced as the salt extraction advances. The increase of dynamic stress corresponds to an increase of seismic activity. Micro-seismic surveying after ore blasting recorded up to a high seismic activity in

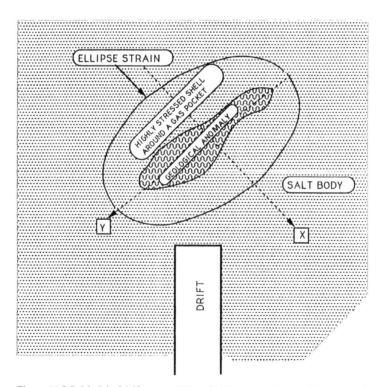

Figure 11.2.7. Model of drift approaching a highly stressed zone of the gas pocket

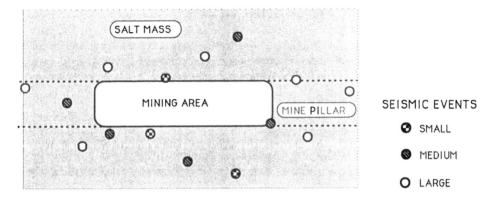

Figure 11.2.8. Model of seismicity distribution around a room after blasting

the mine pillars, almost three times higher than in the floor of the rooms and drifts. Figure 11.2.9 shows the longitudinal velocity of seismic waves which have been propagated along a longer room axis. The changes of the longitudinal velocity are due to salt heterogeneity and the change of strain energy caused by mining operations.[8, 9]

The monitoring of induced seismicity should be an important part of instrumentation in regard to strata control, particularly in the case of possibility of salt and gas outbursts due to the change of strata energy.[9]

The seismic characterization offers the background for a classification of the gas and salt outbursts. Actually, the most realistic classification would be the one which corresponds to the three types of seismicity described earlier.

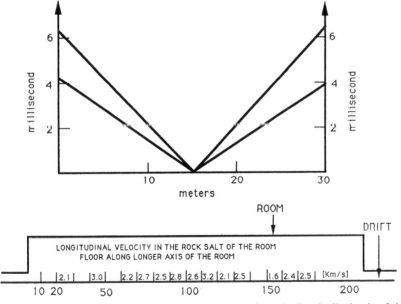

Figure 11.2.9. Changes of velocity of seismic waves along the longitudinal axis of the room floor (Tusanj salt mine)

11.3 DYNAMICS OF SALT AND GAS LIBERATION

The dynamics of salt and gas liberation is not very well known. A certain knowledge could be obtained both by laboratory experimentations and practical phenomenological investigations in underground salt mines. The further consideration of this topic is given as follows.

11.3.1 *Principles of salt and gas liberation*

On the basis of general knowledge of gas hydrodynamics and salt dynamics, the liberation of salt and gas may be analyzed. The strain liberation is more violent if salt dynamic stress and gas hydrodynamic stress are integrated.[10, 11, 12]

Laboratory investigations suggest that dynamics of stress liberation is governed by mass velocity of particles behind the front of the decompressive waves which vary most rapidly up to the pressure drop of $(3.50 - 5.50) \times 10^6$ MPa. However, beyond this stress magnitude, the relative changes are negligible except in the case of some salts of which the pores are significantly deformed before the attainment of high stress concentration. In this case, it gets the greatest increment in the accumulation of strain energy. If the external pressure is further increased, the relative changes in the accumulation of strain energy become less important, which is discussed in Subsection 11.3.2.[12]

The instant relief of the elastic energy and pore pressure causes salt breakage by the mechanics of tensile stress in the decompressive wave, giving a rapid restoration of elastic deformation of the mine structure. The simultaneous gas release increases the volume of rock or salt in a porous medium, and it is followed by the expansion phenomenon, which additionally induces a tensile stress in the salt face area.[11] The underground phenomenological studies indicated that impulsive salt fracturing is controlled by the particular bed separation, which is caused by relief of accumulated strain energy. The salt thrown out is greatly broken up and to some extent pulverized, as discussed in Subsection 11.3.3.[1]

The distinction of gas and salt liberation in regard to types of energy accumulation as described in the previous section, named: active strain energy – active faults; excess strain energy-pockets; change of strain energy – mine workings, is not always an easy task. For example, a certain number of Gulf domes have a large vertical shear structural feature formed during the growth of domes. These structural features represent the zone of interbedded sediments with gas and brines. These zones create difficult mining conditions as experienced in existing mines of salt domes.[13] Certain parts of these difficult zones in some salt mines consist of the shear faults with accumulated seismic strain energy, which could be liberated by its build-up to a critical point or triggered by mining activities in their vicinity. However, the source of gas and salt strain energy liberation for flat and slightly folded deposits is more or less easily recognizable for determination.

11.3.2 *Dynamics of outburst of salt and gas*

The strain energy relief manifested as a salt and gas outburst may occur in all phases of dry as well as solution mining. The strain accumulation results due to factors as described in previous sections. Generally speaking, two technological elements are important for the occurrences of gas outburst: firstly, by salt excavation removal of lateral constraint and changing a stable equilibrium of triaxial compression into an unstable equilibrium of biaxial or uniaxial compression, and secondly, by changes of geometry of the mine

structure in accordance with the method of mining. For example, gas and salt outbursts occur when a room with a large excavated area approaches near enough to an anomalous zone with high gas pressure, that mining stresses by impact could be transferred on it. There is an opinion that the impact of dynamic stresses due to blasting does not trigger outburst, regardless that it occurs after blasting.

The mechanism of energy release could be analyzed by the assumption that the lateral expansion is not in effect and that the deformation coincides with the direction of the mass movement, which is opposite to the direction of propagation of the decompressive wave.[12] The deformations which correspond to directions perpendicular to the direction of motion are equal to zero:

$$\varepsilon_2 = \varepsilon_3 = 0$$

$$\varepsilon_1 \neq 0$$

Further generalized from Hooke's law:

$$\sigma_2 = \sigma_3 = \frac{v}{1-v} \sigma_1$$

$$\varepsilon_1 = \frac{(1+v)(1-2v)}{E(1-v)} \sigma_1$$

where:

σ_1 = stress in the direction of motion;
σ_2, σ_3 = tangential stresses;
v = Poisson's ratio;
E = modulus of elasticity.

To transfer these mechanisms in the dynamic process, the dynamic Poisson's ratio and the modulus of elasticity could be used and are expressed as:

$$v_d = \frac{0.5 - R^2}{1 - R^2}$$

$$R = \frac{V_s}{V_p}$$

where:

v_d = dynamic Poisson's ratio;
V_s = velocities of transverse waves;
V_p = velocities of longitudinal waves.

$$E_d = \frac{\gamma_c}{g} \frac{(1+v_d)(1-2v_d)}{1-v_d} V_p^2$$

where:

E_d = dynamic modulus of elasticity;
γ_c = unit weight of salt;
g = acceleration due to gravity.

Substituting for the strain equation is obtained as follows:

$$\varepsilon_1 = \frac{\sigma_1}{E_d}$$

and further substituting Poisson's number by the following expression:

$$K_\tau = \frac{v_d}{1 - v_d}$$

The relationship between the stress and deformation may be written as follows:

$$\sigma_1 = E_d \varepsilon_1$$

$$\sigma_3 = K_\tau \sigma_1$$

where: K_τ is the coefficient of lateral thrust.

According to the theory of elasticity, the stress arising in deformed media can be calculated by the following equation:

$$\sigma_1 = \frac{\gamma_c V_p u}{g}$$

where: u is the mass velocity of the particles.

The minimum stresses behind the decompressive wave front are σ_{refl}, at which a violent fracture may occur in the salt near their strength within static loading conditions, which is written as:

$$\sigma_{refl} = \frac{\gamma_c V_p u_{cr}}{g}$$

where u_{cr} is the critical particle velocity.

The final consideration should be given to the energy of the salt and gas outburst, which is actually an integration of the potential and kinetic energies:

$$u_{tot} = u_{pot} + u_{kin}$$

Total energy is equal to the potential energy of elastic deformations plus the kinetic energy of the particles moving behind the wave front.

The total energy equation can be composed by the equation for potential energy and by a modified equation of the theory of elasticity for kinetic energy, as expressed by Khanukaer:

$$U_{tot} = \left(\frac{\sigma_1^2}{2E} + \frac{(\gamma_c/g)u^2}{2} \right) 10^{-6} J$$

This equation can be used for the calculation of total energy of the decompression wave per unit volume of the salt if the mass velocity of particles is known.[12]

The underground studies in the salt mines showed that salt and gas liberation from the mining face is due to tremendous stress relief of integrated strata dynamic stresses and gas hydrodynamic stresses. Some cases of history are briefly commented in Section 11.4, which considers outburst effects in some mining operations in Germany and the USA.[11, 13]

11.3.3 *Mechanism of outburst fracturing*

Mechanics of outburst fracturing could be analyzed by stress relief criteria, which have been inferred from observations and investigations of fractures in the field caused by energy strain relief[14] and in the mines caused by outbursts in shear zones.[15] As stated for the outburst types of excess strain energy and the change in strain energy, namely that biaxial or uniaxial compressive stress and tensile stress have been in effect, is exactly the same for the case of relief strain energy. Under these circumstances the geometry of the outburst cavities has a conical or ellipsoidal form, with the shingle like jointing to brittle failure in a biaxial stress field.[15] The fracture surface indicates rock salt mass separation by shear failure with splinters being convexly curved to the axis of cavity.

The energy of salt is more intensive in the case of salt and gas outbursts than in the case of mechanical fracturing, because there is a liberation of a great amount of accumulated strain energy. The fracturing of salt propagates to the critical boundary, with a common geometry of a hemispherical or an elongated cavity with elliptical cross-section (Figure 11.3.1). This conical shape of outburst cavities is caused by the impulsive nature of violent stress release, and also indicates the promotion of continuation of outburst further in solid salt beyond the boundary surface.[11] The diameter of these cavities may range from 1 to 30 m and their height and/or length may range from 1 to 50 m or more.[2] The outburst may suddenly expel a few tonnes to tens of thousands of tonnes of salt and a large volume of gases. Due to differential anisotropy of the salt face, salt fragmentation and breakaway may start deeper in the salt body within stronger salt layers and are accompanied by ejection of nearby soft or loose salt with poor cohesion, which does not give appreciable resistance to the moving

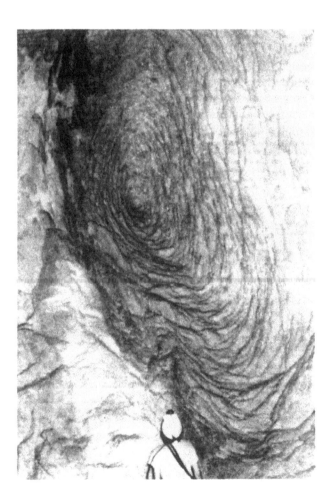

Figure 11.3.1. Typical gas outburst cavity in a Louisiana salt mine (USA)

mass. This explains why the axes of cavities are close to parallel of the salt faces, with their collaring in softer salt layers. Usually, the strong salt is coarsely broken and soft salt pulverized.

The energy sources causing stress relief failure influence the mechanics of fracturing and thus, this criterion must differ from the loading failure. The unloading fracturing follows expansion of the rock mass and proceeds from the place of maximum rebound towards the place of minimum rebound, e.g. free face. Upon completion of the processes of salt expansion and release of potential energy stored in the fault zone by shearing and violent fracturing, primitive stress once more becomes the principal loading force.

The mechanism of fracturing could be represented by a stress ellipsoid (Figure 11.3.2) where two types of failure are inferred, as further discussed.[14]

The shear failures are products of potential stress action, along which outbursted salt has been released from the intact mass. In the half-ellipse axis the failure stresses are:

$$\sigma_n = \sigma_x \cos^2 \alpha + \sigma_y \sin^2 \alpha \text{ (normal stress)}$$

$$\tau_n = (\sigma_x - \sigma_y) \sin \alpha \cos \alpha \text{ (shear stress)}$$

If $x = 2y$, $\sigma_x = 2\sigma_y$ and $\alpha = 20°$, then:

$$\sigma_n = 1.89 \, \sigma_y$$

$$\tau_n = 0.32 \, \sigma_y$$

Most likely, the expressed relationship between shear and principal stresses of a pressure pocket is the initiating factor of a violent shear fracture.

The mechanics of violent and progressive fracturing is governed by overcoming the tensile strength of salt, with its stress concentration further from the free boundary. Although, near each newly formed solid surface, a tensile stress arises that is manifested by repeated sudden failure. The salt fragmentation takes place by loss of adhesion between the

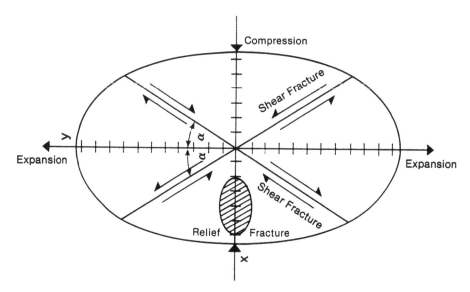

Figure 11.3.2. Model of induced biaxial stress parallel to the mine face (stress ellipsoid)

slab and the solid salt face. The instrumentation indicated that a salt slab moves freely at a velocity equal to twice the particle velocity in the wave ($V = 2\ U$).[6] Relief strain energy fracturing occurs in a number of salt deposits in Central Europe, the Gulf domes (USA) and others.

11.4 OUTBURSTS IN MINING CONDITIONS

The risk of gas outburst is a safety hazard for mine personnel, collapse and damage of mine structure, breakage of equipment, loss of production and others. It is important to know how to identify in advance an outburst-prone zone and to devise procedures for mining through this area so that the danger of outbursts is minimized. In this regard four topics are discussed: characterization of outburst zone, outburst effects on mining workings, control of gas outburst and numerical modelling of outburst.

11.4.1 *Characterization of an outburst zone*

The characterization of the burst-prone zone is generally related to specific geological-structural features of a salt deposit such as a shear zone.[13] The main characteristics of a shear zone related to the salt and gas outbursts in Louisiana domal salt are briefly discussed below.[15]

1. The burst-prone zones are characterized by extensive shearing, tectonic lensing of the included sediments and folding. The geometry of shear zones is highly variable but in the majority of cases they could be approximated as tubular and vertical. A typical shear zone

Table 11.4.1. Characterization of gas outburst zones in Louisiana salt mines (after Kupfer)

Mine	Depth of mining levels	Inclusions	Anomalous (or shear) zones	Depth of outbursts	Remarks
Winnfield (not operating)	Sublevel, at 190 m, flooded. Main level 250 m	Brine, CO_2, 10% anhydrite	Not penetrated or recorded	200-240 m	The five outbursts or 'blowouts' within 150 m of the edge of the dome
Jefferson Island (not operating)	Four levels, deepest at 450 m	Some sediment, brine and gas	At edge and in centre of dome	350-400 m	Several large outbursts near edge shear zone
Avery Island	Three levels, deepest at 275 m	Sandstone and brine	In centre of dome		No outbursts
Weeks Island	Two levels, at 150 m and 250 m	Sediments, some brine and oil, gas	At edge and in centre of dome	220-240 m	Small to medium-size outbursts in shear zones
Côte Blanche Island	One level, at 375 m	Sandstone, gas, black clay	Not detected at edge or in centre, but two sediment-bearing, thin, and persistent zones	360-370 m	Medium-size outbursts follow clay-bearing salt
Belle Isle	One level, at 425 m	Clay-lime sediment, brine, gas	At edge and in centre of dome	390-420 m	Several large outbursts associated with the central shear zone

may be from 3 to 100 m wide, of a very long horizontal extent, and probably of a large vertical extent.

2. The burst-prone zone contains the inclusions of petroleum, brine and various gases. The distribution of inclusions within a shear zone is discontinuous and in places is almost random. Gas and other fluid inclusions in the salt occur in intergranular fractures as well as in intergranular bubbles.

3. Generally speaking, the stress build-up in the pressure pockets is associated with the rock salt porosity and inclusions as well as the depth of mining. The entrapped gas is at least under hydrostatic pressure. The distribution, shape and extent of a pressure pocket (with a potential for generating gas outbursts) are most likely random, as evidenced by a clustering or complete absence of outbursts within short distances.

The summarized characterization of geological-structural features related to the gas outbursts in Louisiana salt mines is tabulated by D.H. Kupfer (Table 11.4.1).[16] The Louisiana domal salt is medium-grained, vertically bedded and occurs in spines surrounded by inclusion-bearing anomalous zones.

It is necessary to point out that the characterization of gas outburst zones presented here was mainly for domal deposits in the USA, however, it could be also applicable for domal deposits in Europe. The characterization of gas outburst zones in flat salt deposits, particularly in the case of the contact of salt beds with rock strata is sometimes difficult. A roof is usually represented by marl, shale or other sediments, in which gas pockets occur. This is the case of the Goderich salt mine in Southern Ontario (Canada), where pressure pockets are located in the roof strata composed by shale (Figure 11.4.1). Characterization of the gas outbursts zone is given by three parameters.

1. The shale strata consist of a single zone or interconnected multiple fractured zones. They are the product of shear deformations.

2. The fractured zone may delineate a gas pocket, which contains mainly methane gas (Figure 11.4.1). The gas under high hydrostatic pressure could be violently and suddenly released when mining working reaches a critical distance.

3. The salt beds in the Michigan basin are approximately 25 m thick, with a tendency to pinch out at the outer perimeters. In this case the actual salt back thickness will decrease and as excavation approaches these perimeters, the protection layer of the rock salt moves

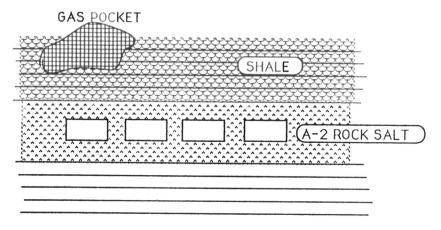

Figure 11.4.1. Pressure pockets in the shale roof strata prone to gas outbursts

towards the shale strata, the pressure pocket is weakened considerably and a gas outburst is facilitated.

It is interesting to note that in all incidents, dripping brine and roof bolts breaking were observed prior to gas outbursts, which is of fundamental importance for its detection in advance.

Generally speaking, the detection and characterization of gas outburst-prone zones is essential for controlled salt mining. The methods of detection that have been tried, with various degress of success, are largely centered on geophysical methods of investigation or monitoring of the gas flow from the rock face or drill holes. Advance drilling for detection and characterization of gas outburst zones is commonly recommended.

The geophysical techniques for detection of the anomalous zones are based on contrasts involving resistivity, sonic velocity and electromagnetic wave propagation. It should be mentioned that geophysical techniques have been rather successful in the characterization of anomalous zones but not so for the prediction of gas outbursts in a short period of time.

11.4.2 *Outburst effects on mining workings*

The gas outburst is triggered when an excavation approaches a pressure pocket containing heterogeneous and anisotropic material. This results in a preferential transfer of mining stress, thus increasing the level and anisotropy of stresses in stiffer rock. In addition, the stored strain energy in the pressure pocket will be higher than that in the surrounding salt, thus increasing the specific capacity of the pressure pocket to outburst.[17]

It is generally recognized that the outburst may occur in the development phase of salt deposits during breakage of virgin ground or in the extraction phase during room excavation. Both of these cases are discussed separately.

1. *The gas outbursts during the development of salt beds* have been observed in the Goderich salt mine, Southern Ontario. The salt extraction is by single level regular room and pillar mining in the stratified salt deposit which leaves a protective rock salt layer towards roof and floor strata. The gas outbursts are related to gas release from pressure pockets due to the close vicinity of roadways excavation, and pressure pockets which are located in the shale formations overlying salt beds. Within three months (1986/87) three gas outbursts occurred, as briefly discussed in the following (Figure 11.4.2).

a) The first outburst occurred at the intersection of the 81 and 16 roadways, followed by large fall of ground. It is not known that the outburst was due to one pressure pocket or several pockets connected by fractures common in the shale roof. It is estimated that approximately 17 000 m³ of gas was released and due to the nature of the roof failure it was estimated that the gas pressure upon release was in the range of 4 MPa of hydrostatic pressure. The outburst extensively damaged ventilation curtains, some almost 2.5 km away.

b) The second outburst and roof fall occurred at the intersection of the 81 and 10A roadways. In this instance, approximately 9000 m³ of gas was released and this outburst damaged more ventilation curtains and windows at the crusher station approximately 300 m away.

c) The third outburst and roof fall occurred at the intersection of 12A and 82. A large emission of gas was detected, but the damage was of a lesser degree.

The gas outbursts have been followed by a methane flash fire which was ignited during its expulsion. Those events are the first gas outbursts recorded in the history of the Goderich mine, from its inception in 1959 to date.[17]

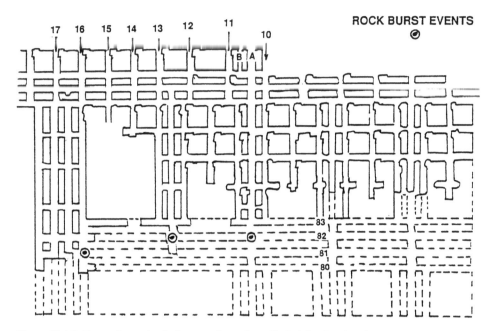

Figure 11.4.2. Gas outbursts in virgin ground openings, Goderich salt mine, Southern Ontario (Canada)

2. *The gas outbursts during the extraction phase* are observed in the large rooms within salt domes of Central Europe (Zechstein's deposit) and North America.

Gas outbursts in salt domes of Central Europe occur when a room with a sufficiently large free face approaches near enough to a horizon of a rock zone prone to outbursts. At this distance, the stress state in the rock salt next to the face changes from a triaxial to a biaxial or uniaxial one. This results in a sudden failure and violent release of gas and fluid from the rock salt face. Under these circumstances, the technology of salt excavation (cyclic or continuous) is unrelated to the occurrence of bursts. A common feature of the outburst phenomenon is the formation of large cavities in the room back and walls due to expulsion of significant amounts of rock salt (50-500 tonnes). Sometimes, the outbursted fragments of salt might fill up mine workings of a length up to 100 m. Gas outbursts sometimes take the form of a chain reaction in which the characteristics of the gassy rock salt are significant. For example, gas outbursts in the Werra district occur in gassy zones that are located in the proximity of young tectonic disturbances.[11]

Gas outbursts of salt domes in North America occur normally in the upper part of the room or in the roof of the room. The room height is between 20 and 30 m. It should be noted that an outburst does not occur in the lower part of the floor of a room. Most likely it is due to some confinement by the broken salt.[15] Underground phenomenological investigations established that when mining excavation approaches the burst-prone salt zone, the stress in the pressure pocket is raised to a level that is several times higher than the lithostatic stress. The outburst is facilitated by stress concentration induced by the mining geometry and the stress increase in the stiffer sediments. The gas outburst is usually triggered when the clastic zone around a mine excavation intersects the pressure pocket with an elevated stress concentration due to mining operations.

The outburst effects on mining workings may endanger a safe mining, and controlled mining has to be implemented, as discussed in the next sub-section.

11.4.3 *Control of gas outburst*

Control of gas outburst has been summarized by M.A. Mahtab for Louisiana dome salt mines, where he suggested two important approaches: identification of burst-prone zones in advance of mining and destressing and degassification ahead of a face, as individually described below.[2]

1. *Identification of burst-prone areas* is a starting measure for outburst control. For example in Louisiana domes, pressure pockets generally tend to be associated with other anomalous features and inclusions found in the edge or central shear zones (see Table 11.4.1). Mapping of these features on a regular basis should be considered essential for identifying potential gas outburst areas.[2]

Drilling-exploration ahead of the face should be carried out to obtain the following data:

   a) Identification of anomalous features;
   b) Information on core disking;
   c) Pressure and flow-rate of gas;
   d) Detection of microseismic activity or change in stress, if required, in highly suspect areas.

The methodology for application of micro-seismic techniques to the study of geologic structure and changes in stress is described by Hardy and Blake et al.

Evaluation of the tendency of a rock mass to outbursts (or degree of risk to outbursts) is a useful idea that was proposed by Gimm and Pforr.[11] They expressed the simple degree of risk to rockbursts as a function of the specific energy released and the compressive strength of rock, both at a given rate of loading.

For the purpose of evaluating the degree of risk of a mass of salt in Louisiana salt mines to gas outbursts, further work is required (in the field and the laboratory) to establish empirical relationships among a group of critical parameters, which should include:

   – degree of elasticity, E_{loading}, $E_{\text{unloading}}$;
   – frequency of disking in core-holes drilled into a face;
   – compressive strength (uniaxial);
   – tensile strength (Brazilian);
   – pressure and rate of flow of gas from boreholes;
   – method of working (configuration, rate of advance, excavation scheme).

2. *Degassification and destressing ahead of a face* is the most important element in the control and prevention of gas outbursts. Degassification reduces the pore pressure in the rock, thereby increasing its strength.[2]

One promising method of degassification that needs to be studied is to drill one or more large, but safe, diameter drainage holes (Barr, 1975, recommends a 25 cm-diameter hole) into the pressure pocket; to drill smaller holes in the face around the drainage holes; and to load and blast the back ends of the smaller holes, a procedure to be called 'shock blasting'. Since this procedure may trigger a gas outburst, due precautions and methodology must be used during drilling, blasting, and degassification. In particular, the location and 'strength' of the shockblast in a given situation, and its requirements on mine safety and mining costs, should be examined in detail.

Shock blasting, though not practised in Louisiana salt mines, has successfully reduced the number of accidents resulting from gas outbursts in coal and salt mines outside the USA.[2] The role of shock blasting is to pre-condition the salt beds, by reducing its modulus of elasticity and strength. The burst-prone zone, with a high strain concentration, by destressing blasting will change the salt to yielding behaviour, which allows the mine structure to fail non-violently.

The transfer of stress from the face to other areas is effected by long-hole drilling ahead of mining, and blasting of the back ends of the holes. To further reduce the risk of outbursts, one or more large diameter (approximately 25 cm diameter) holes can be drilled which act as gas drainage holes. Around the drainage holes, smaller diameter holes are drilled, loaded, and blasted. In qualitative terms, shock blasting is between pre-splitting and pulverizing. In general, the amount of explosives and hole spacing used should be sufficient to fracture the rock, but in such a fashion that fracture density and propagation do not eliminate the rock mass cohesion. A rule of thumb suggested by Jeremic is for the blast to produce aproximately double the number of joints and cracks that were present in the rock. However, this rule may not always apply in most salt deposits where the joints are infrequent.

Shock blasting lowers the level of stress and stored strain energy and changes the failure mode from brittle-elastic to elasto-plastic or plastic. The net result of shock blasting is that the fractured zone yields gradually, rather than failing suddenly.[15]

It should be mentioned that control of gas outbursts as well as high stress concentration could be done by so-called stress relief layouts, such as the 'split panel' by abutment pillar, the 'yielding pillar' concept, the 'pillar notching' into the abutment pillars, as well as 'stress envelope, parallel room and time control stress relief' by Serata.

11.4.4 *Numerical modelling of outburst*

A numerical analysis of the mechanics of gas outbursts in salt deposits has been studied by M.A. Mahtab et al. The results of their work are summarized below.[18]

The model stress analysis have been carried out by finite differences (STEALTH) for the chosen room-and-pillar layout (Belle Isle mine), with the following approximations:

1. Twodimensional, plane-strain idealization is accepted. A longitudinal section through the centre of the room is discretized into a grid of 30×26, finite difference zones (Figure 11.4.3).

2. The depth of the room under consideration is approximately 305 m. It is assured that in situ stress is hydrostatic with a magnitude of 6.9 MPa.

3. The uniaxial compressive strength of the salt determined in the laboratory at 10.5 MPa has been scaled to an in situ strength of 5.8 MPa.

4. It is assumed that the pressure pocket of the model has an equal change of containing gas or liquid under pressure.

5. The Mises criterion for failure of the salt was used.

The properties of the different materials (types 1 to 6) used in numerical model analysis are given in Table 11.4.2.

The room and kerf (undercut) are modelled as air. The salt mass and the pressure pocket ahead of the face are allowed to further weaken after yielding according to the Mises criterion ($\sigma_e \geq C_o$). The weakened region, where the strength of the salt is reduced, or where the strength and shear modulus of the material in the pressure pocket are reduced, is called the burst zone.

A comparison of stress contours (σ_e given in MPa), as illustrated in Figure 11.4.4 and

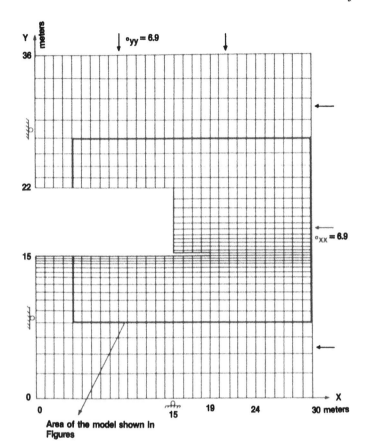

Figure 11.4.3. Finite difference grid for numerical solution

Table 11.4.2. Properties of material types used in the numerical model

Material No.	Name	Unconfined compressive strength 10^6 Pa	Bulk modulus K, 10^9 Pa	Shear modulus G, 10^8 Pa
1	Salt	5.8	1.148	5.3
2	Pressure pocket	5.8	1.148	5.3
3	Air	0.0	0.000	0.0
4	Weakened salt	5.8	1.148	5.3
5	Burst salt	0.58	1.148	5.3
6	Burst pressure pocket	0.0058	1.148	0.0053

11.4.5, indicated the transfer of stress concentration caused by kerf. The plastic zones, as could be expected, will first develop near the edge of the kerf and the top of the face (Figure 11.4.6). For a situation, however, where the plastic salt is not allowed to further weaken, or a weakened pressure pocket is not present near the face, as illustrated in Figure 11.4.7 and 11.4.8, the progress of the plastic zone is limited by the in situ strength of the homogeneous salt.

When a weakened salt zone is modelled ahead of the face together with a vertical pressure pocket extending from the edge of the kerf to several meters beyond the kerf,

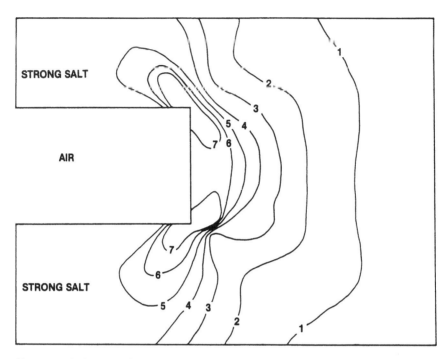

Figure 11.4.4. Contours of stress around a mine heading with a kerf (strong, homogeneous, elastic salt)

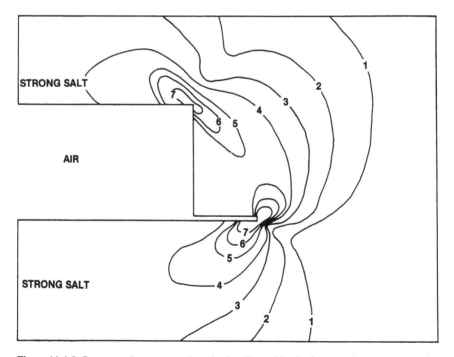

Figure 11.4.5. Contours of stress around a mine heading with a kerf (strong, homogeneous, elastic salt)

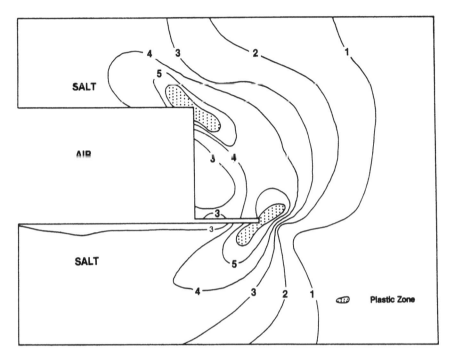

Figure 11.4.6. Contours of stress and plastic zones around a mine heading in modelled salt

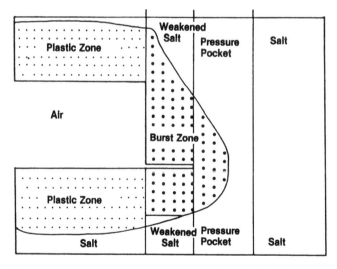

Figure 11.4.7. Plastic and burst zones around a mine heading near a pressure pocket with unconfined floor

plastic zones develop ahead of the face as well as in the roof and floor of the room. The strength of the weakened salt is one tenth of the intiial strength (0.58 MPa) and it is considered to be 'burst salt'.

The weakened volume elements in the pressure pocket, together with the burst salt

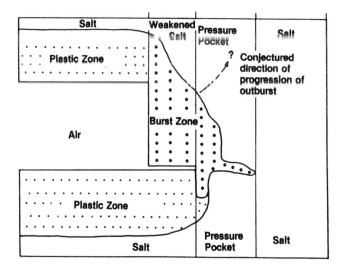

Figure 11.4.8. Plastic and burst zones around a mine heading near a pressure pocket with the mine floor confined by broken salt.

(Figures 11.4.7 and 11.4.8), are preliminary estimates of the zones involved in a gas outburst when broken salt does not cover the floor and when broken salt covers the floor. This indicated that the confined salt below the kerf is not burst-prone.

The numerical analysis of the phenomenon of gas outburst in salt indicated that weakening (fracturing) of the salt and the material in the pressure pocket is located at a small distance beyond the face of a mine heading is a necessary condition for the occurrence of a gas outburst.

The removal of any volume element of the 'burst rock' from the finite difference grid or model has not been carried out. However, the model analysis involving removal of burst rock salt may show a progression of the outburst in a direction that is normal to the 'advancing' boundary of the burst zone shown in Figure 11.4.8.

The model analysis demonstrated the possibility of the evaluation of stress distribution around a mine heading and formation of pressure pockets. The authors suggest that identification of burst-prone zones in advance of mining, destressing and degassification ahead of the heading face is of significant importance for ground control. Degassification of the pressure pockets in advance of the heading face reduces the pore pressure in the rock, thereby increasing its strength.

Finally, a knowledge of gas outburst mechanics in conjunction with mining excavations, beside academic consideration, is also an important engineering factor for the design of safe mine structures integrated with ground control techniques.

REFERENCES

1. Jeremic, M.L. 1985. Coal and gas outburst. In *Strata mechanics in coal mining*, pp. 173-189. Rotterdam: Balkema.
2. Mahtab, M.A. 1981. Occurrence and control of gas outbursts in domal salt deposits. In *The mechanical behaviour of salt*, pp. 775-789. Clausthal: Trans. Techn. Publ.
3. Kupfer, D.H. 1978. Problems associated with anomalous zones in Louisiana salt stocks, USA. In

Proc. of 5th Int. Symp. on Salt, Northern Ohio Geological Society, Cleveland, Ohio, Vol. 1, pp. 119-134.

4. Stojkovic, J. et al. 1987. Salt deposits of Tuzla basin and prospect of future geological exploration. *J. Min. & Geol.* (Faculty of Mining and Geology, Tuzla Univ.), 15/16: 9-18 (in Yugoslav).

5. Jeremic, M.L. 1966. Risk of water penetration in the salt deposits. *Min. & Metall. Bulletin* (Belgrade), 8: 175-182 (in Yugoslav).

6. Kelsall, P.C. & J.W. Nelson 1983. Geological and engineering characteristics of Gulf region salt domes applied to underground storage and mining. In *Proc. of 6th Int. Symp. on Salt*, The Salt Institute, Virginia, Vol. 1, pp. 519-541.

7. Obert, L. & W.I. Duvall 1967. *Rock mechanics and the design of structures in rock*, pp. 344-350, 447-459. New York/London/Sydney: Wilcy.

8. Mandzic, E. et al. 1980. Static and dynamic modulus of elasticity of rock salt, *Bulletin of Min. & Geol. Engng.* (Tuzla Univ.), pp. 111-117 (in Yugoslav).

9. Mandzic, E. & M. Cvetkovic 1970. In situ determination of dynamic modulus elasticity of rock salt from Tusanj mine by micro-seismic method. *Bulletin of Min.* (Institute of Mines, Belgrade), 10 (2): 1-10.

10. Oelsner, C. 1965. *Schlagseismische Untersuchungen unter Tage zur Bestimmung des Gebirges in Situ*, pp. 18-28. Leipzig: Frieberger Forschungshefte, Geophysik, C 178.

11. Gim, W.A.R. & H. Pforr 1964. Breaking behaviour of salt rock under rock burst and gas outbursts, In *Proc. of 4th Int. Conf. on Strata Cont. and Rock Mech.*, Columbia Univ., New York.

12. Medvedev, B.J. & V.V. Oxokin 1973. Mechanics of gas outbursts. *Fiz. Tek. Publ. Raz Polez, Isk.*, 1: 114-117 (in Russian).

13. Kupfer, D. 1967. Mechanism of intrusion of Gulf coast salt, Contribution to Symp. on the Geology and Technology of Gulf Coast Salt, Louisiana State Univ.

14. Jeremic, M.L. 1976. Deformation of a contact-metamorphic rock mass at Lost River, Alaska. *CIM Bulletin*, April 1976: 93-99.

15. Mahtab, M.A. et al. 1984. Controlled mining of salt under gas outburst conditions, pp. 1-18. In *Proc. of 12th World Mining Congr.*, New Delhi, India, November 19-23, 1984, RT-III 3-19.

16. Kupfer, D.H. 1978. Problems associated with anomalous zones in Louisiana salt stocks, USA. In *Proc. of 5th Int. Symp. on Salt*, Northern Ohio Geological Society, Cleveland, Ohio, Vol. 1, pp. 119-134.

17. Salazar, L. 1987. Methane occurrences in outhern Ontario's salt beds. Manuscript, pp. 1-16. School of Engineering, Laurentian Univ., Sudbury, Ontario.

18. Mahtab, M.A. et al. 1983. A numerical analysis of the mechanics of gas outburst in salt. In *Proc. of 6th Int. Symp. on Salt*, The Salt Institute, Virginia, Vol. 1, pp. 549-560.

CHAPTER 12

Mining of moderately thick deposits

Mining of thin and moderately thick salt beds includes all mineralogical types of evaporite deposits, gypsum, trona, rock salt and potash. The mining methods used in the application of rock mechanics to the design of mining layouts for these deposits are similar to those used in the mining of flat coal deposits with thicknesses of between 1.0 and 3.5 m. It is, however, essential to recognize and make allowances for the differences in ground behaviour between evaporite and coal deposits when converting a mining method from one type of deposit to the other.

The full face mining of planar deposits with thicknesses of less than 5 m offers the potential for a high degree of stability for stope structures for mining geometries other than classical room-and-pillar layouts. The key is a combination of a mining sequence and system of pillars which causes the transfer of stress to the pillars to occur in a progressive and uniform manner. The principles of pillar loading and pillar stress concentrations associated with the mining of moderately thick salt deposits are, to some degree, applicable to the mining of thick and inclined evaporite deposits.

This chapter deals with the following mining and rock mechanics topics as pertain to: principal mining systems, design of the mine structure, its stability and its stress analysis.

12.1 PRINCIPAL MINING SYSTEMS

For moderately thick evaporite beds with dips of less than 15°C the choice of layouts must take into account the geometrical configuration and the material properties which make up the roof, pillar and floor strata. The responses of these strata elements to mining should be tested with a geomechanical model prior to any implementation of the mine layout.

Fundamentally, there are three groups of mining methods to choose from for mining a moderately thick evaporite deposit. Each group in turn, however, consists of several mining system subsets which can cause the strata to respond differently to mining.

12.1.1 *Angular pillar mining*

There are a wide variety of angular pillar shapes, either by design or as a result of irregularities in the deposit. Angular pillar mining can be planned as either a one-pass advanced mining system, where the pillars are left in place, or as a two-part advance and retreat sequence, where the pillars are partially or totally recovered in the retreat phase. To a large part, the layout of the angular pillars is influenced by the excavation methods and equipment. Mining can be drilling and blasting (cyclic mining), by road headers, drum miners, etc. (continuous mining).

1. *Square pillars* were used extensively in the past in the mining of all types of evaporite deposits. The common application was in flat bedded deposits with thicknesses up to 2.5 m, and which were sandwiched between rock strata. At the present time square pillar mining has only a limited application. Examples:

a) *Drumbo gypsum mine*, Southern Ontario (Canada), square pillars with a drill and blast mining system are used. The gypsum bed is 1.65 m thick and occurs at a depth of 115 m. It is overlain by 60 m of overburden and 55 m of shale and limestone. Limestone above the gypsum bed is water-bearing. The pillar size is 6 m wide by 6 m long and 1.65 m high (full thickness of the gypsum bed). The rooms are also 6 m wide and 1.65 m high. The extraction ratio is 75%. The exposure of the limestone rock to moisture causes its strength to deteriorate, resulting in rock falls.

b) *Tenneco's trona mine* in Wyoming (USA) is primarily developed for square pillar mining. Some areas have also been mined using rectangular pillar configurations (Figure 12.1.1). Tenneco's mineable ore is mainly in bed 17 at a depth of about 457 m. Bed 17 is characerized by a sharp contact with oil shale at the floor. The roof is capped by impure trona, 1 m thick, that grades into shales and marlstones. Full support is provided by square pillars and an advancing mining sequence is used. Development entries and rooms are excavated by a twin rotor NMS boring machine and Jeffrey Heliminers. The NMS boring machines cut oval openings measuring 2.6 m high by 4.5 m wide. The higher output Heliminers cut openings up to 3.7 m high by 4.6 m wide. Trona from the faces is transported by electric shuttle cars to feeder breakers that feed onto the mine belt haulage system. Roof support is by 1.6 m resin bolts on 1.3 m centres.[1]

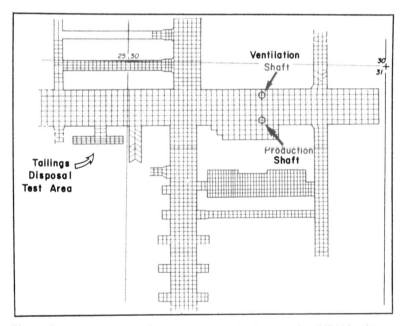

Figure 12.1.1. Development of Tenneco's trona mine in Wyoming (USA) by the square pillar mining layout

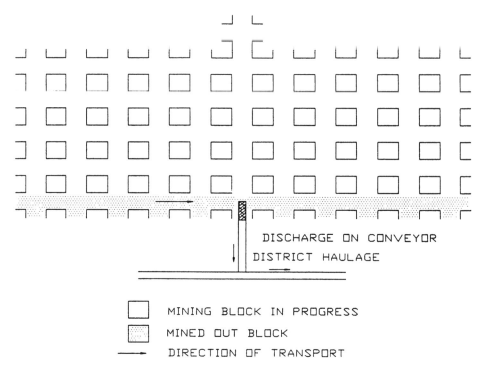

Figure 12.1.2. Layout of a square pillar, room-and-pillar operation in a mining block at a German potash mine

c) In *German potash mines* a room-and-pillar system is used, where the potash bed is divided into block units. Square support pillars are formed by a system of multpile entries connected by crosscuts.[2] Mining is by drilling and blasting. Drilling of the salt face is done with two types of drills. The first drills large diameter cut holes either with a single auger drill (diameter, 420 mm) or with three parallel auger drills (diameter, 280 mm). The length of the auger rods is 7.5 m. The second drill type drills the blast holes. The drills are mounted on the power trucks. Figure 12.1.3 shows the sequence of room excavation, with a lead 5.5 m wide by 2.5 m high advance heading over 10.0 m ahead of the room face. The final room size is 20.0 wide by 2.5 m high, with an arched shape to the roof and walls. The pillars are 40.0 m wide by 40 m long by 2.5 m high (Figure 12.1.3).[3]

2. *Rectangular pillars* have also been applied to the mining of evaporite deposits in a close parallel to square pillar mining. It is currently limited generally to drill and blast operations. Examples:

a) *The Caledonia gypsum mine*, Southern Ontario (Canada), mined with rectangular pillars. The gypsum bed is 2-3 m thick and occurs at depths of 25-30 m. The stratum above the gypsum bed is composed of shale beds up to 6 m thick. The shallow position of the mineable bed and the influence of weathering limited the unconfined compressive strength of the gypsum to less than 25 MPa. The immediate roof of the gypsum bed is 1.8 m thick. The rectangular pillars were 3.6 m wide by 7.2 m long and the room width 6.0 m (Figure 12.1.4). Blasting had an adverse effect n roof stability, particularly where the roof bed thinned by inducing bed separation.

b) *FMC's Westvaco trona mine*, Wyoming (USA) uses rectangular pillars in conjunc-

PLAN

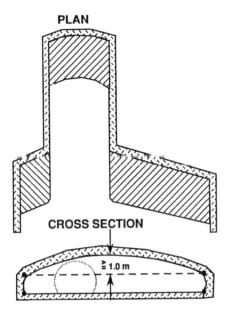

CROSS SECTION

≧ 1.0 m

Figure 12.1.3. Plan and cross-section of room excavation using an advancing slot sequence (German potash mine)

tion with drilling and blasting. A typical mining layout with mine entries is presented in Figure 12.1.5. The mining equipment consists of a Joy 3.6 m bar cutter, a Joy 4 m auger drill, and Eimco power truck blaster, a Joy 14 BU loader, NMS torkar shuttle cars and Joy single boom roof bolters. The mining sequence is to load ore, both the top, cut a slot in the face, drill holes in the face, and blast (shoot) the face. The sequence begins at the right side and moves to the left side of each block. It is repeated as the face advances. Roof bolting is on 1.2 m centres using fully grouted resin rebar bolts, 1.5 m long. Where unstable ground is encountered, every other row of bolts is strapped with a channel, 2 m resin bolts are used in intersections.[4]

3. *Diamond shaped pillar mining* is a variation of square pillars mining systems. Examples:

a) *FMC's Westvaco trona mine*, where pillars are angled up to 30° to allow easier LHD equipment access. Two types of continuous miners are used to excavate rooms and entries.

– *Drum continuous miners* are used in areas that are considered to have a short to medium-life (ten to eighteen months). The mining is by a two-phase advance and retreat sequence. It is therefore essential that parts of the development remain in good condition for safety, haulage, and air requirements. These sections are referred to as submains. A typical drum miner panel can have from five to eight rooms (Figure 12.1.6). Drum miners are mainly used in the advance phase, but may also be used for the retreat depillaring phase when ground conditions permit to do so.

– *Borer miners (double rod headers)* are used for main line entries (life of more than fifteen years) with an enlarged room-and-pillar structure. Borer miners are also used for retreat mining pillar extraction.

b) The *Drumbo gypsum mine* used angled openings at 30° to accommodate LHD equipment movement. Mining by drilling and blasting has been abandoned in favour of continuous mechanized mining. With continuous miners, it is possible to leave a protective gypsum layer 0.35 m thick towards the roof strata. Under these circumstances the mineable

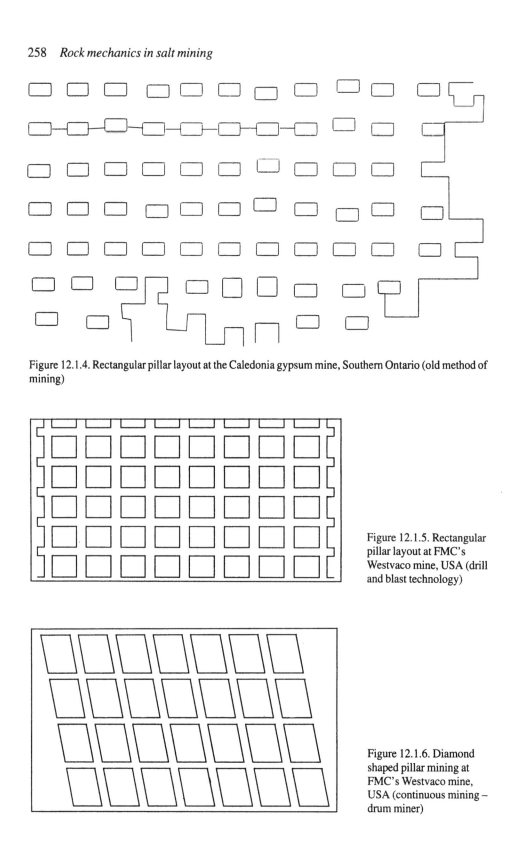

Figure 12.1.4. Rectangular pillar layout at the Caledonia gypsum mine, Southern Ontario (old method of mining)

Figure 12.1.5. Rectangular pillar layout at FMC's Westvaco mine, USA (drill and blast technology)

Figure 12.1.6. Diamond shaped pillar mining at FMC's Westvaco mine, USA (continuous mining – drum miner)

thickness of the gypsum bed is restricted to 1.36 m. The protective layer seals moisture in the roof to limit limestone deterioration and reduce the numbers of rock falls.

4. *A Chevron room-and-pillar layout pattern* has an application in either deep salt mines or very shallow mines. In the first case the strata pressure is very high as for example at Saskatchewan, where the depth of mining is 1000 m or more. In the second case, it is used where weathering has weakened near surface deposits. For example, this layout pattern occurs in Southern Ontario.

Potash deposits in the Saskatoon area contain multiple clay bands in the roof which in conjunction with a high strata stress causes unstable roof conditions.[5] The fish bone room-and-pillar pattern provides a partial extraction mining system within a pattern of small pillars and rooms excavated at an angle of 30° (Figure 12.1.7). The inter-room pillar

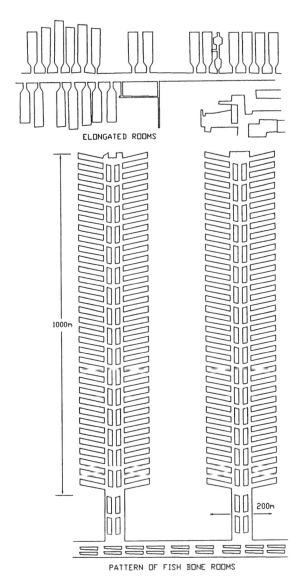

ELONGATED ROOMS

1000m

200m

PATTERN OF FISH BONE ROOMS

Figure 12.1.7. Comparative layouts of elongated rooms and the fish bone room pattern (Saskatchewan)

Figure 12.1.8. A fish bone room-and-pillar pattern (gypsum mine in Southern Ontario)

width is up to 3 m, and the room width up to 10 m. The small inter-room pillars are in a yielding state and cause stress to be transferred to surrounding strata. The layout of a large elongated room (total extraction) and the alternative fish bone pattern rooms (partial extraction) is illustrated in Figure 12.1.7. From the rock mechanics point of view, the fish bone pattern offers more stable ground conditions, particularly where the mining is in a retreat sequence towards the access haulage way.

Some very shallow gypsum underground mines in Southern Ontario changed over from drilling and blasting to continous mechanized mining. This was accompanied by a change in the mine layout from rectangular pillars to some variation of the fish bone pattern. The use of continuous miners permitted leaving a safe protective layer of gypsum against the weak roof, a feature that was impossible to accomplish with blasting, due to the tendency for the gypsum to overbreak the roof contact. The fish bone pattern consists of pillars up to 3.3 m wide and rooms are up to 13.2 m wide, driven at an angle of 30° on both sides of the axial entry (Figure 12.1.8).

12.1.2 *Long pillar mining*

The long room-and-pillar mining system has been developed in the deep potash mines in Saskatchewan. The objective of the long pillar mining system was to maximize potash ore extraction without compromising ground stability.

The thickness of the overburden over the potash deposits in Saskatchewan is in the range of 950 to 1050 m and consists of the normal sequence of limestones, dolomites, sandstones and shales typical to the prairie regions of Western Canada and the United States. Of particular concern is the overlying, water-bearing, Dawson Bay limestone formation. This is separated from the mining horizon at the Rocanville mine, for example, by a salt back which varies in thickness from 30 m down to about 18 m. This plate must provide a barrier against the intrusion of brines from the Dawson Bay formation into the mine workings. Locally the Dawson Bay formation is known to host a large, and re-chargeable, reservoir of brines Surface drill stem tests have measured formation hydraulic pressures in the order of 7.9 MPa.

Other factors which influence long pillar mining layouts include the mechanical properties of potash ore (low compressive and shear strengths) and the plastic flow characteristics of the salts.

Access to the deep Saskatchewan potash deposits is provided by pairs of shafts spaced at centres of between 100 m and 225 m. From the shafts, main haulages are driven to open the 1500 m wide by up to 1500 m long panels for development. Panel access is provided by four panel entries: two for fresh air and return air, one for vehicle transport, and one for the panel conveyor. Panel development consists of a three entry system in which the entry height depends on the potash mineable thickness (2-4 m), and width is usually up to 15 m (Figure 12.1.9). All development workings are driven with two to four rotor continuous miners of various manufactures.

The room-and-pillar development within a panel depends on the chosen long pillar system layout and on the projected ground conditions. The length of the development workings depends also on the opening spacings.

Over the past two decades in which potash has been mined in Saskatchewan a number of long pillar mining layouts have been developed.[6]

1. *A uniform room-and-pillar system* layout is represented in Figure 12.1.10. It consists

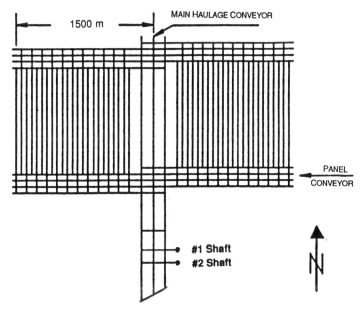

Figure 12.1.9. Access to potash by twin shafts and three main haulages; panel development is also shown

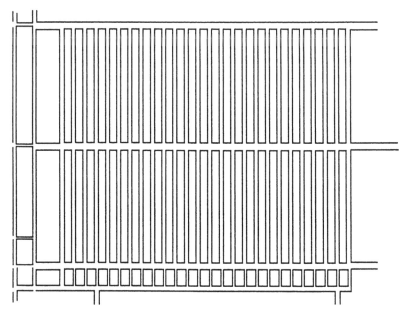

Figure 12.1.10. The uniform room-and-pillar system

of a uniform pattern of 20 m wide rooms separated by 33-36 m wide pillars. The extraction ratio is 33%. It has been successfully used in several Saskatchewan potash mines.

2. *Single rooms with stiff pillars*. This was the first production long pillar layout at several Saskatchewan potash mines. In the Rocanville potash mine, the rooms were 20 m wide with pillars of 25 m wide. Experience showed this layout to be overdesigned and it has been abandoned. It was found that rooms with a span of 20 m would stand unsupported, and that leaving stiff pillars between them as a reinforcement was not necessary.

3. *Two-room groupings with yielding pillars* are composed of three separate structures: rooms 20 m wide, a yielding pillar of 8 m wide left between the rooms, and strong abutment pillars of 36 m wide (Figure 12.1.11). The overall extraction in a two-room grouping is slightly higher than in other yielding pillar patterns: 47.5%.

4. *Three-room groupings with yielding pillars* share the same three structural units as the two-room grouping, except that there are two yielding pillars and the abutment pillar widths increase to between 55 and 60 m. The room widths remain at 20 m, the yielding pillars at 8 m. The extraction ratio is 46% (Figure 12.1.11).

5. *Four-room groupings with yielding pillars*. This is illustrated in Figure 12.1.11. The dimensions of individual structural units are similar to or the same as in the three-room groupings. The room width in some mines is 20 m, rather due to technological convenience than to ground stability requirements. A 20 m room width is formed by three passes of a Marietta 780 continuous miner. Room widths of 20 m and more generate a larger stress envelope over the room back, transferring loads higher up into the salt back.[7] The extraction ratio is up to 50%.

Some potash mines in Saskatchewan use a long pillar mine layout with five-room groupings and yielding pillars. At the Potomac mine, the extraction panels are 1220 m wide and 1400 m long (Figure 12.1.12). The rooms are 14.6 m wide, 5 m high and 1220 m long. The yielding pillars are 6 m wide and the abutment pillars 53.5 m wide. Rooms and pillars

2 - ROOM GROUPING

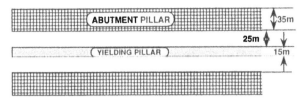

0 ROOM ONOUГIIO

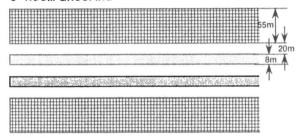

4 - ROOM GROUPING

Figure 12.1.11. The multi-room groupings with yielding pillar systems

run the full panel length of 1220 m. The extraction ratio is approximately 49%.

The rooms which are cut to a 5 m height are done in a four-pass configuration, utilizing the Marrieta 890-aw4 mining machine. The first two passes (top lift) are mined 2.7 m high (Figure 12.1.13). An extensible conveyor is installed on the first pass and a cross conveyor is used to feed onto the belt on the second pass. The remaining two passes (lower lift) are cut 2.4 m high. The extensible conveyor is relocated to the bottom lift during the initial bottom lift pass.

12.1.3 *Longwall mining*

Longwall mining of potash beds is practised in Europe, including the USSR. The first longwall face was set up in France in 1948, and was then adapted to the more regular deposits of Spain in 1963 and the USSR in 1972. Longwall mining was a complete departure from the normal room-and-pillar operations used in evaporite deposit mines with flat and moderately thick beds. Success depended on close cooperation between the mine operators and mine equipment manufacturers. This allowed equipment, designed for

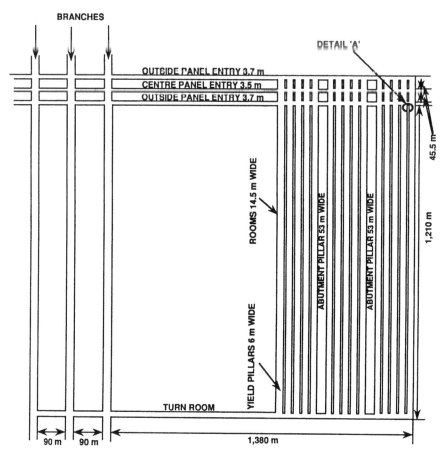

Figure 12.1.12. The layout of five-rooms groupings with yielding pillars

longwall coal mining, to be modified for use in the very different strata conditions and ground behaviour associated with the mining of salt. Longwall salt mining, at the beginning, followed the coal mining practice of face undercutting followed by blasting. Loading was done manually or with a scraper-hoist system. Developments in mechanized longwall coal extraction were immediately transferred to salt mining operations.

Longwall mining has been mainly applied to stratified potash beds, where the strata gradient could be as steep as 35°. The mineable seam thickness is between 1 and 3.5 m (thin and moderately thick potash beds). The main limitation of the application of longwall mining to potash deposits is the presence of water-bearing horizons, as is the case for Saskatchewan potash. Roof caving, as longwall mining progresses, will propagate fracturing in the strata above which can intersect water-bearing horizons and make channels for water to flood into the mine workings.

The following examples of longwall potash mining are based on Herget's paper 'Longwall mining and potash'.[9]

1. *Longwall mining of potash in France* is conducted by Mines de Potasse d'Alsace in the upper Rhine valley. The company produces 55 000 tonnes of potash ore per day from three operating mines. The evaporite strata are post-Eocene in age and dip steeply to the

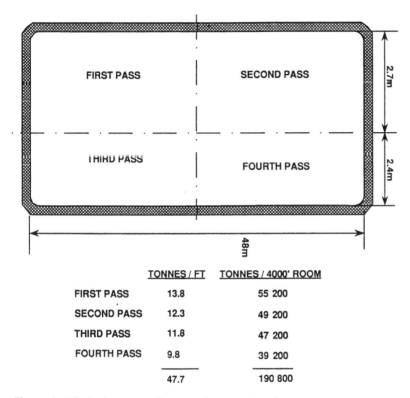

	TONNES / FT	TONNES / 4000' ROOM
FIRST PASS	13.8	55 200
SECOND PASS	12.3	49 200
THIRD PASS	11.8	47 200
FOURTH PASS	9.8	39 200
	47.7	190 800

Figure 12.1.13. Cutting profile of a room with a marrieta miner

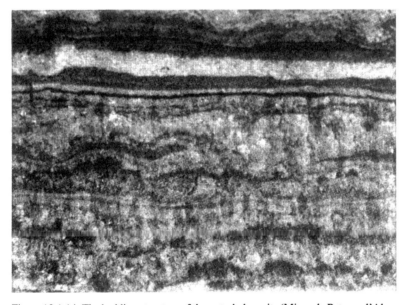

Figure 12.1.14. The bedding structure of the potash deposits (Mines de Potasse d'Alsace, France)

north. The mineral worked is sylvinite and its lies in two seams 21 m apart at a depth of 450 m to 1100 m. The lower bed thickness varies from 2.4-5.0 m, with a K_2O content of 14-20%. The upper bed is 1.2-2.1 m thick and contains 22-25% K_2O. The potash deposit is interbedded with marl anhydrite, and rock salt (Figure 12.1.14). At a depth of 1100 m the ambient temperature can reach 54°C due to a very high temperature gradient. The temperature limits underground mining to a depth of 1200 m (56°C). Below this depth, potash will be exploited by solution mining.

The longwalls are operated in the retreat mode. Twin galeries (4 m each with a 3 m pillar) are developed over the full length of the panel. Panels are usually 220 m wide and are developed for a 5000 m length at a depth of about 600 m. At a depth of 1000 m, twin drifts are developed for shorter distances because the stand-up time is less than one year. Observations have shown that twin drifts with a yield pillar are more stable than single openings. No barrier pillars are left and the recovery is close to 85%.

The shearer is floor mounted on three rollers and skids. It will cut in one direction only. On the return, picks are maintained and hydraulic support is moved ahead. Hydraulic support systems have 90 chocks of 4×120 t capacity. To move the shearer ahead, one of the twin drifts is widened by blasting (Figure 12.1.15).

Caving occurs immediately behind the hydraulic supports. Occasionally an overhang develops for a distance of up to 20 m for a mined seam height of 2 m. Generally the strata break up into relatively small blocks and the broken material lies immediately against the vertical gob shields. If the mined seam is 1.2 m high, closure of the beds is predominantly by flexure.

Some 200 m of rock salt are in the immediate roof and no serious water problems have

Figure 12.1.15. Advancing with a three-drum shearer

been encountered. The risk of inflow increases when cutting to the edge of the ore body (faulting and thin salt cover) or if by chance an old exploration drill hole is intersected. Radar will not work in the strata for brine detection because marl and clay beds reflect radar impulses.

The upper seam is mined ahead of the lower seam. In one area the lower seam had been mined ahead of the upper seam. When upper seam mining moved across the old mining face of the lower seam, deflection of beds and bed separation were observed. A hand could be put between the beds, but no fragmentation or caving was visible when mining 25 m above the lower, mined-out seam.

Maximum subsidence on the surface is 85% to 90% of the mining height and occurs about one year after mining. On the surface the outline of the 400 m radius shaft pillars can be seen clearly as a 3 m drop. Periodic level surveys are carried out and a predictive subsidence model has been established.[9, 10]

2. *Longwall mining of potash in Spain* is carried out by the company Potasas de Navarra near Barianin (Pamplona), located in the south-east corner of the Bay of Biscay in the Pyrenees. The potash deposit is of the same age as the ones in France (post-Eocene) but the geothermal gradient is less severe than in the upper Rhine valley. The potash bed is faulted and displaced laterally 600 m. Overburden is about 800 m thick. The composite salt bed sequence is nearly 15 m thick and from roof to floor comprises a bed of carnallite, 1.0-2.0 m thick, which is separated from a 2.0 m sylvinite seam by beds of rock salt and marl of 0.25-1.9 m thick. Below the sylvinite is a rock salt bed, 9.0-10.7 m thick, which forms the working floor.[8]

A 180 m longwall face is operated in the 2-2.5 m thick potash seam. Six men produce up to 300 t/h of potash (Figure 12.1.16). Overall recovery is 80%. The strata do not cave readily. After three passes of the shearer, holes are drilled (6 m at an angle of 60°) into the overhang to blast it down. The longwall faces are Anderson Strathclyde equipped.

Development roadways having a carnallite roof exhibit some problems because the salt chips, moves and bulges readily between bolts. This requires extensive roof maintenance which is carried out every two years (Figure 12.1.17). Some of the development is laid out in the footwall salt to avoid the problem of operating below a carnallite roof.[9, 11].

The presence of a very extensive water-bearing stratum above the salt deposit does not prevent longwall mining operation. The thick sandstone capping stratum holds a large quantity of water which does not seriously penetrate the mine openings. There are several layers of impervious beds of clay and marl between the water-bearing stratum and the roof of the salt deposit. The cracks or fissures in the overlying stratum induced by longwall mining are rehealed in the clay and marl beds which, in contact with water, expand in volume so rapidly that they effectively seal the beds. Water from the water-bearing sandstone strata cannot penetrate the mine operation.

3. *Longwall mining of potash in the former USSR* has been carried out in the Soligorsk basin situated about 150 km south of Minsk. There are two mineable potash beds. The no. 1 seam is located at a depth of 420-440 m and has a thickness of 1.7 to 3 m. The no. 2 seam is located at a depth of 600-650 m and has a thickness of 2.3-4.8 m. The potash deposits are nearly free of faults and have a regular flat dip of between 3 and 4°. Above the potash seams, there is an impervious rock salt deposit, 200-300 m thick, interbedded with shale and marl. The evaporite-bearing strata belong to the Upper Devonian period. The salt beds are interbedded by clay layers of 1 to 3 cm thick at spacings of 5 to 15 cm.

Initial mining was carried out with a room-and-pillar layout. There were relatively high ore losses and many roof falls. Studies in 1972 indicated that the roof problems were due to

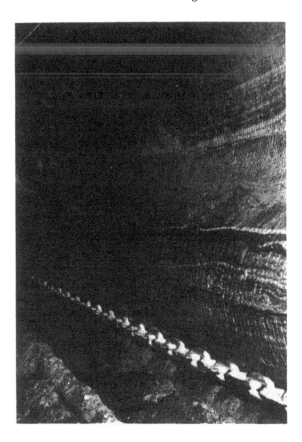

Figure 12.1.16. Longwall face (Minas Potasas de Navarra, Spain)

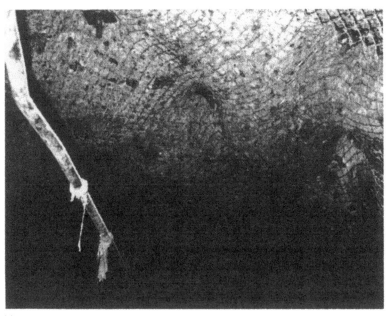

Figure 12.1.17. Roadway roof instabilities due to carnallite exposure above the opening

stable pillars punching into the roof strata. With the introduction of fully mechanized cutting machinery, a stepwise reduction of pillar widths was carried out in seam no. 1 until a yielding pillar layout was achieved with rooms of 5.3 m wide, a pillar width of 1.20 m, and a room height of 2.7 m.

With these dimensions, the pillars start breaking up when the third room has been cut. After four or five rooms have been cut, the pillars collapse and the footwall and hanging wall strata converge significantly. With an overall span of between 100 and 150 m, complete closure is achieved. This approach is used only in seam no. 1 at a depth of 420 to 440 m. The overall recovery is 76% (Figure 12.1.8).

Longwall mining is carried out in both seams no. 1 and no. 2. Seam no. 2 has a centre portion with a low potash content. To increase the KC1 content of the broken ore, longwall mining has been carried out simultaneously in two slices by retreat and a layer of a low potash grade (0.7 m) is left between them as waste (Figure 12.1.19). The upper face is ahead of the lower face by a distance of 4.5 to 6.0 m. The roof stratum breaks up readily beyond the self-advancing supports. The method is comprehensively described in coal mining literature.[13]

Before a decision was made to go ahead with longwall mining, models of equivalent materials and trial stopes were set up to look at the potential for water inflows from water-bearing strata above the mining horizon. In situ measurements of permeability, geophysical measurements and extensometer measurements indicated that the fracture zone above the longwall would be limited to a height of 8 m above the upper longwall face. This indicates that a ratio of 1:4 exists between mining height and thickness of fractured roof strata.

It is interesting to note that the surface subsidence velocity is about twice (4.5 mm/d) as high in comparison to that above room-and-pillar mining (2.5 mm/d) with stiff punching pillars.[8, 12]

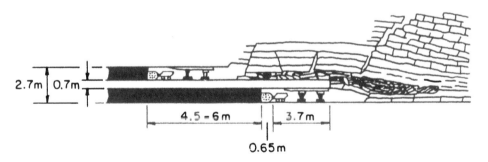

Figure 12.1.18. Yielding pillars (Soligorsk mine, former USSR)

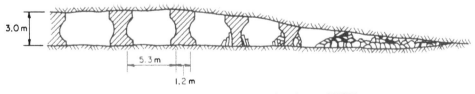

Figure 12.1.19. Multiple slice longwall mining (Soligorsk mine, former USSR)

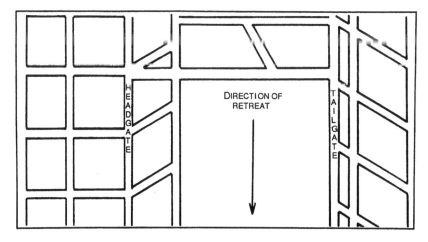

Figure 12.1.20. Longwall layout (FMC's Westvaco mine, Wyoming)

4. *Longwall mining of trona* is used in FMC's mining operation. Longwall mining began early in 1981, as an experimental method to determine if trona could be mined economically and safely at a higher level of ore recovery. Panel length is about 1 km (Figure 12.1.20).[4]

Longwall mining is still a relatively new mining method at FMC Wyoming and, as might be expected of any new mining system, it has gone through many growing pains to reach its present state. Longwall mining offers the opportunity to achieve outstanding production levels.[4].

The advantages that longwall mining offers over other forms of mining are a very high extraction rate and the elimination of bolting requirements in the longwall. Disadvantages are much higher roof support costs for the access openings than is normally required, coupled with a large capital investment for the longwall equipment.[4]

12.2 DESIGN CONSIDERATION OF MINE STRUCTURES

The design of salt pillars involves the determination of an adequate strength of pillar for the calculated load or average stress. The average stress computation used in this section has been mentioned elsewhere in sections dealing with pillars and abutment pillars. Basically, pillar strength depends on the strength properties of the pillar material, the pillar material homogeneity and on geometrical factors related to the pillar size and shape. Many formulas have been developed to calculate pillar strengths, particularly for coal mining. Here, as will be the case in the next two chapters of mining methods, formulas will be described which are suitable for the design of salt or evaporite pillars.

12.2.1 *Average stress in pillars*

The most common and simplest approach to estimating pillar loads is through the tributary area theory which depends on a number of simplifications.[13] In many cases it errs on the side of caution by overestimating pillar stress.

Room pillars have an open face on all four sides and thus are uniaxially loaded by the superimposed overburden (Figure 12.2.1). The average stress in pillars with an adequate size in relation to mine depth to not fail, is calculated on the assumption that they uniformly support the entire load of the overlying strata above both the pillars and the mined-out areas. The effects of deformation and failure of the roof strata are disregarded.[14] The following formulae simplify the task of using the tributary area theory to calculate average pillar stresses for different pillar geometries for flat bedded deposits.

1. *Rectangular pillars* are common for salt mining. They are shown in Figure 12.2.2. The pillars uniformly support the entire load overlying the pillar and one half of the mined-out area on either side of the pillar.

$$\sigma_v = \frac{\gamma h}{WL} (W + w)(L + l)$$

$$\sigma_v = \gamma h \left(l + \frac{w}{W}\right)\left(l + \frac{l}{L}\right)$$

$$\sigma_v = \gamma h \frac{A_t}{A_h}$$

or

$$\sigma_v = \gamma h \frac{A_t}{A_t - A_e}$$

$$\sigma_v = \gamma h \frac{1}{1 - R_e}$$

where:

γ = unit weight of overburden strata;
h = height of overburden;
W = least width of pillar;
L = length of pillar;

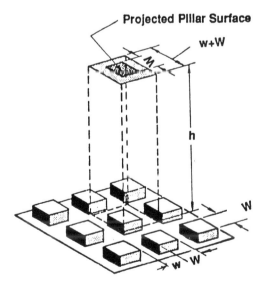

Figure 12.2.1. The tributary area concept of loading of pillars

w = least width of excavated area;
l = width of excavated rooms cross cutting the pillars;
A_t = tributary area;
A_e = extracted area;
A_h = pillar area;
R_e = recovery factor.

The load equations show that the average pillar stress depends upon the ratio of the total area mined out to the total area remaining in the pillars, e.g. The recovery factor.

2. The layout of *square pillars* is given in Figure 12.2.3. The vertical stresses in the pillars are calculated by modifying the above equations by setting ($W = L$) and ($w = l$).

$$\sigma_p = \gamma h \left(1 + \frac{w}{W}\right)^2$$

or

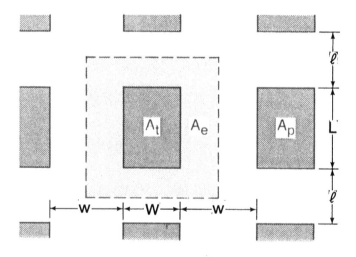

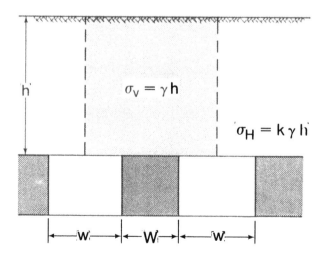

Figure 12.2.2. Rectangular pillar loading by tributary area theory

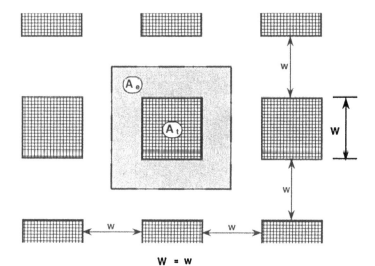

Figure 12.2.3. Square pillar loading by tributary area theory

$$\sigma_p = \gamma h \, \frac{1}{1 - R_e}$$

3. *Long pillars* are typical for the potash flat bed mining, particularly in Saskatchewan (Canada). A portion of a layout is shown in Figure 12.2.4. The calculation of the average pillar stress by the tributary area theory is based on a unit pillar length (Figure 12.2.4).

$$\sigma_p = \gamma h \left(1 + \frac{w}{W} \right)^2$$

or

$$\sigma_p = \gamma h \, \frac{1}{1 - R_e}$$

Long pillars can be designed so that the average pillar load is insufficient to fail them (uniform room-and-pillar mining) or that it is sufficient to fail them (multi-room groupings with yielding pillars).

12.2.2 *Average stress of abutment pillars*

The concept of abutment pillar design was introduced in underground mining long ago to improve ground conditions in ore extraction areas. The concept is based on the stress diversion from the mining extraction panel onto the abutment pillars. It is accomplished through the use of inter-room yielding pillars which can only support a limited load.[14] The magnitude of the load resting on the abutment pillars depends on the shape and size of the arch of equilibrium above the mining panel. For the practical purpose of the load calculation on abutment pillars, an arch of equilibrium is approximated by a trapezoidal shape.[13]

Abutment pillars in flat evaporite deposits with limited bed thickness are used for both room-and-pillar systems utilizing yielding pillars and for longwall systems with roof caving. The extent of roof caving is limited by the size and shape of the equilibrium arch.

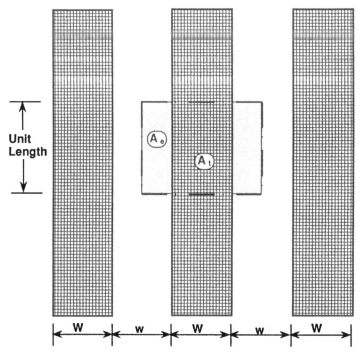

Figure 12.2.4. Long pillar loading by tributary area theory

Abutment pillars serve two ground control purposes: to absorb stress from the extraction panel, and to control the vertical extension of the fissured area. The latter is important as the intersection between the fissures and an aquifer could be detrimental to mine production. Leaving salt in abutment pillars lowers the overall mine recovery. Yet, in the case where abutment pillars are not introduced, losses might be higher due to severe ground problems.

Stress is diverted from an extraction area onto the abutment pillars in the following manner (Figure 12.2.5).

1. Abutment pillars are overdesigned so that they can support the overburden load without failure (stiff pillars).

2. Interroom extraction pillars are underdesigned so that they cannot support the overburden load without failure (yielding pillars).

3. As the rooms are mined out, the load above the panel is slowly transferred onto the abutment pillars through the action of pillar yield.

The stress diversion is described by the arching action of the roof strata above a mine panel. The yielding pillars carry only the load of the rock below the arch. The rest of the overburden load is shifted onto the abutment pillars. The dead load on the yielding pillars is uniformly distributed where the roof remains intact, i.e. The roof and floor slowly converge without failure.

The average abutment pillar stress can be calculated on the basis of the pillar layout and the arch geometry. Roof fissuring is controlled by compressive stresses at the excavation corners and by tensile stress at the centre of the excavation. The angle of inclination and

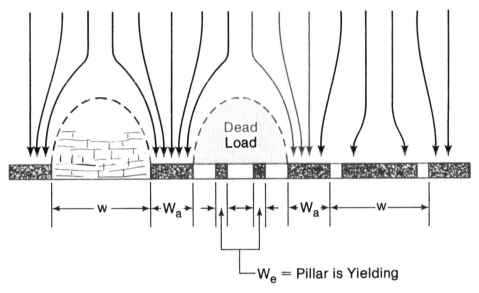

Figure 12.2.5. Stress transfer to abutments due to pillar yield

propagation of failure laterally and vertically due to tension depends on the rock lithology and strength. Two types of arch geometry are suggested, each defining average stress on the abutment pillars.

1. *The triangular type of arch* is approximated for relatively homogeneous low strength material above the extracted panel (Figure 12.2.6). The average stress per unit pillar length can be calculated using the following equations:

$$\sigma_p = \sigma_{p1} - \sigma_{p2}$$

$$\sigma_p = \gamma h \, (w_o + W) - \frac{w_o \times w_o \tan \beta}{4}$$

$$\sigma_p = \gamma h \, (w_o + W) - \frac{w_o \tan \beta}{4}$$

where:

h_o = height of stress relieved ground;
γ = unit weight of overburden strata;
h = mine depth;
w_o = panel width;
W = abutment pillar width;
β = angle of break

2. *A trapezoidal type of arch* forms when the sedimentary strata are heterogeneous and relatively strong (Figure 12.2.7). The calculation of the average stress in this case is derived as follows:

$$\sigma_p = \sigma_{p1} - \sigma_{p2}$$

$$\sigma_p = \gamma h \, (w_o + W) - \gamma h_o \, (w_o - h_o \cot \beta)$$

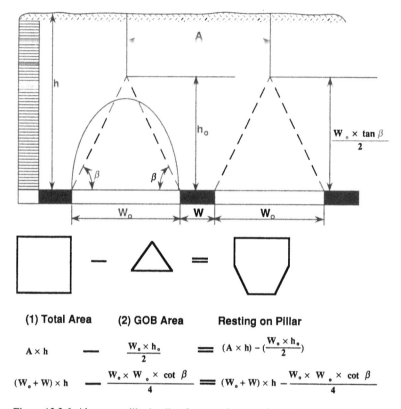

(1) Total Area   **(2) GOB Area**   **Resting on Pillar**

$$A \times h \quad - \quad \frac{W_o \times h_o}{2} \quad = \quad (A \times h) - (\frac{W_o \times h_o}{2})$$

$$(W_o + W) \times h \quad - \quad \frac{W_o \times W_o \times \cot \beta}{4} \quad = \quad (W_o + W) \times h - \frac{W_o \times W_o \times \cot \beta}{4}$$

Figure 12.2.6. Abutment pillar loading for an arch approximated as a triangle

$$\sigma_p = \gamma \, [h(w_o + W) - h_o \, (w_o - h_o \cot \beta)]$$

where h_o is the height of caved gob cavity.

The trapezoidal type of arch is more common in underground mining than the triangular one. The relationship between the abutment pillar and panel widths is important because it controls the arch height and the potential to intersect an aquifer. In mine design this factor has to be taken into consideration when choosing abutment pillar and extraction panel widths.

12.2.3 *Geometrical factors of pillar strength*

The geometrical factors for salt pillar design have primarily been studied with models of equivalent materials. There are two main parameters.

1. *The size effect* or the concept of critical size. For a cubical sample of coal or ore, the strength decreases with increasing sample size until it becomes constant from some critical sample size onward. However, in the case of cubical salt samples the opposite effects occur and strength increases with increasing sample size until it becomes constant (Chapter 9). Figure 12.2.8 shows the functional relationship between size and strength for coal and salt. It can be seen that for salt cube sizes up to 20 cm, the strength of the sample is increasing until it becomes constant. For coal cube sizes up to 40 cm the sample strength decreases

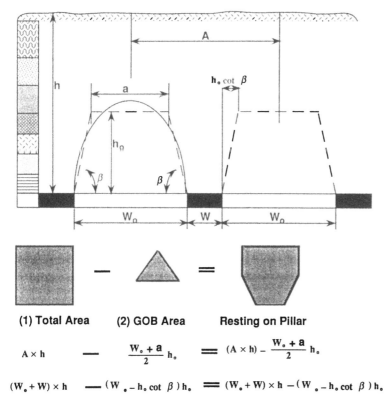

Figure 12.2.7. Abutment pillar loading for an arch approximated as a trapezoid

until it too approaches a constant condition. For this reason, in salt pillar design the size effect characterization is not required as in the case for coal and hard rock pillars. It is common practice to measure strength values for salt pillar design on large salt samples (cube size 10-20 cm) where the size effect of the sample size is eliminated.

2. *Shape effects* or the concept of pillar slenderness is the second element of pillar strength. Pillar models of different slenderness ratios have been cut from blocks of salt brought from operating mines. The following discussion on the shape effects is based on the research carried on for salt mined at the Werra district (Flache deposit, Germany).[15,16,17]

The salt model pillars were tested in a loading frame, as illustrated in Figure 12.2.9. The testing was carried out for models of mono-mineral salt pillars[15] and of bi-mineral salt pillars.[16] The model salt pillars were sandwiched between rock salt plates which uniformly transferred loads from the 6000 tonnes loading machine to the models, and simulated underground pillar loading conditions better than would be the case for models sandwiching between metal plates.[17]

The shape effects on the strength of monomiral model pillars was studied by Uhlenbecker for two types of salt ores: hard salt and carnallite.[15] The testing of model pillars clearly indicated the different ultimate strengths of each salt type in regards to pillar slenderness ratios (Figures 12.2.10 and 12.2.11). In addition, the relationship between model pillar deformations at failure and the pillar slenderness factor varies considerably for these two salt mineral compositions (Figure 12.2.12). It can be seen in the ultimate strength

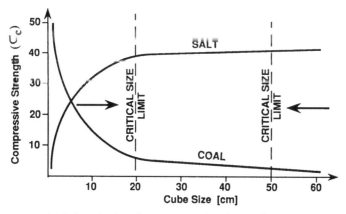

Figure 12.2.8. Sample size effects on salt and coal strengths

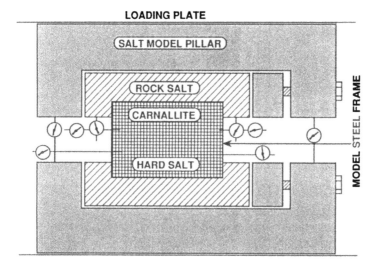

Figure 12.2.9. Model loading set up for salt pillars

diagrams that model pillars of hard salt have a higher bearing capacity than model pillars of carnallite. However, the deformation of hard salt pillars at failure is higher than that for carnallite, which suggests that hard salt pillars support load to their larger deformation limits.

The strength of bi-mineral pillar models was also studied by Uhlenbecker in regard to the distribution of carnallite and hard salt and in regard to the shape of a model pillar composed of layers of carnallite (½ model high). The model studies indicated that with a decrease in the hard salt content in the pillar, its strength decreased accordingly to the degree shown in Figure 12.2.13. Further modelling investigations with respect to the shape effects of pillars with 50% of hard salt and 50% of carnallite showed that their bearing capacity is between the bearing capacity of mono-mineral hard salt pillars and mono-mineral carnallite pillars (Figure 12.2.14).

In addition, the bearing capacity of salt pillars of varying slenderness ratios has been

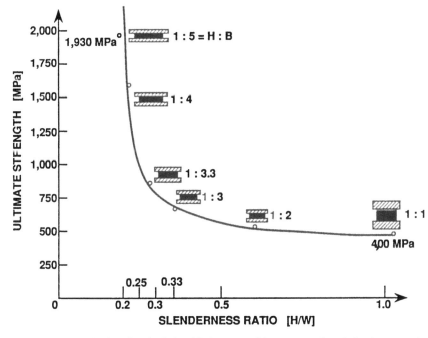

Figure 12.2.10. The functional relationship between ultimate strength and slenderness ratio for model pillars of hard salt

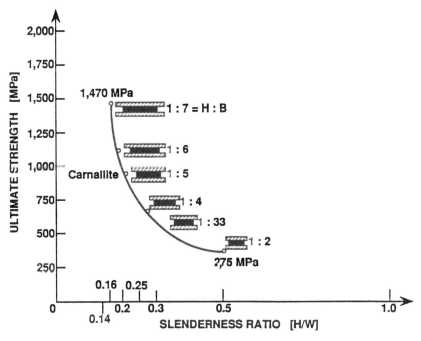

Figure 12.2.11. The functional relationship between ultimate strength and slenderness ratio for model pillars of carnallite

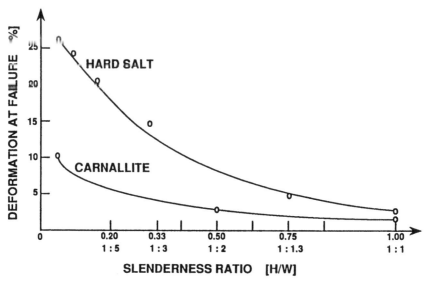

Figure 12.2.12. The relationship between deformation (failure) and slenderness ratio for hard salt and carnallite model pillars

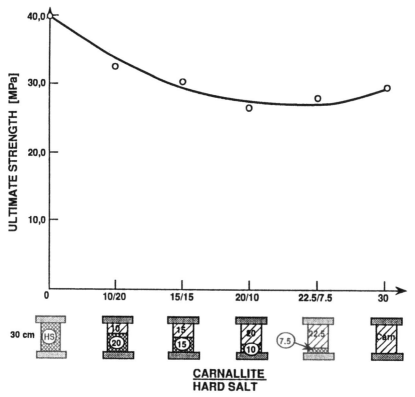

Figure 12.2.13. Bi-mineral model pillars; the relationship between ultimate strength and content ratio of carnallite/hard salt

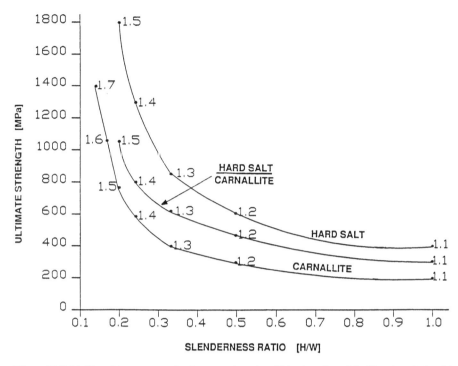

Figure 12.2.14. The ultimate strength of mono-mineral and bi-mineral model pillars in relationship to the slenderness ratio

studied in underground mines. The pillars studied have been sized in three groups (Figure 12.2.15). In the mine studies, the pillars have the same height but different widths. This differs from model pillars which were tested as a constant width and a varying height. The constant height of the mine pillars was governed by the thickness of the potash bed. The test pillars underground were monitored for stress concentration, and showed the same trends as the model pillars. The diagram relating ultimate strength and slenderness ratio could not be made, because none of the underground pillars failed.

12.2.1 *The pillar strength*

Pillar strength formulas which are generally applicable to coal pillars are not directly applicable to salt pillars, because of the differences in salt properties and behaviour in comparison to coal.[18]

From the research conducted for the room-and-pillar mining of thin to moderately thick salt deposits it has been found that one strength formula which was developed from coal mining experience and is widely accepted in hard rock and non-metallic mineral mining, is also applicable to salt:

$$\sigma_p = K \frac{W\alpha}{H^\beta}$$

where:

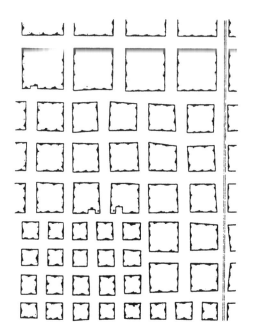

Figure 12.2.15. Layout of underground pillars of three different slenderness ratios (H/W 1:4, 1:6, 1:8)

K = the compressive strength of a cubical sample of the pillar material with a size between 10 and 20 cm for salt and potash pillars;

α, β = constants expressing the shape effect, obtained from the diagram of the relationship between ultimate strength and slenderness ratio of the salt in question;

W = pillar width, given by the mine layout;

H = pillar height, the mineable thickness of the salt bed.

The only major difference is that for salt the exponents α and β are near equal, eliminating the size effect on strength. For coal and hard rock applications the ratio of $\alpha/\beta \cong 0.7$ adds a size effect component to the pillar strength calculation.

For the calculation of pillar bearing capacity up to ultimate strength of Saskatchewan potash deposits, the foregoing pillar formula is simplified by setting α and β constants even and equal to 0.5:

$$\sigma_p = \sigma_c \left(\frac{W}{H}\right)^{1/2}$$

where σ_c is the compressive strength of potash ore determined in the laboratory for cubic specimens 10 to 20 cm to a side.

The safety factor is calculated by the relationship between the ultimate pillar strength (σ_p) and average pillar stress (σ_V):

$$F_s = \frac{\sigma_p}{\sigma_V} > 1 \text{ load-bearing pillars}$$

$$F_s = \frac{\sigma_p}{\sigma_V} < 1 \text{ yielding pillars}$$

For salt pillars the same safety factors should be applied as for coal and hard rock pillars,

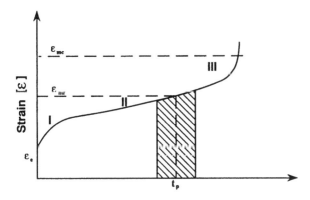

I   **Transient creep**
II **Steady state of Creep**
III **Accelerating Creep**

Figure 12.2.16. The creep curve
for predicting long-range strain in
salt pillars

where $F_s = 1.6$ is required for long-term supporting pillars, $F_s = 1.3$ for intermediate-term supporting pillars, and $F_s = 1.1$ for temporary pillars.

It has been found that this pillar design formula for thin and moderately thick beds can also be applied to the design of salt pillars in thick salt beds.

C.D. daGama discussed in his paper[19] that where the compressive strength design concept was applied to salt pillars with heterogeneous structures, the calculations should be based on the strength of the least resistant component.

12.2.5 *The strain criterion of pillar design*

An alternative approach to salt pillar design is the strain criterion, which may be superior to the strength criterion. The knowledge of the long-term behaviour of salt is essential because salt exhibits viscoelastic behaviour. It undergoes increasing deformation with time under a constant load, and at some point in time the constant load overcomes the pillar strength. Failure of most salt pillars occurs when strains reach the accelerating phase of creep. The dimensions of a salt pillar can be established by means of predicting long-range strain (Figure 12.2.16) and designing pillar with an adequate safety factor such that the critical strain level is never reached within the required time frame.

$$F_s = \alpha \frac{\varepsilon_{mc}}{\varepsilon_{mr}}$$

where:

α = numerical coefficient $\alpha \leq 1$;

ε_{mr} = maximum strain for failure under rapid uniaxial compression;

ε_{mc} = maximum strain in a creep leading to rupture.

The creep curve can be obtained using finite element methods and average pillar stress, as calculated by the tributary area theory. The strain criterion, as daGama stated, accomodates the creep properties and the relative proportions of the salt types included in the pillar. A conservative approach considers the pillar to be composed totally of the softer element.

The allowable strain for the design of pillars (and the corresponding time frame) is obtained by an appropriate selection of the safety factor F_s and the coefficient α.

12.3 STABILITY EVALUATION OF MINE STRUCTURES

Stability analysis in salt mining in the last two decades has been particularly extensive for bedded and moderately thick deposits, such as the potash deposits in Saskatchewan. As a result, several concepts of ground stabilization in salt mining have been developed, primarily by S. Serata.[20, 21] Others have also contributed to this field.[6, 22, 23, 24, 25]

12.3.1 *Strata stabilization by stress control system*

This subsection is written exclusively on the basis of Serata's published data.[20, 21] He first defined the nature of roof failure problems in bedded deposits, establishing that failure is a time-dependent mechanism dictated by specific ground conditions. When the ground starts to fail, no conventional strata support could prevent large rock falls. The planes of rock separation continue to propagate deeper into the roof strata. The pattern of failure depends on three rock mechanics factors, namely: stress distribution and concentrations, rock behaviour and properties, and the geometry of the opening. A typical example of massive roof failure is shown in Figure 12.3.1. It shows the five general stress zones which might be found around a failing underground mine excavation. They are brittle-fracture, yielded, plastic, elastic and stress-relieved zones.

Serata proposed three principal methods of stress control that covers a wide range of problems associated with failing roofs and heaving floors.

1. *The stress-relief method* is based on the phenomenon that narrow rooms experiencing

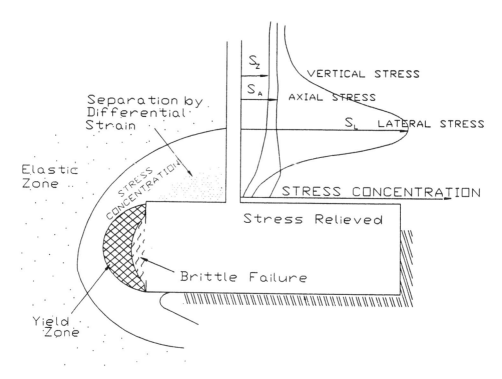

Figure 12.3.1. General pattern of roof failure in bedded strata

extensive roof failure and floor heave from high horizontal stresses. They can be stabilized by increasing the room width, not reducing it. The greater the room width, the larger the decrease of stress relief created in the ground immediately above and below the opening due to the circular expansion of the stress envelope (Figure 12.3.2). The stress-relief method is suitable only for uniform ground conditions, where there are no sensitive separation seams (i.e. clay) immediately above or below the opening. It would not be suitable for potash horizons to contain several clay seams as interbeds. When water-bearing formations overlie a mine opening, two factors must be carefully balanced. One is room width, because an excessive room width creates tension resulting in deep tension cracks in the roof. The other factor is pillar width, determined by an amount of load which has to support.

Widening rooms has greatly improved roof and floor stability in the potash mines of Saskatchewan.

2. *Parallel-room method* is used for stress control in weak ground which has a tendency for bed separation, usually along clay layers. It requires the creation of a room parallel to the first excavated failing room, with a narrow yielding pillar between the two (Figure 12.3.3). By this method the second room is protected while the first continues to deteriorate. Protection of the second room is provided by a secondary stress envelope which developes around the entire structure of the two-room system. Inside the stress envelope, the stress-relieved ground is formed immediately above and below the second room (Figure 12.3.3). This process can be repeated several times. The process of expanding the number of parallel rooms, however, cannot be continued infinitely. A limit is imposed by the collapse of the secondary stress envelope in the overburden formation and/or excess stress built up in the ground during the new excavation. Such limitations can be predicted effectively from a knowledge of the overburden strata and rock properties.

3. *The time control method* is used for stabilizing failing ground. In the case of extremely weak ground, the stress control technique by the parallel-room method is

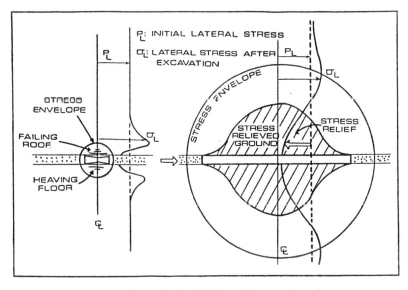

Figure 12.3.2. Stress-relief methods of stabilizing underground openings

PARALLEL-ROOM METHOD

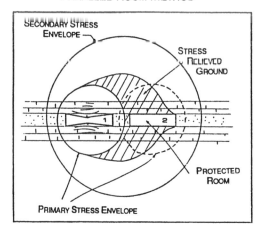

REPEATED PARALLEL-ROOM METHOD

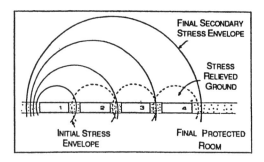

Figure 12.3.3. Parallel-room method of stabilizing of mine openings and the repeated application of parallel-room excavations

insufficient. One form of extremely weak ground is characterized by slab buckling, and is commonly found to occur in the northwestern part of the Saskatchewan potash deposit. In ground which fails by slab buckling, the strength of the roof is dictated by the strength of clay seams and other discontinuities. In general, slab buckling is not restricted to deep deposits, because the failure can occur at any depth regardless of the strength of the rock itself. Once sequential roof failure by slab buckling starts, it is impossible to stop by artificial support work. Traditional methods, such as decreasing the room size, slow down roof failure but do not stop it. That is where the time control method was developed, because the other control strategies did not work. It is capable of making stable entries in the an area where normal entries collapse. It involves a specific time-controlled sequence in the excavation of a group of rooms, as illustrated in Figure 12.3.4.

The first step of the procedure is to cut two rooms sufficiently far apart to allow individual primary stress envelopes to form. The spacing between the openings should be designed to cause strain-hardening to occur. The spacing, however, should not be so close that the stress envelopes overlap to cause a tensile stress build-up in the ground between the openings, which inhibits strain-hardening. Strain-hardening of potash involves squeezing the grains to such an extent that the grains become locked together. The strength is increased four to five times the original strength. Finally, the protected opening is made through the strain-hardened ground. The two pillars which are created are made to fail quickly upon excavation of the third opening. In the process of failing, the three primary stress envelopes surrounding the individual rooms are transformed into a single secondary stress envelope. The formation of the secondary stress envelope counteracts the failing process of the small pillars resulting in a stable equilibrium. This secondary stress envelope stabilizes the centre room and at the same time substantially improves the stability of the two outside rooms. Most of the stress is distributed outside the secondary stress envelope onto the abutment pillars.

A variation of the time control method is used at the Potomac potash mine (Saskatchewan). In the first step, instead of driving just two initial openings in the ground, three are

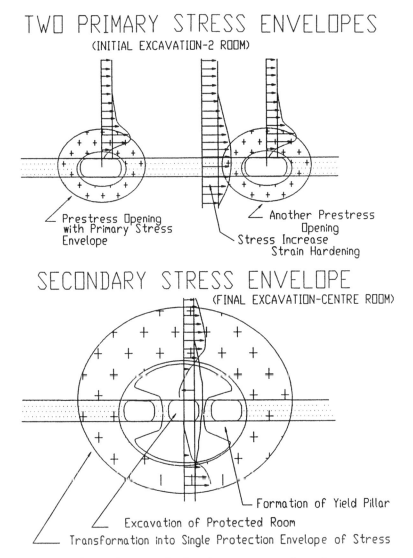

Figure 12.3.4. Development of twin primary stress envelopes (outer two rooms excavated) and development of the secondary stress envelope (centre room excavated)

STEP 1

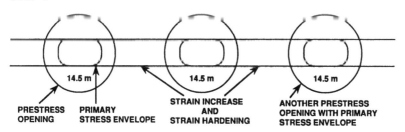

PRESTRESS PRIMARY   STRAIN INCREASE   ANOTHER PRESTRESS
OPENING   STRESS ENVELOPE   AND   OPENING WITH PRIMARY
                         STRAIN HARDENING   STRESS ENVELOPE

STEP 2

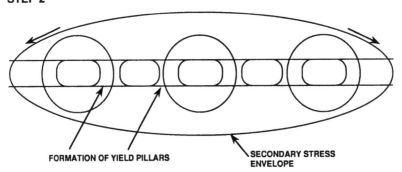

FORMATION OF YIELD PILLARS   SECONDARY STRESS
                                        ENVELOPE

Figure 12.3.5. Application of Serata's time control method for stress relief at the Potomac mine (Saskatchewan)

driven (Figure 12.3.5). Thay are again driven far enough apart to allow for individual primary stress envelopes that are not overlapping and yet close enough to allow for overloading of the ground between them. After approximately one week of overloading, the potash becomes strain-hardened to a strength of four to five times that of virgin potash. In the case of Potomac ore this amounts to a strength of 72 MPa.

The second step involves driving two rooms in the strain-hardened ground between the previously driven openings. Driving the two rooms causes the small, 3.3 m wide, newly created pillars to fail quickly. In the process of failing, the five primary stress envelopes surrounding the individual rooms are transformed into a single secondary stress envelope surrounding all five rooms. The formation of the secondary stress envelope counteracts the failing process of the small pillars. It stabilizes the protected rooms as well as improves the stability of the three previously driven rooms. Stress is distributed, outside of the protective secondary stress envelope, onto the 50 m wide abutment pillars on either side of the five-room group.

The Serata stress control technique was not required for panel entries. The 45.4 m wide pillars were able to support the load of the overlying rock, without inducing failure to these entries.[24]

12.3.2 *Comparison of stress controlled systems*

A study of ground stability has been carried out for fifteen years at the Rocanville Division of the Potash Corporation of Saskatchewan Mining Ltd. Here, the top of the prairie evaporites occurs 920 m below surface, with the potash ore elevation at about 940 m. The ore itself can be described as a potash-enriched salt bed, between 2.1 m and 2.7 m thick, with an average grade of 22% K_2O equivalent. A basic consideration of strata stability is the overlying water-bearing Dawson Bay limestone formation. It is separated from the mining horizon by a salt-shale back, between 18 m and 30 m thick. The salt-shale bed is

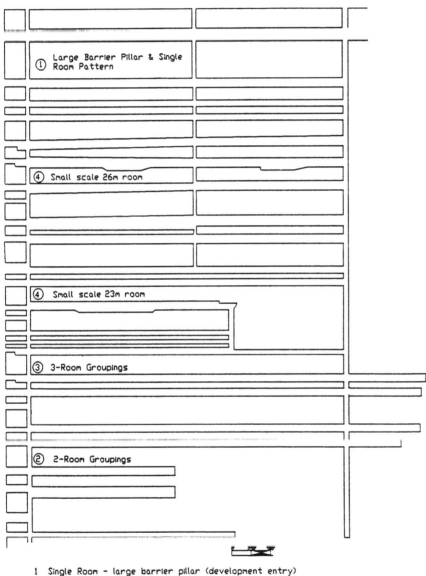

1   Single Room - large barrier pillar (development entry)
2   2-Room groupings - 15m yield pillars
3   3-Room groupings - 7,5m yield pillars
4   Small scale Room - 26m and 23m width

Figure 12.3.6. Various layouts of test mining patterns at the Rocanville mine (Saskatchewan)

relied upon to provide a barrier against brine penetration from the Dawson Bay formation into the mine workings. The uniaxial compressive and shear strengths of the ore are 15.9 MPa and 4.8 MPa respectively. The ore and the salt cap exhibit viscoelastic flow under pressure.

Mining trials of a number of layouts have been carried out, as described by Molavi and Wooley[6] (Figure 12.3.6).

1. *The uniform room-and-pillar system* was briefly described in subsection 12.1.2 as a long pillar mining method. The layout is illustrated in Figure 12.1.10. Field data indicate that the uniform room-and-pillar system provides a more even subsidence of the upper strata and as a result the chances of fissuring to the water-bearing Dawson Bay formation are reduced. The room convergence rate in this type of panel is dependent upon the composition of the ore. For example, panels with a carnallite content up to 9% have a closure of up to 70 cm in five years, at which point in time massive roof falls have occurred. In other panels, with more competent ground, closures of up to 65 cm occur before there are massive roof falls and the time frame is eight years. In Figure 12.3.7, curve 1 shows the large amount of closure in weak carnallite ground prior to a roof fall. Curve 5 in the same figure represents the behaviour of a point in the middle of the room-and-pillar panel. Visual inspection indicated a normal, initial primary state, and shearing of the corners resulting in thin slabbing of the corner of the room. A secondary shear separation, however, followed this with cracks initiating from the corners and at an angle very close to the stress envelope boundary (approximately 45°) and propagating towards the centre span of the room. This caused a dooming effect which resulted in fairly massive repeated failures, a common phenomenon in rooms in weak strata.

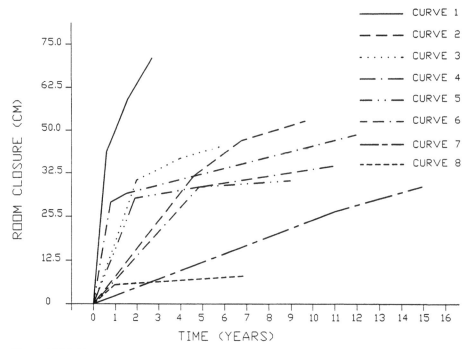

Figure 12.3.7. Room convergence deformation, measured at mid span (Rocanville mine, Saskatchewan)

Curve 2 in the graph of Figure 12.3.8 is from the instrumentation at a station placed in a 38 m pillar and 26 m room area. The pillar is classified as stiff, and it is exhibiting progressive deformation with time as do a majority of the long pillars in the Rocanville potash mine.

2. *Single rooms with large stiff pillars* were mentioned in the sections on long pillar methods. Field data and visual observations suggested that the first 23 m production rooms with 36 m pillars were wasteful. Confidence was gained that a span of 23 m would stand without support. The rooms in this area exhibited less closure than in other systems. Four years after cutting, the closure measured about 25 cm. The relaxed zone reached to about 15 m into the wall.

Single rooms with large barrier pillars, as shown in Figure 12.3.6, proved to be very stable and were used for mainlines and panel entries. Mainline entries consisted of three, 13.5 m wide rooms separated by two 114 m wide pillars that were isolated from production panels by 152 m barrier pillars. Curves 7 and 8 in Figure 12.3.7 indicated the accumulated closure common to single rooms with large pillars.

Closure deformation of the development entries showed that the ground is fairly competent, as is evident by closures in the range of 50% of that in rooms in production panels. There are no signs of instability in mainline entries unless they are in geologically disturbed ground. The local carnallite content is around 9%. The reloading effect due to mining extraction has not been recorded or observed.

Curve 4 in Figure 12.3.8 is from a monitoring station located in one of the main entries. There is relatively little pillar expansion.

3. *Two-room groupings with yielding pillars* were tried in four panels. Two pillar widths were used. The first, a 15 m yielding pillar, which later proved to be too wide to yield effectively, was left between the two rooms. The second, a large 36 m wide pillar separated

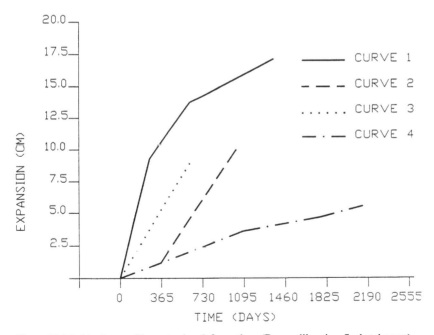

Figure 12.3.8. Maximum pillar extrusion deformations (Rocanville mine, Saskatchewan)

the paired room groupings to limit the effect of ground reactions and to reduce the surface subsidence (Figure 12.3.6). The overall extraction is slightly lower than in other panels. The factors considered to have contributed to the inability of the 15 m pillars to yield, are the very competent ground both above and below the mining horizon coupled with the increased pillars. Apart from localized primary shearing limited to within 0.75-1.0 m of the corners, the rooms are very stable. A room closure of approximately 0.5 m has occurred over a period of twelve years at mid span and 0.3 to 0.38 m beside the pillars. The closure rate of 1.25 cm per year is higher than usual for rooms of this age. The surface subsidence over these panels is in the range of 50 to 200 mm. Curve 6 in Figure 12.3.7 shows the amount of closure in a room in this system compared to closures in rooms with different configurations.

4. *Three-room groupings with yielding pillars* were mined over ten years ago in two panels. In one case two 7.5 m wide pillars were left between 20 m rooms and in the other case 12 m wide pillars separated the rooms (Figure 12.3.6). The pillars separating these groupings are in the range of 55-60 m wide.

Visual inspection indicates that the pillars separating rooms are yielding rather than punching into the roof or floor. There is some roof slabbing around the corners. Slabs formed by the mechanism of primary shear separation fell about five years after excavation. There has not been a massive roof collapse covering an entire room span. The total mid span closure to date is around 0.40-0.45 m. At the ribside it is between 0.25 and 0.38 m. Roof slabbing is more pronounced beside the yielding pillars. The surface subsidence over these rooms is not distinguishable from the other panel.

Part of the 75 m total span is supported by the two 7.5 m yielding pillars. The balance is supported by the 55 to 60 m pillars. The system, in relatively competent ground, is stable to date. Utilization of this system is not recommended for situations with geological anomalies and carnallitic ground. Curve 4 in Figure 12.3.7 compares the closure of this style of room development to other patterns. Curve 1 in Figure 12.3.8 shows the great amount of lateral pillar expansion in one of the pillars.

5. *Large trial rooms* of 23 m and 26 m width were excavated on a very small scale, as shown in Figure 12.3.6. The experiment was carried out to establish if 26 m wide rooms were stable and could replace 20 m wide rooms for reasons of economy. The data collected in two years have generally verified their stability for both short-term and long-term loading time. Curve 3 in Figure 12.3.7 shows room closure of a 26 m wide room in comparison to other opening widths. Curve 2 in Figure 12.3.8 indicates a fairly high pillar expansion, but the consequences are inconclusive.

Molavi and Woolley concluded that the most troublesome factor to long-term stability is the presence of carnallite. Short-term stability (two to three years) of rooms within a panel (regardless of geometry) has proven to be excellent. Mainline entries have successfully been isolated from each other and the production panels by leaving barrier pillars 225 m and 150 m respectively. The 7.5 m wide yielding pillars in the three-room groupings have yielded. Pillars of 9-16 m which were originally believed to be yielding, no longer appear to be yielding. The two-room groupings appear to be a stable design. However, a uniform room-and-pillar design has been chosen for production purposes. It provides a very uniform deformation of the overlying strata, with a slightly larger extraction ratio. Logistically, it is easier to set up a standard pattern of uniform distances between rooms from the surveying and production standpoint.[6]

12.3.3 *Instrumentation of mine structure stability*

Hebblewhite et al.[23] carried out and interpreted a simple instrumentation program at the Boulby potash mine, in North Yorkshire, England (Cleveland Potash Ltd). The mineable potash seam is 3.4 m thick and is part of the middle evaporites (Zechstein). The roof stratum over the mineable potash bed is an undulating potash zone with many rolls and troughs. There is a major roof control problem. The lower section of the mineable potash bed is gneissose potash, the upper section is a secondary potash, very rich in sylvinite. The floor stratum is a massive halite bed, 40 m thick, with a planar horizontal contact (Figure 12.3.9).

Two mining methods have been employed at the Boulby mine. In the first, panels separated by large stable confined core barrier pillars, were mined by drilling and blasting in a conventional room-and-pillar configuration. The second method for more uniform parts of the deposit used a long pillar mining system consisting of narrow five-headings panels. Rooms were mined in a strict sequence to cause yield to take place in the narrow pillars and outer sacrificial headings to cause a major stress distribution to the abutment pillars. The central heading was then driven through the stress-relieved zone where stable conditions were expected (continuous mining technology). This is referred to as either the yielding pillar technique or the pressure arch method and was developed in the English coal field over thirty years ago. It is known in Canada as the stress relief system.[25]

The instrumentation consisted of roadway convergency measurements with convergence monitors, tape extensometers, multiple anchor extensometers, borehole wire extensometers, and level surveys. Extensometer anchors were installed to depths of up to 27 m. Each borehole could accomodate up to four anchors. Stress measurements were conducted at one site in the mine using a Talbott strain cell. Field stresses were measured by the overcoring technique using doorstopper strain cells above the mineable potash bed.

$$\sigma_1 = 30 \text{ MPa (vertical stress)}$$

$$\sigma_2 = \sigma_3 = 115 \text{ MPa (horizontal stress)}$$

The low ratio of horizontal to vertical stress seems uncharacteristic for an evaporite deposit, but the strata distortion in the overlying shale could prevent time-dependent stress redistribution to a hydrostatic stress state.

1. *Roadway stability* was monitored in openings of 3.0-3.4 m high and 4.5-8.0 m wide,

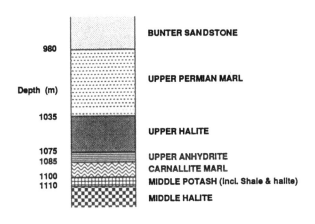

Figure 12.3.9. A simplified geological section of the Boulby potash mine (North Yorkshire, England)

the normal width being 6.0 m. All roadways were driven with a rectangular profile. Although roadway stability must be considered in conjunction with the associated pillar and panel stability, a number of general points were made from an assessment of the convergence and roof extensometer results:

a) Roads exhibited overall stability when at least a 2 to 3 m thick competent potash or halite roof was left below the weak shales or marl as a protective layer.

b) A number of roadway intersections exhibited stress relief in the roof and floor zones. This resulted in high initial closure rates, but within 30 days, closure rates decreased to less than those of the surrounding roadways which exhibited very stable ground in the vicinities of these intersections. Figure 12.3.10 is an isometric sketch showing roof instrumentation at the conveyor road intersection. It indicates that within 75 days, the floor closure rate in the centre of the intersection was down to 0.17 mm/day while 20 m along the main heading the closure rate was 0.32 mm/day. This area has 6 m thick potash and a competent halite. A similar site, with an immediate shale roof, had within 100 days ceased to maintain the stress redistribution. Closure rates increased prior to major rock falls.

c) In all cases where roadways have remained stable, deformation rates have continued to decrease, even after 800 days. This cannot be directly related to a primary creep since it is associated with a changing loading condition as the stratum continues to relax. It does appear, however, that for any roadway where roof strain or closure holds at a fixed rate, instability and eventual failure will result.

d) Figure 12.3.11 summarized the convergence monitoring conclusions. The shaded envelope represents an area within and below which closure rates are indicative of roadway stability over several years. Monitored roadways where roof falls occurred have produced

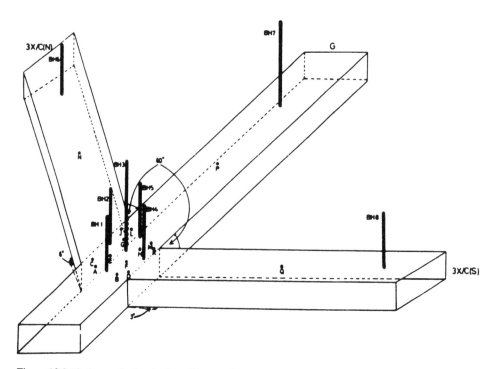

Figure 12.3.10. Isometric sketch of road intersection roof instrumentation

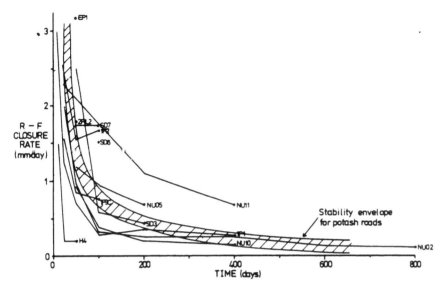

Figure 12.3.11. Plots of roadway closure rates

closure rates above this envelope. The upper limit of the closure rate envlope, on the basis of all results, has been chosen as a reasonable criterion for assessing potential roadway instability. The limiting curve may be approximated by the equation:

$$c = 34.81 \, t^{-0.785}$$

where:

      t = time since excavation (days)

      c = roof to floor closure rate in mid span (mm/day).

2. *Confined core pillars* used in conventional room-and-pillar mining along the 60 m square pillars in the shaft area were instrumented with borehole extensometers, generally to a depth of 6 m at mid-pillar height. It was found in every case that deformation measured in the first 6 m of the pillar edges accounted for just in excess of half the total wall to wall closure measured. This implied that pillars of 3.4 m height have a yield zone around the triaxially confined core extending more than 6 m into the sides of the pillar. At one pillar it was found that, initially, 65% of the wall to wall closure was a result of induced lateral tensile stress and fracturing within the first 6 m of the pillar. This percentage, however, dropped to an average 45% by 250 days and then remained relatively stable.

Figure 12.3.12 is a plot of the average strain rate (microstrain/day), measured over the first 6 m of a 60 m pillar, plotted against time. The strain rate is plotted on a logarithmic scale. The results, over the first 200 days, can be approximated by a non-linear relationship giving rise to the following expression for average strain over the 6 m zone:

$$\varepsilon = 1.136667 \, (1 - e^{-0.01125 \, t})$$

The results shown in Figure 12.3.13 were obtained by combining the estimate of strain with the component of wall to wall closure from the adjacent roadway. It shows the equivalent average strain rate across the central 48 m core of the 60 m pillar over a time period of 400 days. The minimum value of 5 microstrain per day after 150 days indicates that, even for

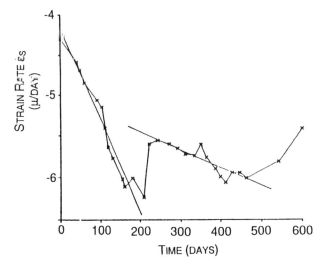

Figure 12.3.12. Average strain rate vs. Time for pillar

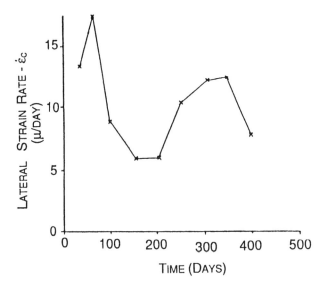

Figure 12.3.13. Lateral strain vs. Time for confined core pillars of 48 m and 60 m wide

pillars of this size, time-dependent deformation exists across the entire pillar.

The results for the 60 m pillars showed that after about 150 days, a constant width of confined core existed to maintain stability. The results, however, in Figures 12.3.12 and 12.3.13 both indicate an abrupt change in behaviour after 200 days. This change was also apparent in all closure readings in the vicinity of the site. The fact that the strain rate in the central core of the 60 m pillars more than doubles, indicates that a fairly widespread load increase had occurred in this area of the mine.

Subsequent development of production panels generated 40 m and 25 m pillars. An instrumentation program was implemented for monitoring through the full pillar widths. Figure 12.3.14 shows a comparison of strain rates against depth for 60 m, 40 m and 25 m potash pillars in the centres of wide panel areas where virtually the full tributary area load

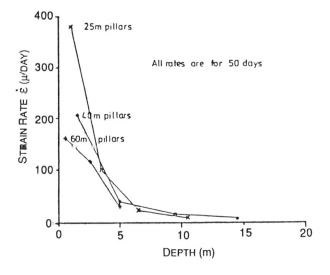

Figure 12.3.14. Pillar strain
rate along the longitudinal
axis

could be expected to be carried by the pillars. In the case of 40 m and 25 m pillars, significant strain rates extend through to the centres of the pillars, although at a greatly reduced level. This implies that the pillars possess a confined core, even though there is significant lateral and hence vertical deformation of the pillars over their full width. It could be concluded that a 25 m wide pillar should be considered an absolute minimum for long-term pillar stability, in the light of the fact of the 10 m yield zone along each side. Instrumentation of 40 m halite pillar and 40 m potash pillar showed rock salt pillars to be significantly more rigid than potash pillars, as shown in Figure 12.3.15.

3. *Yielding pillars* have been experimented with to determine their characteristics.

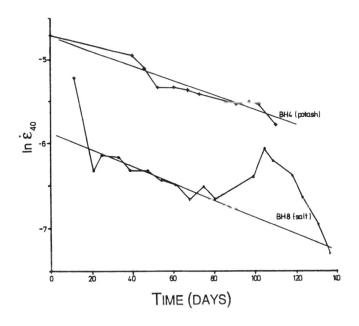

Figure 12.3.15. Potash
and rock salt pillars,
deformation vs. time for
40 m pillars

Figure 12.3.16 illustrates a mining area where 11 m and 12 m pillar widths were created by splitting a 46 m pillar with three rooms. Extensometers and convergence meters indicated that the initial 10 m yield zone is:
- in the solid ribside;

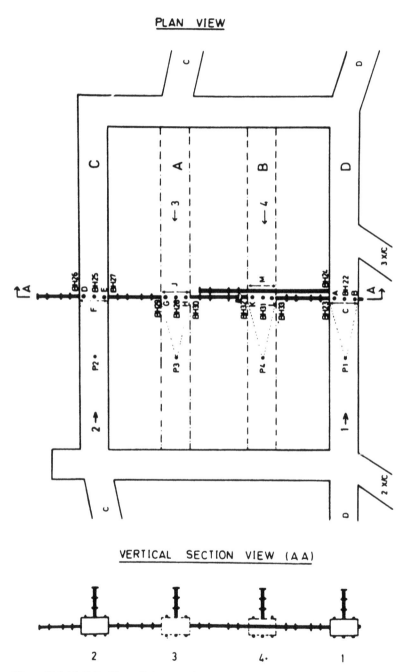

Figure 12.3.16. C-D pillar splitting experiment (no. 1 panel)

- in the 46 m pillar between C and D;
- in the solid rib beyond C;
- in the 30 m pillar between A and D.

The excavation of road A has induced strain throughout the 30 m width of the A-D pillar, though the centre appears well confined as indicated by the reduced strain rates. However, the 11 m A-C pillar is in state of yield over its entire width. This is also evident in the subsequent A-B and B-D pillars with very large strain rates, particularly adjacent to B, the last road to be driven. Although these pillars yielded and the strain rate decreased slightly, there was no significant stress relief to the central roads and no stabilization of the yielding pillars to indicate a redistribution of stress onto the adjacent abutments (stress relief mechanism). Two reasons could be given for this lack of stress relief. First, the pillars were too wide to yield quickly enough to force stress redistribution. Secondly, the shale present in the immediate roof in this area could have caused a breakdown of any stress redistribution mechanism which might have been set up. Similar behaviour has been observed over intersections where local geological factors seemed to play a major role in overriding yield-induced stress relief.

Narrow pillar experiments with panels of four headings mined in sequence with 8 m yield pillars between rooms indicated stress relief providing greater stability. The overall conclusion on the use of yielding pillars was that pillar size, panel width, sequence of mining, and geology were all of paramount importance. Pillar widths must be restricted to less than 8 to 9 m to induce sufficient yield to cause stress redistribution. Pillar sizes larger than this, but smaller than the 25 m minimum for triaxially confined pillars, were found to be too stiff to yield sufficiently for stress relief but not large enough to be stable. They tended to attract excess load and hence caused very bad roadway conditions. Pillars smaller than 8 m tended to be too weak to carry even the load of the immediate roof strata. They deform greatly, producing concentrated zones of excessive subsidence above the seam.

Under the prevailing conditions, panel widths in excess of 60 m were found to be too wide for stress relief techniques to work, even with correct sequencing and suitable geological conditions.

Hebblewhite et al. concluded that the results have indicated the benefits which can be gained from using simple instrumentation. The behaviour of potash ore and the surrounding strata has been quantified to form an empirical model for the prediction of creep deformations in roadway backs and pillars. The instrumentation program improved the understanding of the effect on stability caused by the raodway geometry, local geological anomalies, pillar yield zones and confined cores, and the mechanism of stress relief with yielding pillars. Instrumentation and supporting mathematical and laboratory studies can assist in the design of mine layouts with proper roadway, pillar and panel geometries that take into account in situ conditions.

12.4 STRESS ANALYSIS OF MINE STRUCTURES

The stress analysis discussions which follow are related to room-and-pillar mining systems in flat and moderately thick salt deposits. The analysis for longwall mining is identical to that used for coal mining, and for which there is extensive published literature, including author's book *Strata control in coal mining*. The stress analysis topics covered included theoretical analysis, physical model analysis, and mathematical model analysis.

12.4.1 *Theoretical analysis of structure*

The basis for this subsection is the published study 'Method of correlation of laboratory test with in situ criterion', written by Mraz and Dusseault.[26]

1. *Geometrical correlation.* In a circular opening, or in a segment of an 'equivalent' mine opening, the radial flow of material is inversely proportional to the distance of the element of volume from the centre of curvature (Figure 12.4.1). Assuming that volume is conserved, the annular layer of material containing the element of volume will change its thickness by ε_d as it moves toward the centre.

$$\frac{\pi\alpha}{180}\,rD = \frac{\pi\alpha}{180}\,(r - \varepsilon_r)\,(d + \varepsilon_d)$$

$$rd = rd + r\varepsilon_d - d\varepsilon_r - \varepsilon_r\varepsilon_d$$

The product $\varepsilon_r\varepsilon_d$ is very small and can be taken equal to zero. Thus, the creep ε_d in the element of volume equals:

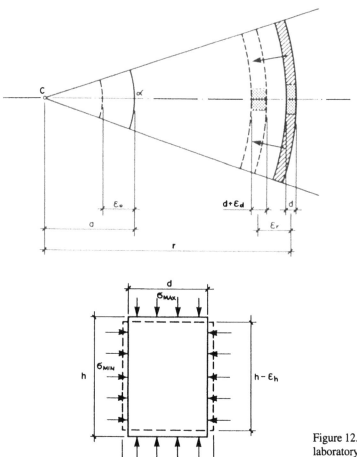

Figure 12.4.1. Correlation of laboratory specimen results to in situ properties

$$\varepsilon_d = \varepsilon_r \frac{d}{r}$$

The term 'creep' represents the velocity of plastic deformation in units of length per unit of time. Similarly, because the circular section contains the same volume of material:

$$\frac{\pi\alpha}{180} r\varepsilon_r = \frac{\pi\alpha}{180} a\varepsilon_0$$

$$F_{,} \equiv F_0 \frac{a}{r}$$

From the foregoing equations

$$\varepsilon_d = \varepsilon 0 \frac{ad}{r^2}$$

where:

ε_0 = creep rate at wall opening (one half of total creep);
ε_d = creep of element of volume;
a = radius of curvature of 'active opening';
r = radial distance from centre of curvature;
d = dimension of element of volume.

The standard laboratory sample is cylindrical and in the triaxial test cell it can creep into the fluid-filled annulus in all directions, as follows (Figure 12.4.1):

$$\frac{\pi d^2}{4} h = \frac{(d + \varepsilon_d)^2}{4} (h - \varepsilon_h)$$

$$d^2 h = d^2 h + 2\, dh\varepsilon_d + \varepsilon_d^2 h - d^2\varepsilon_h - 2\, d\varepsilon_d\varepsilon_h - \varepsilon_d^2\varepsilon_h$$

Neglecting multiples and powers of ε_h and ε_d, we get:

$$\varepsilon_h = \varepsilon_d \frac{2h}{d} = \varepsilon_s$$

where ε_s is the creep of laboratory sample.

On the other hand, the in situ element around a long opening is laterally restricted and therefore will only creep in the direction of the centre of curvature:

$$dh = (d + \varepsilon_d)(h - \varepsilon_h)$$

And after neglecting small multiples, we get:

$$\varepsilon_h = \varepsilon_d \frac{h}{d}$$

In the in situ element all dimensions are taken to be equal and therefore:

$$F_h = F_d$$

By combining strain equations ε_d and ε_h we get:

$$\varepsilon_s = \varepsilon_o \frac{2\,ah}{r^2}$$

and at $r = a$ (at the wall of the underground opening):

$$\varepsilon_s = \varepsilon_o \frac{2\,h}{u}$$

where h is the length of laboratory sample.

The following general conclusions can be drawn from the above:

a) The ratio of in situ creep to laboratory creep is proportional to the ratio of the radius of curvature of the equivalent opening to twice the length of the sample. The term 'equivalent opening' is described in the following section.

b) The magnitude of creep in the wall of an underground opening is indirectly proportional to the ratio of radial distance from the centre of curvature to the radius of the equivalent opening (Figure 12.4.2). This statement is supported by underground measurements.[27]

A typical example of the geometric correlation follows:

$\varepsilon_o = 0.05$ mm/day – one half of total creep in the underground opening;

$h = 100$ mm – length of laboratory specimen;

$a = 1000$ mm – radius of curvature of equivalent underground opening;

$\varepsilon_s = 0.05 \times 2 \times 100/1000 = 0.01$ mm/day.

2. *Plastic deformation.* The solution for a thick-walled cylinder in a generalized, perfectly plastic material takes this form:

$$\sigma_r = \frac{K}{\tan \rho} \left[\left(\frac{r}{a} \right)^n - 1 \right]$$

$$\sigma_t = \frac{K}{\tan \rho} \left[\left(\frac{r}{a} \right)^n \frac{1 + \sin \rho}{1 - \sin \rho} - 1 \right]$$

where:

σ_r = radial stress;

σ_t = tangential stress;

r = radial distance;

a = radius of opening;

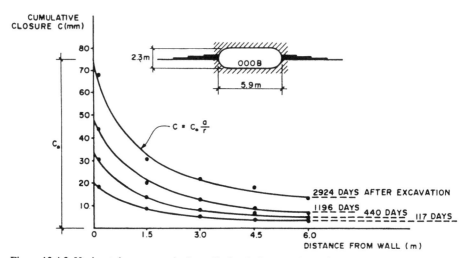

Figure 12.4.2. Horizontal movement in the wall of a virgin ground opening

$$n = \frac{2 \sin \rho}{1 - \sin \rho}.$$

A minimal creep in virgin ground openings converges to zero, and to material within the yielding zone is in the state of an ideally plastic substance. The equations of radial and tangential stress assume the form:

$$\sigma_r = 2 \, K ln \frac{r}{a}$$

$$\sigma_t = 2 \, K \left(1 + ln \frac{r}{a} \right)$$

This was demonstrated by direct measurements of radial stress in a stabilized, five year old virgin ground opening (Figure 12.4.3).

3. *Equivalent opening.* It has been repeatedly observed that in a deep salt rock continuum, an 'active' or equivalent opening develops around the mine opening. Because the radius of curvature of the equivalent opening is an important parameter in stress equations, it is necessary to know the shape of the equivalent opening.

The equivalent opening is an ellipse circumscribed to the mine opening (Figure 12.4.4). The equivalent opening of a square hole has a circular cross-section. In a rectangular opening (Figure 12.4.5) this would result in an envelope $B_1 D_1$ with a tangent line t_1 at 45° to the horizontal axis. However, at point T_A the radius of curvature changes. The common radius can be determined as:

$$a_2 = \frac{1}{2} \sqrt{W^2 + H^2}$$

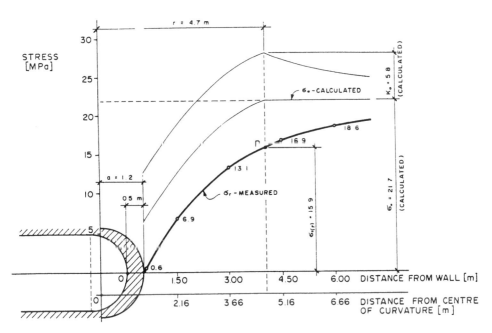

Figure 12.4.3. Radial stress in the pillar of a virgin ground opening

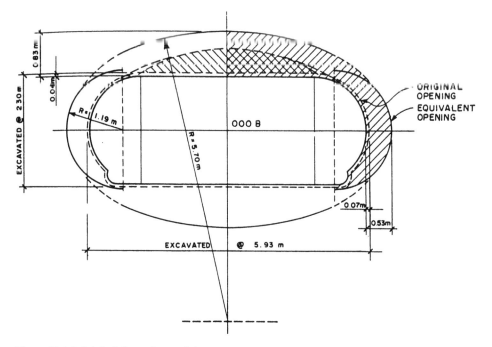

Figure 12.4.4. Original dimensions and the equivalent elliptical opening, eight years after excavation

The slope of the tangent line t_2 is:

$$\tan(180° - \gamma_2) = -\frac{H}{W}$$

$$\tan\gamma_2 = \frac{H}{W}$$

which defines the ellipse B_2D_2 having the following dimensions:

minor semi-axes:

$$b_2 = \frac{H}{\sqrt{2}}$$

major semi-axis:

$$s_2 = \frac{W}{\sqrt{2}}$$

radius of curvature at wall:

$$a = \frac{H_2}{W\sqrt{2}}$$

It has been observed by Mraz that the equivalent opening applies with reasonable accuracy to virgin ground conditions in a homogeneous body of salt rock in deep mines. However, due to initial stress relief creep and in particular to subsequent reloading as a result of ore extraction, the dimensions of the equivalent opening change, as shown in Figure 12.4.5. As more material adjacent to the wall fails due to reloading, the equivalent opening changes to

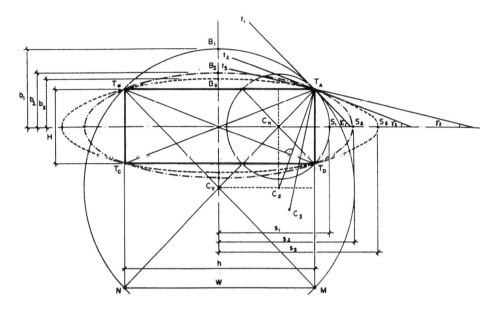

Figure 12.4.5. Derivation of the equivalent (active) opening

B_3D_3 with the tangent line at point T_A flattening. This reduces the radius of curvature.

In the case of plane strain, the total unsupported load in any direction is equal to the product of concentric pressure and that perpendicular to the section of the opening. Hence, we can postulate:

$$\tan \gamma_3 = \frac{\sigma_o H}{\sigma_M W}$$

$$C_M = \frac{\sigma_M}{\sigma_o}$$

$$\tan \gamma_3 = \frac{\tan \gamma_2}{C_M}$$

where:

σ_o = original geostatic pressure;

σ_M = average pressure at the core of mining pillars.

From the geometry of the tangent line:

$$s_3 = \frac{W}{2}\sqrt{1 + C_M}$$

$$b_3 = \frac{H}{2}\sqrt{1 + \frac{1}{C_M}}$$

From the geometry of the ellipse:

$$a_3 = \frac{b_3^2}{s_3} = a_M$$

$$a_M = \frac{H^2}{W} \cdot \frac{\sqrt{1+C_M}}{2 C_M}$$

where a_M is the radius of curvature of equivalent opening adjacent to mine pillar.

These equations can be used to calculate the influence of a mine opening on the stability of the adjacent mine pillars, assuming that the rock salt bed is homogeneous and without clay interlayers. It should be concluded that Mraz and Dusseault have demonstrated the validity of their theoretical analysis and its application to the solution of practical underground mining problems. Their present studies and ongoing work strengthen this conclusion.

12.4.2 *Physical model analysis*

The following physical model analysis of evaporite mining is based on the published study 'Physical modelling of roof strata in potash mines of Saskatchewan' by Jeremic and Farah.[28] The observed failure mechanism in modelled roof strata has also been observed in Saskatchewan mines, and in evaporite mines of Southern Ontario.

1. *Physical modelling of mine opening roof strata.* The relevant variables for destructive model testing, where the effect of body forces can be neglected and where equality of strain in the prototype and model is required, are the compressive (C_o), shear (S_o), and tensile (T_o) strength, the modulus of elaslticity (E), Poisson's ratio (v), applied load (L), and the geometric dimensions respectively. For our purpose, the stress, geometric dimensions, load and specific gravity scales will bear the following relationships:

$$\xi = \rho \lambda$$

$$\psi = \rho \lambda^3$$

where:

ξ = stress scale;
ρ = material density scale;
λ = geometric scale;
ψ = load scale.

It is not always possible to find or design a model material to satisfy all the above scaling requirements. In such cases the quantities most relevant to the mode of failure should be satisfied. Generally speaking, Poisson's ratio has the least effect on the modeling process.[29] In the experiments reported in this work, the most important parameter is the inter-bed shear strength. Once this strength is exceeded, roof beams start to sag and eventually fail.

Physical models of roof strata are composed of multiple bed arrangements where the main structural defects are the sedimentary planes. Models were made of Plaster of Paris, with a compressive strength of 5 MPa, a modulus of elasticity of 3 GPa, Poisson's ratio of 0.24, and a tensile strength of 0.63 MPa. Two sets of models were constructed. The first with internal reinforcement made by braided wire grouted with epoxy resin. The second, without reinforcement, was anchored just above the upper sedimentary plane so that no suspension effects from the overlying layer were present. The layout of the support is given in Figure 12.4.6.

To simulate uniform vertical loading of roof strata, a plank of timber of 488 mm long, 100 mm wide and 100 mm thick was placed on top of the model. The load was applied by a universal testing machine. Although this arrangement is not the most ideal loading method, resources and time limitations did not permit the use of more elaborate loading techniques

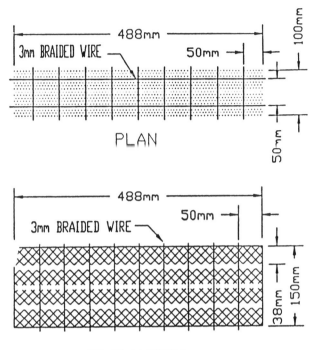

Figure 12.4.6. Layout of the internally reinforced physical model

such as membrane bags filled with water or other materials.[30] Calculations based on the assumption of stacked beams show that the timber plank carries about 95% of the bending moment.

The simulation of the lateral load was achieved by having the roof of the model compressed between two plates of the loading machine, as shown in the photographs (Figure 12.4.7).

The modelled roof layers of the mine openings experienced several modes of failure with the following factors influencing the failure modes the magnitude of the stress field, the direction of the stress relative to the bedding planes, the geometry of the openings, the material strength properties, the density of the discontinuities and the method of its reinforcement.

2. Gravitational roof failure. Critical gravitational stress is developed when rock separation is induced by differential strain along an arch of balance above the mining area. A vertical stress is developed by the rock beam's weight because the lateral stress of the elastic zone is relieved.[31]

The bedded nature of the rock cupola and the character of its loading produce a unique failure mode.

It is assumed that a small amount of frictional resistance (10 kPa approximately) exists between the beams since both the coefficient of friction and normal stresses are relatively low. The value of shear resistance used in this work is reasonable and comparable to the values measured by Brandis et al. in studies on the shear behaviour of joints using direct shear tests on replicas cast from natural surfaces.[32] Thus, the beams act as a composite beam until the frictional resistance is overcome by horizontal shear. This occurs at a

 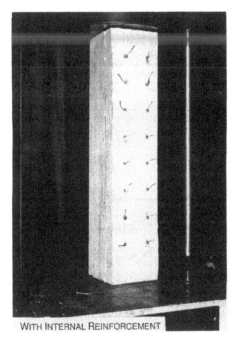

WITHOUT INTERNAL REINFORCEMENT   WITH INTERNAL REINFORCEMENT

Figure 12.4.7. Axial loading simulation of lateral stress in the model strata roof

calculated applied load of 840 N. The tensile stress is 70 kPa, well below the tensile strength of 630 kPa.

The shear stress along the bedding planes is the critical parameter in the modelling process. It signifies the transformation of the system from one with a composite action to one of stacked beams without resistance along the bedding planes.

This shear stress is used as the basis for the stress scale $\varepsilon = S_{op}/S_{om}$. If the geometric scale is taken as $\lambda = 20$, and the density scale as $\rho = 2.5$, then the stress scale is $\xi = 50$. This gives the shear stress along the prototype bedding planes as 0.5 MPa which is of the correct order of magnitude. Note that the corresponding mine opening would be 9 m wide and the thickness of the beds 0.75 m. These values are within the range existing in the evaporite mines of Canada.

As the load increases, the tensile stresses increase. This causes cracking, separation and sagging of the beams and subsequent failure. For the model, this occurred at 15 kN. Analyzing the system as stacked beams, gives a maximum tensile stress of 610 kPa, close to the tensile strength of the model material of 630 kPa. The results of these studies can be summarized as follows:

a) Sagging of the mine roof, which can be approximated as a beam in bending. The intensity of the mine roof deformation is also a time-dependent phenomenon.

b) Rock separation occurs along the bedding planes (Figure 12.4.8). Maximum separation is at the bedding bed closest to the opening roof.

c) As sagging and bed separation progresses, the lateral tensile strength of the bed is exceeded. Cracks perpendicular to the bedding beams are initiated and propagate (Figure 12.4.9).

d) The fracturing of a multiple bed beam is followed by rock falls. The fractured rock

PHYSICAL MODEL

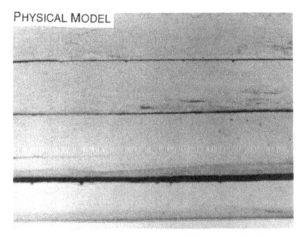

MINE OPENING

Figure 12.4.8. Strata sagging
and separation along bedding
planes (model and mine opening)

blocks rotate to a limited degree, developing kink bands and these can be seen in the
physical model (Figure 12.4.10).

e) The final stage is a collapse up to an arch of equilibrium above which the stratum has
developed a time-dependent stability. This type of roof failure occurs commonly in the
mine roadways but not in rooms. Bolting support of the roof strata could prevent bed
separation and by this means prevent roof falls.[34]

3. *Lateral stress failure.* Underground observations and model investigations indicate
the presence of fractures in the roof strata which are the product of high lateral stresses
acting approximately parallel to the bedding planes.

Under axial loading, beams are expected to support the applied load equally. The failure
load causing the beams to separate and approach buckling failure was 34 kN. Due to the
method of loading, the beam can be approximated as a column fixed at one end and free at
the other. The failure load using the Perry-Robertson formula[33] is 7 kN. This is close to the

PHYSICAL MODEL

MINE OPENING

Figure 12.4.9. Tensile failure
normal to the roof beam

value of 8.5 kN which caused failure of the individual beams. The formula mentioned above assumes that the column has both an initial curvature and a small initial eccentricity due to the modelling and loading. Note that the load scale $\psi = \rho\lambda^3 = 20\,000$, which results in a prototype lateral failure load of 1.4×10^5 kN. Theoretical calculations show that the prototype failure load is in the range of 0.56×10^5 kN to 1.95×10^5 kN, depending on the assumed boundary conditions.

Failure is characterized by the development of shear fractures, which are initiated both along the bedding planes and within the beds (Figures 12.4.11 and 12.4.12).

Roof failure results in small to large rock falls, and is more common in large rooms in

Figure 12.4.10. Strata beam failure by the development of kink bands

thicker seams. Shear failure planes in the roof strata usually are shallow. Bolting of the rock beds could provide adequate ground control.

4. *Failure of reinforced structure*. Experience and model tests have demonstrated that bolting of the roof strata can be a valuable tool in strata control. Dead load and lateral compressive model tests show that for reinforced beds, failure occurs at higher applied loads than for models without reinforcement.

Model tests indicate that the increased stability of internally reinforced bedded strata is due to a resistance to dilation between the bedding planes.[34]

Two different failure modes of reinforced models have been observed with the mode of a function of the loading conditions:

PHYSICAL MODEL

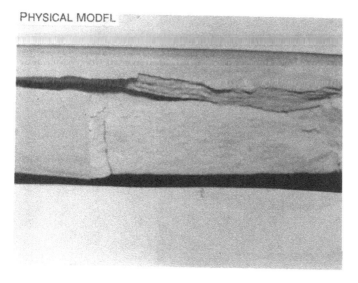

MINE OPENING

Figure 12.4.11. Pattern of fracturing during shearing along bedding planes

a) Separation of the bedding planes is initiated above the reinforced linear arch by gravity. A fall of the entire reinforced structure can occur. Shear stresses develop at the beam ends and set into action a yield process which continues until shear failure is complete (Figure 12.4.13A).

b) The reinforced roof structure becomes stiffer and under high lateral stress stores some strain energy. In such a situation the reinforced linear arch can suddenly fail and fall to the floor (Figure 12.4.13B). This results in the formation of a large cave in the back of a mine opening.

Physical model investigations as well as mining observations have indicated that these two types of roof falls can be controlled to some degree. In the gravity load case, some bolts

PHYSICAL MODEL

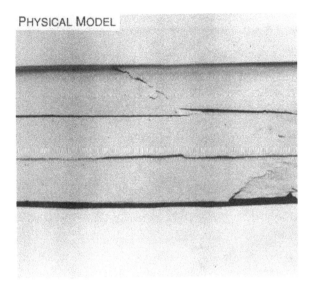

MINE OPENING

Figure 12.4.12. Development of shear failure at an angle to the bedding planes

should be installed at an angle to resist the formation of shear planes. In the second case, the stiffness of the reinforced roof strata should be decreased so that yielding deformation will develop.

It is concluded that the use of physical modelling for an analysis of the stress at failure is useful and simulates observed in situ behaviour.

12.4.3 *Mathematical model analysis*

Mathematical model analysis is widely used for all types of mining to determine the stress distributions displacements, and the timing and nature of ground failures.

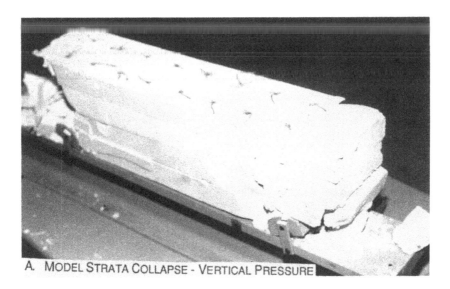

A. MODEL STRATA COLLAPSE - VERTICAL PRESSURE

B. MODEL STRATA COLLAPSE - LATERAL PRESSURE

Figure 12.4.13. Models of reinforced roof falls as a function of loading conditions

The mathematical model used in the analysis which follows was a boundary element model. It divides the boundary surfaces of the openings into discrete elements. The behaviour of any point in the solid about the openings is the sum of the influences of each element. The computer program that was utilized is the University of Toronto's two-dimensional elastic plane strain program 'EXAMINE', developed for use on an IBM compatible PC. The program produces a complete diagram of major and minor stress concentrations as well as safety factor distribution. The model analysis was carried out at the Geotechnical Research Centre of Laurentian University as a part of a research project on potash mining in Saskaktchewan. Two layouts were modelled.

1. *A five-room grouping* was analyzed for a layout of rooms of 12 m wide with yielding pillars of 6 m wide. The five-room grouping was modelled in two steps. In step one, the middle and outer rooms were excavated. In step two, the two remaining rooms were excavated in the strain hardened ground.

An analysis of the major principal stress distribution plot (Figure 12.4.14) for the five rooms shows that immediately adjacent to the first room and to the fifth room the stress concentrations in the abutment pillars were fairly high, ranging from 30-45 MPa, but still not high enough to cause failure. As you move away from these openings, further into the abutment pillars, the stresses begin to decrease to the range of 15-30 MPa. For the three yielding pillars the stress zones were quite high, ranging from 45-60 MPa. The reason for this is that the stress zones are overlapping (the stresses around one opening interact with the stresses related to the adjacent opening), plus the fact that the model does not let the yielding pillars fail. This plot agrees with the calculations for the load on the yielding pillars prior to yield in the earlier sections.[36]

An examination of the minor principal stress plot (Figure 12.4.15) shows the minimum stress increasing away from the openings. Also there is some tensile stress acting in the pillars.

Looking at the factor of safety plot for the rooms (Figure 12.4.16), one can see that it is highest at about 16 m into the abutment pillars and ranging from 4-5. It indicates that the abutment pillars will support the load of the overlying rock that is distributed to the abutments when the yielding pillars begin to fail. The factor of safety that was previously calculated using the formula for the abutment pillars is not in the same range (2.75) because

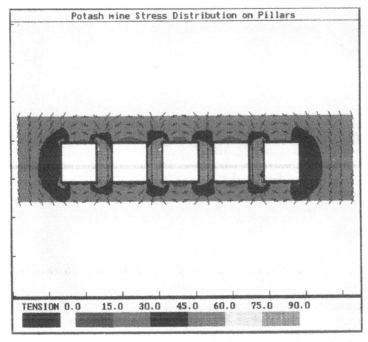

Figure 12.4.14. Major principal stress distribution (σ_1) in and around a five-room grouping (cross-section)

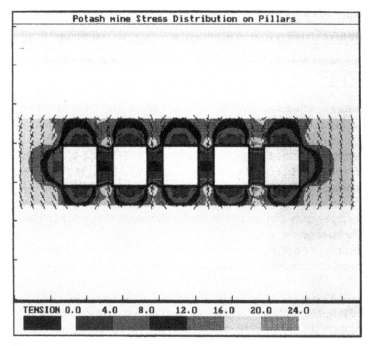

Figure 12.4.15. Minor principal stress distribution (σ_3) in and around a five-room grouping (cross-section)

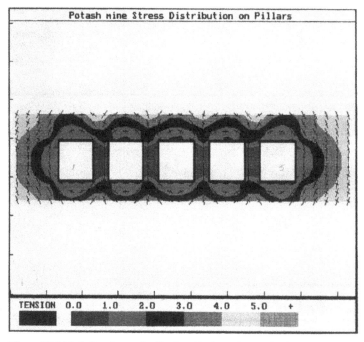

Figure 12.4.16. Safety factor distribution (F_S) in and around a five-room grouping

the formulas assume that the load distribution in the abutment pillars is uniform. The boundary element plots, however, correctly calculate the stress distributed in the abutment pillars. That is, the close one gets to the opening, the more the stress in the abutment pillar lowers the safety factor. From about 16 m from the last opening to about 36 m into the abutment pillar the factor of safety is very high (4-5). Beyond 36 m the proximity of the next five-room group begins reducing the safety factor. From the factor of safety plot one can see the trend of the factor of safety decreasing towards the opening. Immediately adjacent to the two end rooms (1 and 5) the factor of safety is between 1 and 2, which means that the pillar could be on the verge of failing.

2. *A uniform room-and-pillar layout* has also been modelled. For the model the widths of the rooms and pillars are 25 m.

An analysis of the major principal stress distribution plot (Figure 12.4.17) indicates critical stress concentrations limited to around the corners of the rooms. The model used rectangular openings and the stress distribution plots are based on square corners. However, continuous miners cut rounded corners. The stresses concentrated around the sharp, angular corners in the boundary element plots may not be present in situ. The plot indicates low stress distributions in the immediate roof and floor of the rooms.

Also from the minor principal stress plot (Figure 12.4.18) it can be seen that nowhere the tensile strength of potash was exceeded. The very slight tension in the floor and roof of the rooms on the west side is more likely a result of the model rather than a real stress distribution.

The safety factor plot (Figure 12.4.19) shows its lowest value (between 1.0 and 2.0) around the room corners and walls. The geometrical distribution of the minimum safety

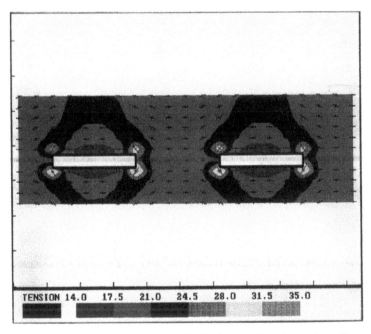

Figure 12.4.17. Major principal stress distribution (σ_1) in and around a uniform room-and-pillar layout (cross-section)

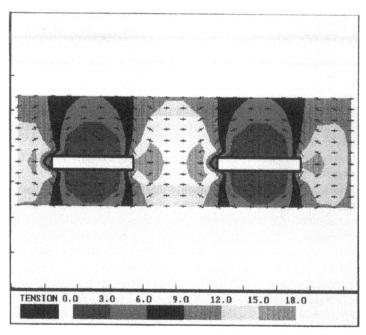

Figure 12.4.18. Minor principal stress distribution (σ_3) in and around a uniform room-and-pillar layout (cross-section)

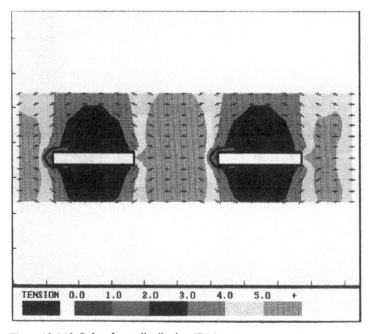

Figure 12.4.19. Safety factor distribution (F_S) in and around a uniform room-and-pillar layout (longitudinal section)

factor indicates that oval geometry of the in situ room walls could eliminate these potentially unstable zones.

12.4.4 *Rock mechanics analysis*

A full rock mechanics analysis is comprehensive. It involves a geological investigation, the testing of the geological materials, analytical analysis of test results, and either a physical model analysis or a mathematical model analysis. It is used to assess the stability of existing underground structures and to produce the engineering elements for the design of new mine layouts. Such a study was undertaken by Fairhurst et al. in the publication 'Rock mechanics studies of proposed underground mining of potash in Sergipe, Brazil'.[36] The following summary is from this publication.

The strength properties determined in the laboratory are given in Table 12.4.1.

The minimum potash ore recovery at a depth of 1000 m in Saskatchewan is 35% suggesting that it should be possible in Sergipe, at an average depth of 600 m, to extract 55% using similar mining methods

$$33\% \times \left(\frac{1000}{600} = 55\% \right)$$

The concept of mining at Sergipe requires leaving a protective layer of potash ore in the room floor to prevent a tachyhydrite flow into the room or to avoid excessive floor heave. A minimum room width of 4 m was considered desirable for continuous mining technology. It was conservatively assumed that the tachydrite behaves essentially as water, exerting a constant uplift pressure equal to the lithostatic pressure at depth on the base of the sylvinite bed.

Assuming a mining depth of 600 m, the stress field would be as illustrated in Figure 12.4.20. The maximum shear stress on the floor beam is:

$$\tau = \frac{15 \times 4}{2\,t}$$

or

$$t = \frac{30}{\tau}$$

$$t = 30/\tau$$

where:

τ = shear strength of the sylvinite;
t = thickness of sylvinite protective layer.

From Table 12.4.1 it can be seen that the shear strength is 18 MPa. Setting τ = 15 MPa results in a protective layer with a minimum thickness of 1.7 m to resist shearing. If the applied shear stress is limited to 60% of the shear strength (i.e. a safety factor of 1.6), then the required thickness of the protective potash layer is approximately 3 m.

The calculated beam formula of maximum flexular tensile stress should not exceed the 15 MPa lateral field stress value in an assumed 4 m thick roof beam or in the floor protective sylvinite beam with a uniform 15 MPa upward load.

The mine layout has been modelled using the two dimensional Stealth finite difference program. It can accommodate viscoelastic material and assumes plane-strain conditions. The modelled mining geometry and loading conditions are shown in Figure 12.4.21. The finite difference grid was made up of three different layers of halite and sylvinite/carnallite. The finite difference analysis indicates several important points with respect to mine stability and the capabilities of this model code.

a) The stress distribution within pillars can be defined even after considerable creep has taken place. Figure 12.4.22 shows the pattern of stress distribution after creep for one year. With this numerical technique it is possible to evaluate mine pillar stability over time.

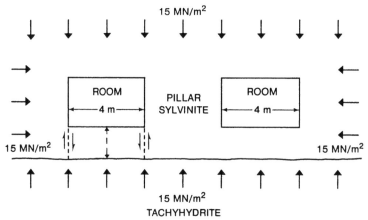

Figure 12.4.20. Cross-section of a room-and-pillar layout with a sylvinite protective layer

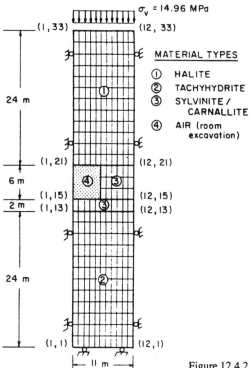

Figure 12.4.21. Finite different grid for numerical modelling

Table 12.4.1. Strength properties of evaporite beds

Mining designation	Type of evaporite bed	Max. thickness (m)	Compressive strength σ_c (MPa)	Tensile strength σ_t (MPa)	Shear strength T (MPa)	Normal stress σ_N (MPa)	Elastic modulus E (MPa)	Poisson's ratio v
Roof	Halite	25.0	33.1-152.8	0.15	–	–	–	–
Ore	Sylvinite	4.5	32.2-143.6	0.19	18.0	15.0	10 000	0.36
Ore	Carnallite	15.0	13.5- 61.7	–	–	–	–	–
Floor	Tachyhydrite	20.0	3.5- 5.3	–	–	–	1000 5000	0.49

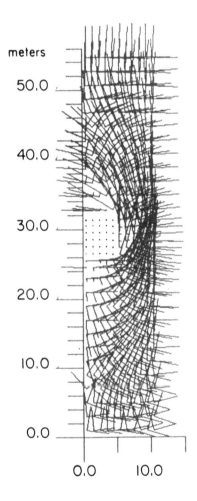

meters

50.0

40.0

30.0

20.0

10.0

0.0

0.0   10.0

Figure 12.4.22. Stress distribution around a room after creep for one year

b) The analysis suggests that stress relief occurs both in the halite roof and in the tachyhdrite floor. Also, most of the roof-floor convergence results from the high stress concentration in the pillars, which are represented by sylvinite.

c) The computer analysis was carried out to obtain the creep behaviour of the respective evaporite materials. A strain hardening form of creep (i.e. creep strain rate decreasing under

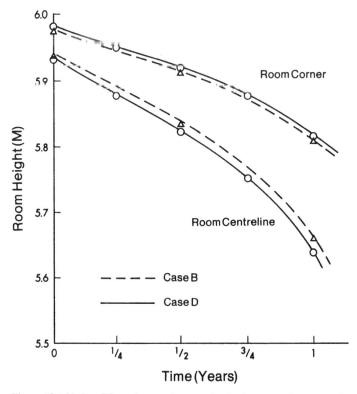

Figure 12.4.23. Roof-floor closure of a room for the first year after excavation

constant stress) was used in the analysis. It shows that room closure continues at a more or less constant rate, with some acceleration towards the end of the period (Figure 12.4.23). This result corresponds to the constant rate of room closure observed in underground mines and shows that constant closure is not inconsistent with the strain hardening creep behaviour measured in laboratory testing.

The study demonstrated the possibility of forecasting time-dependent stress distributions above and below the rooms and pillars, a key factor for strata control in evaporite underground mines. The authors recommend using the method only as an aid in assessing the stability of an actual mine situation. The validity of the results is in question until they can be verified by underground observations.[36]

Finally, it should be commented that the Stealth program was probably more suitable for modelling yielding pillars and associated stress transfers than the EXAMINE program used in the previous subsection.

REFERENCES

1. Delling, D.R. 1985. Tenneco's Green River project. *Min. Engng. ASME*, October 1985: 1197-1200.
2. Michalke, E. 1980. Optimization of a mining block in a flat deposit. *Kali und Steinsalz*, Bd. 8, Heft 2: 50-55 (in German).

3. Hofmeister, W. 1986. Drilling and blasting techniques in potash mines. *Kali und Steinsalz*, March 1986: 223-232 (in German).
4. Post, L. N. 1985. FMC's Westvaco soda ash operation uses a variety of mining techniques. *Min. Engng. AIME* 37 (10): 1200-1204.
5. Moore, G.W. & R. Gauthier 1983. Long-range planning in Saskatchewan potash mines. *Potash technology*, pp. 91-97. Toronto/New York: Pergamon Press.
6. Molavi, M.A. & M.J.S. Woolley 1986. A case study of pillar design, pp. 1-23. Contribution to *CIM Rock Mechanics and Strata Control Committee Workshop on the Role of Pillars in Soft Rock Mining*, at Saskatoon, Saskatchewan, November 1986.
7. Fuzesy, A. 1982. *Potash in Saskatchewan*. Saskatchewan Energy and Mines, Report 181, pp. 1-32.
8. Gerne, J. 1983. Longwall extraction of potash minerals in Europe. In *Potash technology*, pp. 73-79. Toronto/New York: Pergamon Press.
9. Herget, G. 1986. Longwall mining and potash. *CIM Bulletin* 79 (887): 96-100.
10. Bodu, M. 1981. Full or selective extraction at the potash mines of Alsace. *Industrie Minerale – Les Techniques* 1-12, January 1981: 339-346 (in French).
11. Carrasco, G.J. 1979. Explotacion por tajo largo a gran profoundidad, pp. 13-20. In *Proc. of 10th World Mining Congr.*, Instanbul.
12. Gimm, W. & H. Pforr 1976. Kammerbau mit nachgiebigen Pfeilern und Strebbruchbau im Soligorsk. Kalibergbau-methodik und bisherige Ergebnisse bei der Einführung verlustarmer Abbauverfahren. *Neue Bergbautechnik* 6/7: 673-676.
13. Jeremic, M.L. 1985. (1) Principal mining systems of slice mining, and (2) Stability of stope pillar structure, In *Strata control in coal mining*, pp. 354-367 and 244-261. Rotterdam: Balkema.
14. Jeremic, M.L. 1982. Internal structure of coal seams and mechanical stability. *CIM Bulletin*, May 1982: 71-75.
15. Uhlenbecker, F.W. 1971. Uhlenbecker: Gebirgsmechanische Untersuchungen auf dem Kaliwerk Hattorf (Werra-Revier). *Kali und Steinsalz*, Bd. 5, Heft 10: 345-359.
16. Uhlenbecker, F.W. 1974. Neuer Forschungsergebnisse in der Gebirgsmechanik aus dem Salzbergbau. *Kali und Steinsalz*, Bd. 6, Heft 9: 308-315.
17. Walther, C. 1974. Walther: Verteilung vertikaler Spannungen in Pfeilermodellen aus Hartsalz. *Kali und Steinsalz*, Bd. 6, Heft 10: 356-361.
18. Singh, R.N. & M. Eksi 1987. Empirical design of pillars in gypsum mining using rock mass classification systems. *J. Mines, Metals & Fuels* (India), 35 (1): 16-23.
19. daGama, C.D. 1979. Rheological behaviour of heterogeneous salt rocks. In *Proc. of 4th Int. Congr. on Rock Mech.*, Montreux, Switzerland, Vol. 1, pp. 107-115.
20. Serata, S. 1971. The Serata stress control method of stabilizing underground openings, pp. 99-119. In *Proc. of the 7th Canadian Rock Mech. Symp.*, March 1971.
21. Serata, S. 1976. Stress control technique – An alternative to rock bolting. *Min. Engng.*, May 1976: 51-56.
22. Mackintosh, A.D. 1977. Strata control in a deep Saskatchewan potash mine, pp. 1-15. In *Proc. of 6th Int. Strata Cont. Conf. Banff*, CANMET, Canada.
23. Hebblewhite, B.K. et al. 1979. The design of underground mining layouts for a deep potash mine on the basis of rock mechanics investigations. In *Proc. of 4th Int. Congr. on Rock Mech.*, Montreux, Switzerland, Vol. 2, pp. 219-226.
24. Sattler, A.R. & C.L. Christensen 1984. Measurements of very large deformations in potash salt in conjunction with an ongoing mining operations, pp. 485-495. In *Proc. of 1st Conf. on the Mech. Behav. of Salt*, Penn State Univ. Trans. Techn. Publ.
25. Alder, H. et al. 1949. Yield pillar technique in deep coal mining. *Engng. Magazine* (Great Britain).
26. Mraz, D.Z. & M.B. Dusseault 1983. Method of correlation of laboratory test with in situ conditions. In *Proc. of 6th Int. Symp. on Salt*, The Salt Institute, Virginia, Vol. 2, pp. 259-269.
27. Mraz, D.Z. 1980. Plastic behaviour of salt rock utilized in designing mining method. *CIM Bulletin*, March 1980: 111-116.
28. Jeremic, M.L. & A. Farah 1990. Physical modelling of roof strata in potash mines of Saskatchewan, pp. 1-7. In *Proc. of 1st Int. Work. on Scale Effects in Rock Masses*, Loen, June 7-8, 1990.

29. Obert, L. & W.I. Duvall 1967. *Rock mechanics and the design of structures in rock*, pp. 387-401. New York: Wiley.

30. Potts, E.L.J. et al. 1979. The evaluation of the design criteria for underground roof strata considered as linear arch structure. In *Proc. of 4th Int. Congr. on Rock Mech.*, Montreux, Switzerland, Vol. 2, pp. 531-538.

31. Jeremic, M.L. 1985. Deformation and failure of coal structures. In *Strata mechanics in coal mining*, pp. 173-218. Rotterdam: Balkema.

32. Brandis, et al. 1981. Experimental studies of scale effects on the shear behavior of rock joints. *Int. J. Rock Mechs.*, (Pergamon Press), 1-21.

33. Ryder, G.H. 1963. *Strength of materials*, p. 250. London: Cleaver Press.

34. Gilani, A.J. 1982. *Rock mechanics evaluation of cable bolt application at underground mine*, p. 58. Engineering Thesis, Laurentian Univ., Sudbury, Canada.

35. MacNeil, K. 1989. The mining of stratified potash deposits in Saskatchewan by the long room and pillar method. Manuscript, p. 31.

36. Fairhurst, C.M. et al. 1979. Rock mechanics studies of proposed underground mining of potash in Sergipe, Brazil. In *Proc. of 4th Int. Congr. on Rock Mech.*, Montreux, Switzerland, Vol. 1, pp. 131-138.

CHAPTER 13

Mining of thick deposits

Mining of thick salt deposits is mainly undertaken for rock salt beds, but in some cases also for potash. The most suitable mining method of thick salt deposits is room-and-pillar mining, which can offer stability of large rooms. Fundamentally, the room-and-pillar method is typified by large and moderate room excavations where mining can be implemented by either single-level or multiple-level layouts. The room-and-pillar mining methodology of thick salt deposits bear only a remote similarity with room-and-pillar methods applied in coal mining and hard rock mining, except for the case of slicing excavations.

From the point of view of the mining systems and rock mechanics analyses, primary consideration is given to the geometry of the mining layouts. They may vary considerably. For example, the room height could be in the range of 8 m to 35 m and may be rectangular or trapezoidal. The mining considerations and rock mechanics and ground control aspects are discussed in a similar manner as in the previous chapter on moderately thick salt deposit mining. Four basic topics are represented.

13.1 PRINCIPAL MINING SYSTEMS

The fundamentals of mining of thick salt deposits (domes, folded structure, thick beds) are based on the general concept of room-and-pillar mining. The classification of this mining method relates to the geometry, structural elements (strike and dip), size of the salt body and its geomechanical environment. At this point, the classification considers several principal systems of room-and-pillar mining which are individually described and characterized by a particular mining geometry.

13.1.1 *Square pillar mining*

The room-and-square-pillar mining is typical for flat, thin or moderately thick sedimentary deposits, as discussed in the previous chapter. The square pillar mining had been introduced in some number of salt bedded deposits in which the thickness was greater than 10 m. However, due to the introduction of continuous mechanized mining technology, as well as roof instability and difficult ground control, square pillar mining in some number of evaporite mines has changed over to more suitable mining methods.

Square pillar mining has been practised in the Goderich rock salt mine in Southern Ontario. The salt deposit of this mine is massive and tabular of considerable lateral and vertical extent, where the rock salt is relatively uniformly distributed and parallel to the planes of stratification. The salt bed is 25-30 m thick, and occurs at a depth of 575 m below Lake Huron. The salt bed dips at 1° and is hosted by shaly dolomite. The salt deposit lends itself to room-and-pillar mining. The square pillar mining or the so-called original mining method has the following layout:

- square pillars: 45 m wide by 45 m long by 13 m high;
- rectangular rooms: 14 m wide by 52 m long by 13 m high;
- extraction ratio: about 55%.

Figure 13.1.1 illustrates the same layout as described above, but with the room-and-pillar dimensions listed below:

- square pillars: 64 m wide × 64 m long × 13 m high;
- rectangular rooms: 18 m wide × 64 m long × 13 m high;
- extraction ratio: also about 55%.

It should be noted that the mine layout provides a larger pillar (182 m wide × 182 m long) in

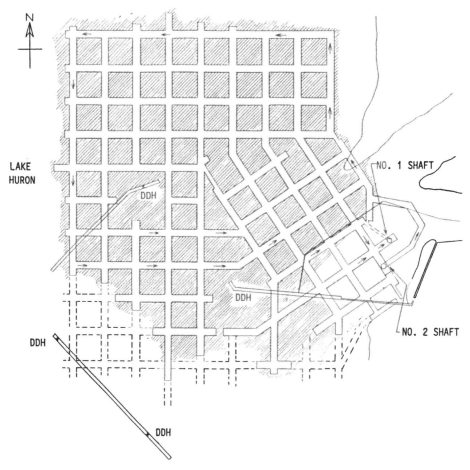

Figure 13.1.1. Original mine layout of room-and-square pillar configuration (Goderich, Southern Ontario)

the area where the exploration diamond drill holes are located. The minimum distance between the edge of the pillars and the diamond drill hole must be 45 m. Therefore small rooms of 14 m wide could be driven in the protection pillar but the advance must not be great enough to break the safety minimum distance, as stated above.

In the Goderich salt mine experienced rock falls are due to action of high horizontal stresses. To increase stability of the roof of the rooms, a new room-and-pillar mining system has been devised and is discussed in the next subsection. It should be noted that the rock salt excavation of both systems is by cyclic mining, e.g. drill and blast technique, which is described together with the current mining system.[1]

13.1.2 Ribbon pillar mining

This mining method could be applied in the moderately thick salt beds as well as thick salt beds. This mining method more and more replaces square pillar mining due to the possibility of the introduction of continuous mechanized mining technology as well as much better ground control.

The ribbon pillar mining at the Goderich salt mine in Southern Ontario entails the layout of a three-room entry system. These rooms are to be stress-relieved, which involves the layout of three rooms, leaving small yield pillars between them. This mine layout implies the time stress controlled technique where the stresses are diverted into large abutment pillars which are left between each set of rooms, as discussed in subsection 13.1.3. The stability of the openings depends on the size of rooms and pillars, as well as the time delay between excavation of the inside and outside rooms.[2]

At the present time, at the Goderich mine the change over from the square pillar mining to ribbon mining is shown in Figure 13.1.2. There are two layouts of ribbon pillar mining, namely the three-entry system and the four-entry system. The dimensions of the mine structure are:
- ribbon pillars: 15 m or 18 m wide × 13 m high;
- rooms: 15 m or 18 m wide × 13 m high;
- abutment pillar: 60 or 70 m wide × 60 or 70 m long × 13 m high;
- the length of ribbon pillars and rooms depends on the panel size;
- extraction ratio: up to 65%.

In the three-room entry, the two outer rooms are mined first and a lapse of approximately one month is required before the commencement of the centre room. After practising three-entry ribbon pillar mining, a four-entry system was introduced for two basic reasons:

1. to supply adequate room for passage of men, equipment, excavation of rock salt and ventilation;

2. to maintain ground control stability of the main entry that will last for the estimated life of the mine.

It should be noted that some number of pillars in this mine have the notch to decrease a total roof area at the entry intersections.

The three-room entries are excavated by full face ground breakage and broken salt removal. In the mining operation, the procedure is to first undercut the salt face with a universal cutting machine. This cut provides relief during the blasting operation and at the same time gives a relatively smooth floor. The cut is 12.7 cm wide and about 3.7 m deep (Figure 13.1.3). It is estimated, based upon the efficiency of the universal cutting machine, that the time required to cut a face width of 18 m is approximately three hours.[3]

A four-drill jumbo is used to drill the face. Two set-ups are required to cover the full face.

Figure 13.1.2. Mine layout
for three-room entry and
four-room entry systems

Also these jumbo drills are double-decked in order to reach and drill holes at the higher sections. The drill jumbos are rubber tired allowing for quick movement from face to face.

To drill and blast the advances into the salt, some form of cut pattern is usually required. A conventional burn-cut stope pattern would be adequate to fragment the salt efficiently. The faces are drilled by auger-type drills of 4.5 m long with carbide tips. The time estimated to drill a complete face of required 127 holes, is three hours.[3] The approximate volume of salt to be blasted per room is calculated to be up to 850 m³, which corresponds to approximately 1800 tonnes. The loading of broken salt is by gathering-arm loaders, or by rubber-tired front end loaders (Figure 13.1.4). The muck, loaded in 27 ton trucks, is hauled to a hopper set-up. The hoppers provide an even feed to the conveyor belt which carry rock salt ore to the crusher installation.

13.1.3 *Angular room mining*

Angular room mining is carried out with a number of variations expressed by different sizes, dimensions, angular shape and layouts of the room-and-pillars. This mining system, however, has a common factor which is exhibited by ascending mining. The principles of rectangular room mining are discussed in the case of history of the Borth salt mine in Germany, as described in Geyer's published papers.[4, 5]

The geological section of salt deposits at the Borth salt mine shows an upper salt bed,

Figure 13.1.3. Universal cutting machine for stope face undercut

Figure 13.1.4. The rock salt muck after full stope face blasting (room 13 m high)

composed of almost pure halite, with a mineable thickness up to 20 m (Figure 13.1.5).

The old stoping mining method is illustrated in Figure 13.1.6 with four phases of the mining, namely: undercut, widening undercut and roof slashing, salt extraction by two breasts, salt mining with four breasts and room roof slashing (high wall). The dimensions of the elements of the rectangular mine structure are:
- undercut slot: 2 m wide × 2 m high × < 600 m length;
- slashing of undercut for 9 m on either side of the slot;
- slashing of the undercut back is 2 m high;
- the breasts are 2 m high × 3.5 m long × 20 m wide;
- the slashing of the roof is 2 m high.

The new rectangular room mining is a result of the redesign of the old mining method mentioned before. The redesign mining method has three principal phases of rock salt extraction (Figure 13.1.7), as is briefly outlined below.

1. *The undercut phase* is the most complex one, because it consists of three sub-phases:

a) The slot drift is simultaneously excavated by two slices with two road headers. Two slices drift excavation has a size of 4.4 m high × 4.4 m wide.

b) The initial undercut is open from the slot drift. The face is 14 m wide × 7 m high (excavation by drilling and blasting).

c) The slashing undercut results in its final 24 m width.

The drilling of the undercut is done parallel to the length of the room (initial undercut) or across the width of the room (slash of undercut). Drill holes are inclined 15-20° to the room floor. The drill rounds measure up to 9 m in length.

2. *High cut* is as simple phase of the ascending room excavation. The drilling is done by drill holes, 5.5 m apart (horizontally) and with 40° inclination. The drilling is done in the back of the undercut. The slice of ore taken by this drilling and blasting is 9 m thick. The drilling and blasting is done from the room floor, not from the muck (Figure 13.1.8).

3. *Back slash* is the last phase of stope extraction and is carried on from the muck pile by a pair of small boom jumbo drills. The back slash is 2 m high and, forming the room, 18 m high. The roof is reinforced by rock-bolting by two bolting machines which are behind the drilling machines. The installed mechanical rock bolt is 1.5 m long. The distance between

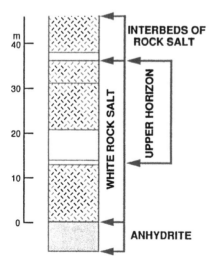

Figure 13.1.5. Geological section of Borth rock salt mine (Germany)

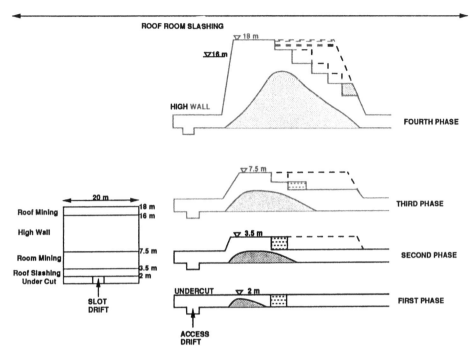

Figure 13.1.6. The phases of open room extraction of the Borth mine (old mining technology)

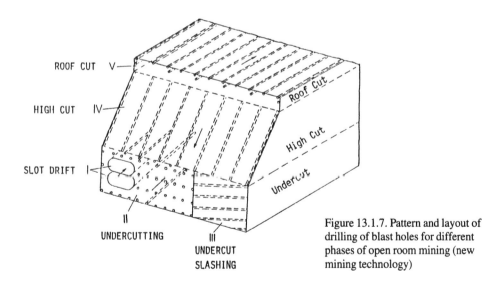

Figure 13.1.7. Pattern and layout of drilling of blast holes for different phases of open room mining (new mining technology)

the top of the muck pile and the room back is constantly kept at 3 m so as to provide enough working space for the mechanized equipment of the mining operation.

The blasted rock salt is loaded by a mucking machine and transported by a 18 tonnes rubber tire haulage truck to the crusher. From the crusher the transport is by a belt conveyor (90 cm wide) to the shaft loading pocket.

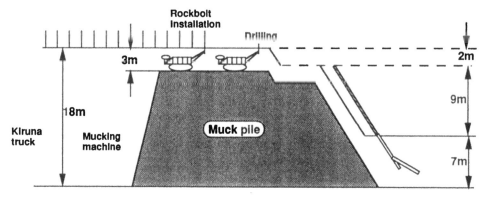

Figure 13.1.8. The phases of open room extraction and muck transportation at the Borth mine (new mining technology)

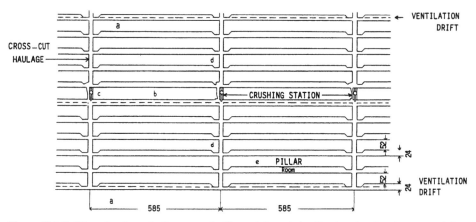

Figure 13.1.9. The mine layout of room-and-pillar mining at the Borth mine for the new mining technology

The mine layout of the rock salt panel is given in Figure 13.1.9, showing the ventilation drifts and the main conveyor drift with the crushing station. the dimensions of the room-and-pillar structure are:
– room: 24 m wide by 18 m high by 585 m long;
– pillar: 52 m wide by 18 m high by 585 m long.
Angular room mining of thick rock salt beds is extensively applied (among other countries) in Poland, Romania (Figure 13.1.10) and Germany.[6, 7, 8] In all cases, the old mining method has been changed to a more mechanized and productive one, as in the case of the Borth salt mine in Germany. The technological improvements of the mines do not permit a selective salt extraction, where low quality salt and waste partings could be manually separated and left in the mined-out rooms. However, the new methodology offers improved safety because room reinforcement is integrated into the mining methods, and rock falls from the roof are thus controlled.

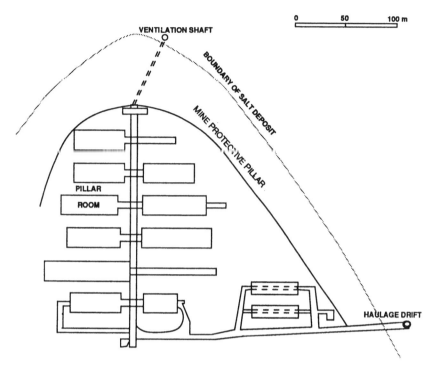

Figure 13.1.10. Mine layout of room-and-pillar mining at the Muresh salt mine in Romania (level 110 m)

13.1.4 *Trapeze room mining*

This mining system had been developed and implemented in Romania.[8] The access from the level haulage to the stoping area is by the winzes (blind shafts). The final depth of a winze has to correspond to the level of the room floor. The room extraction is in descending order by slicing. From the winze a short cross-cut (up to 10 m) is driven to make an access to the room development workings, which consist of the room overcuts and elongated drifts as illustrated in Figure 13.1.11. As may be seen, the development of the rooms is at their top, because the mining is from top to bottom. The drifts are 12 m wide and 2.5 m high. The room overcuts are indicated as A, B, and C in Figure 13.1.11. The distance from their centre line varies from mine to mine. For example, at the Dez salt mine the distance between the centre line is 86 m but at the Slavic mine and other mines the distance between the centre lines is 100 m. At the Slavik Mine, after completion of the overcut, the wooden gallery is installed around the wall of excavation. As the excavation of the room progresses deeper and deeper the stability of the roof of the stope is controlled from the gallery. After room completion the roof stability is also continuously controlled from the gallery, which became accessible by fire fighting telescopic loaders. In other Romanian salt mines with trapezoidal rooms the wooden gallery at the stope back is not installed, because this type of ground control is very expensive. It could be done much cheaper and more efficiently with other methods and techniques of ground control.

The rock salt extraction is by simultaneously slicing. The height of the individual slice varies from mine to mine between 3 and 4 m. In Figure 13.1.12 drilling of an individual

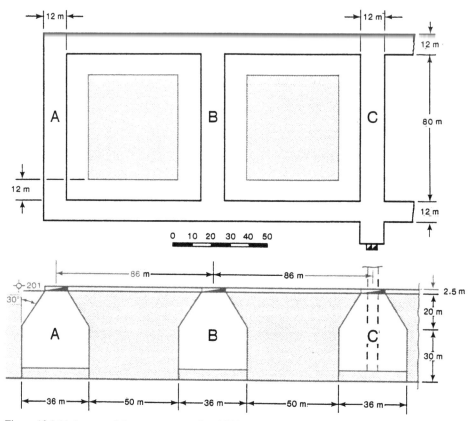

Figure 13.1.11. Layout of the trapeze room (level 200 m) at the Dez salt mine (Romania)

slice is shown. The face of the slice is parallel with the width of the room and its direction of mining is along the length of the room (Figure 13.1.13). After each slice inception the loading platform in the winze is lowered to the level of the next slice, because the salt from the room is lifted by the winze to the level haulage above. However, with implementation of mechanized mining the salt ore is removed from the working face by the loader with gathering arms and loaded on the combi trucks, which will dump ore in one winze used as a mill hole. The salt ore is then delivered to the main level below and hauled to the main shaft and by the skip brought to ground surface. In the case of mechanized salt mining, the trapezoidal rooms are mines by individual descending slicing, one slice after another.

The upper portion of the trapeze rooms have an inclination of 60°. The final width of the rooms is 34 or 36 m, which depends on the width of 10 or 12 m of the central drift which overcuts the room. The height of the conical portion of the room is 20 m, with a total depth of 34 m to 50 m where the height of the room overcut is not taken into account. The pillar dimensions are 50 m and 80 m long (Figure 13.1.11). This clearly shows that more salt is left in the pillar than is extracted by room, what also may be the case with angular room mining. The salt recovery, however, is higher in angular room mining than in trapeze room mining. The height of the angular rooms is in the range of 10 m to 20 m, which limits the time duration of salt extraction, but may be greatly offset by the increase of the room length, which is not the case with trapeze rooms.

Figure 13.1.12. The descending slice extraction by drilling and blasting at the Dez salt mine (Romania)

Figure 13.1.13. Ensemble of trapeze room at the Slanic salt mine (Romania)

It should be noted that mechanized trapeze room mining offers large production with high productivity (OTM 100 tonnes). The ventilation is satisfactory because all rooms are connected among themselves. The ore recovery by the trapeze room system is up to 30% and rather low when compared to angular rooms (up to 60%).

13.1.5 *Room longhole mining*

Room longhole mining has been considered in the Denison-Potocan potash mine in New Brunswick, Canada. The typical extraction panel is developed by a three-entry system, as illustrated in Figure 13.1.14. The shape and size of the panel depend on the form and regularity of the potash ore body. The three-entry system is chosen so that stress will be diverted from the drifts in the surrounding strata. This diversion is achieved by the concept of yielding pillars, as discussed in a previous section. The dimensions of the entries and pillars between them are:
  – drift: 11 m wide and 3.6 m high;
  – pillar: 2 m wide and 3.6 m high.
The drifts are driven at the upper part of the potash body and their roof is at the contact between the potash and the salt beds.

The open rooms are developed oblique to the mining panel and are mined out with the following dimensions:
  – room: 30-35 m wide, 5-40 m high (depends on the thickness of the ore body) and 600-1000 m long (depends on the panel size);
  – inter-room pillars: 60 m wide or 3 room widths, between 5 and 40 m high (depends on the thickness of the potash bed) and 700-1100 m long (depends on the panel size);
  – barrier pillars: 40 m wide, located between panel entries.
The conceptual layout of the rooms is also given in Figure 13.1.14.

The stope extraction is from roof to floor of the potash body. Two techniques of potash extraction are in effect. In both of them it is first required to excavate a so-called top slice of 30 m in length, which corresponds to the length of the stope. The height of the top slice depends on the angle of inclination of the beds and their regularity in regard to thickness, but should be less than 5 m.

1. *The technique of bench mining* is used in ore bodies less than 20 m thick. It is similar to room mining by slicing in descending order. The benching is done by continuous miners, AM-100 or 1012 Marietta miners. The height of the individual bench is less than 5.0 m, with the direction of mining along the length of the room. From the top slice the rest of the ore below is mined out by three benches with slice thicknesses of less than 5 m.

2. *The technique of longhole mining* is used in ore bodies more than 20 m thick. It is similar to the bulk mining in hard rock but with particularities dictated by potash ore. In this case, the stope is developed by two slices, namely: the top slice (at the roof of the ore body) and the bottom slice or undercut (at the floor of the ore body). The bottom slice is divided by a pillar into two sections, as illustrated in Figure 13.1.15. Each undercut has a length of 12 m and the pillars between them are 6 m wide. The top slice and bottom slice are 4-5 m high, and have a width up to 30 m, which corresponds to the room width. Longhole drilling is done by a drilling machine mounted onto a truck and drilled at an 80° angle of inclination (Figure 13.1.16). The drilling of blast holes is in the pattern of 1.20 m × 1.55 m. The diameter of the blast holes is 5 cm and they are charged by an ANFO explosive. The drilling and blasting starts from one end on the stope and retreats towards the other. One blast yields from 1500 tonnes to 5000 tonnes of the broken potash ore.

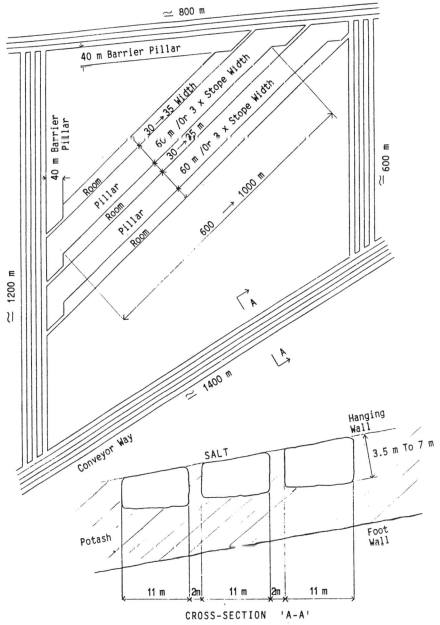

Figure 13.1.14. Development of mining panel by three-entry system with layout of the room-and-pillars

The ore recovery by both techniques is approximately the same, with an extraction ratio in the range of 45%. In the case of longhole mining with backfill, the ore recovery would be greater, because of the possibility of secondary ore recovery from the pillars between filled stopes. Actually, the sequences of drilling and blasting as well as of backfilling would be

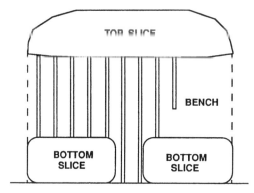

Figure 13.1.15. Cross-section of the bulk mining stoping with two undercuts and lay-out of the long blast holes

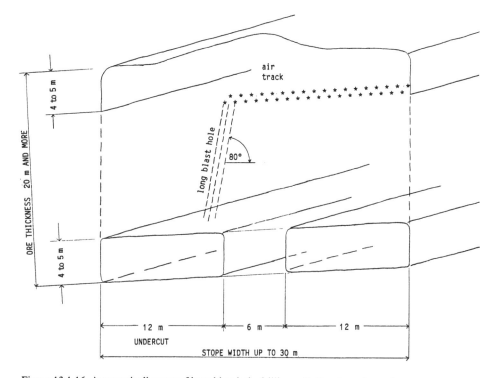

Figure 13.1.16. Axometric diagram of long blast hole drilling with the stope dimensions

similar to bulk mining. Most likely mining with backfill will facilitate larger room extractions by leaving smaller inter-room pillars. The potential of this mining method is significant for large and thick salt deposits and its rock mechanics must be investigated because it is a key factor for greater recovery and ground stability.

13.1.6 *Multi-level room mining*

The multi-level room-and-pillar mining system differs from previously mentioned mining methods only by mine layout but not by techniques of salt extraction. For this reason, in this subsection only layouts of this mining system will be discussed. The mining is by multi-level layouts and by bi-level layouts.

Basically three layouts are in use for multi-level mining and they are influenced by geological-structural features of the salt deposits.

1. *Mining of homogeneous rock salt blocks* mainly corresponds to salt domes with huge blocks of mineable rock salt. This type of mining is practised in some number of European salt mines. The multiple-room mining has been applied also in Poland for thick rock salt bodies, for which a 3D layout is illustrated in Figure 13.1.17. The underground mining is carried out by two principal techniques of rock salt exploitation: dry mining and solution mining of the rooms (discussed in Chapter 15). The layout of room and pillars for both technologies is similar, namely:
  – inter-chamber pillars: 10-40 m wide by 80-100 m long by 8-10 m high;
  – sill pillars: width and length depend on the size of the mining rock salt block; the height is 5-10 m;
  – room: 20 m wide by 80-100 m long by 8-10 m high.
The schematic layout of multiple-room mining is illustrated for the Solno-Inowroclaw salt mine (Figure 13.1.18). This mine applies both principal techniques for salt exploitation from the rooms (dry mining and solution mining). The formation of the skeleton pillar structure is by single-level extraction in descending order. The mining follows the rule that next level extraction will start after completion of mining the level above.[6]

2. *Mining of rock salt blocks interbedded with waste partings* mainly corresponds to intensively folded deposits. The layout of the multiple-level room-and-pillar structure is constrained by the internal geological structural feature of the mining block. For example, the Tusanj salt mine in former Yugoslavia, which is approximated as a lens of a maximum thickness of 500 m and lateral extension of 600 m, is folded and refolded with waste partings (marl and anhydrite), with thicknesses varying from several centimeters to over tens of meters. For delineation of the mining blocks, the inclusion of thick waste parting is

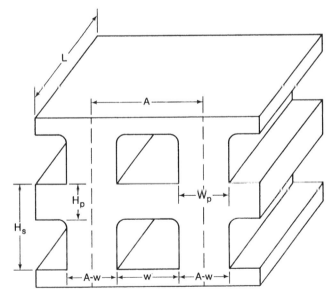

Figure 13.1.17. Block diagram of the skeleton pillar structure formed bi-level mining

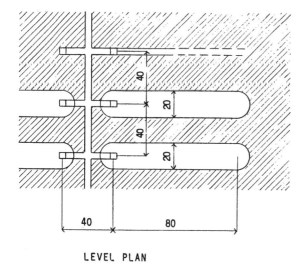

LEVEL PLAN

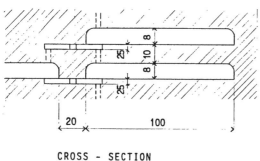

CROSS - SECTION

Figure 13.1.18. Schematic layout of multi-level mining at the Solno-Inowroclaw salt mine

permitted under the condition that they will be left in the pillars, as illustrated in Figure 13.1.19. Multi-level room mining in the Tusanj salt mine is laid out within the mining block at depths between 440 m and 500 m. The mineable portions of the mining block are only the locations of highly deformed and recrystallized rock salt of over 89% NaCl.[9]

The length of the rooms depends on the folding characteristics of salt deposits in their direction, and it could be from 50 m to 200 m. The room is 10 m wide and 10 m high and the inter-room pillars are of the same size (10 m by 10 m). The height of the sill pillar is also 10 m. Figure 13.1.19 shows how the mine layout is influenced by the existence of the thick marl bed within the mine block. Under these geological circumstances, where waste was to be left in the pillar, the extraction factor drops below 30%.[10]

3. *Mining of multiple potash beds* is implemented in some salt mines of Europe and North America. The multi-level room mining is strictly controlled by stratified geological structures, similar to the case of multiple seam coal mining.

The bi-level mining system in potash has been briefly described by Schmitke for the case of PCS Mining, Cory Division, in Saskatchewan (Canada).[11] The mine extracts ore predominantly from the A zone of the Patience Lake member of the prairie evaporite, more than 1000 m below surface. The prairie evaporite is composed of alternating beds of halite and sylvite having an average total thickness, at the Cory mine of approximately 200 m. The mining horizon is situated near the top of the formation. Below the A zone are several

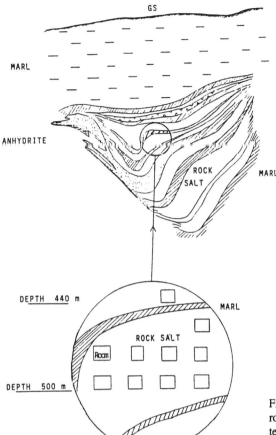

Figure 13.1.19. Layout of the multi-level room-and-pillar mining constrained by internal geological structure of mine block (Tusanj rock salt mine, former Yugoslavia)

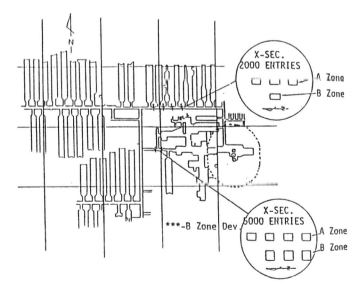

Figure 13.1.20. Layout of the bi-level potash mining by the room-and-pillar system (PCS Mining, Cory Division, Saskatchewan)

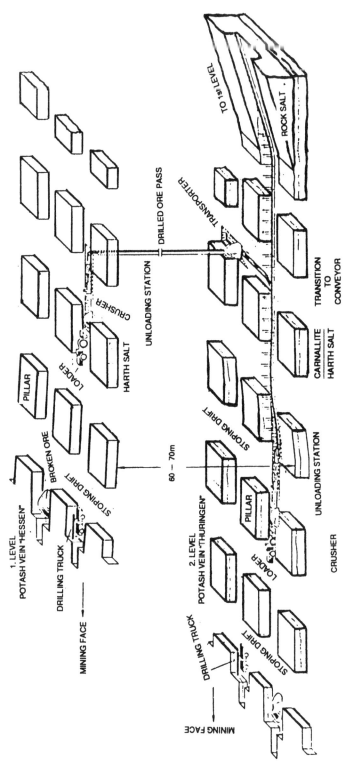

Figure 13.1.21. The block diagrams of multiple potash bed mining (Flash potash mine, Germany)

other potash beds, with the B zone presently being mines below existing A zone workings. The two mining horizons are separated by a distance of approximately 7.5 m.[10]

Figure 13.1.20 illustrates the location of the B zone development with respect to the overall mine layout. The B zone development below the 200 series three-entry system consists of a single, 12 m wide entry directly below the existing A zone centre entry. The B zone development below the 5000 series entry system consists of three 9 m wide entries (separated by two 9 m wide pillars) directly below the existing A zone four-entry system.[11]

The mining heights of the A zone and the B zone 12 m wide entry are approximately 3.4 m. The eventual height of the three 9 m wide entries in the B zone will be about 6 m. The age of the entries in the A zone, prior to the development in the B zone, ranges from eight to twelve years, the oldest being the 2000 series entry system.[10]

The system of multiple-bed mining has a geometrical similarity with the multiple-level room mining. However, the height of the sill pillars between rooms is predetermined by the thickness and properties of the salt strata sandwiched between two potash beds. This can be seen more clearly in the case of the Flash potash mine in Germany. The mining of the potash is by the room-and-pillar method with square pillars. The vertical distance between two levels of room extraction is between 60 and 70 m, as illustrated in Figure 13.1.21. The extraction recovery factor is between 35 and 45%. The technique of potash extraction is by continuous mechanized mining, which is a rubber-tired operation. The main haulage of the potash ore is by belt conveyors.[12]

The system of multiple-bed mining in regard to structural requirements is the same as the two previously described systems of mining. For example, the inter-room pillars have to be exactly centred below each other in order to the load to be uniformly transferred from the upper to the lower pillar regardless of the vertical distance between two or more levels of mining, as discussed in the next sections.

13.2 DESIGN CONSIDERATION OF MINE STRUCTURES

The stability of the mine structure in the first place depends on the adequate design of the rooms and the pillars. The key factor of stability is that the average stress of the pillar is below its strength. The average pillar stress is mainly composed of two factors, namely: depth of the pillar (weight of overburden) and recovery factor of the ore (transferred stress from mined-out rooms). It is obvious that the increase of the mine depth and the increase of the size of the room relative to the pillar also will increase the average pillar stress. The main consideration in this section is given to a determination of the average pillar stress, for the principal mine structure of thick salt deposits as discussed in the previous section. Four pillar structures are considered as follows.

13.2.1 *Loading concept of square pillars*

Square pillars are not common structures in underground mining of thick evaporite deposits. They are used, however, primarily for mining of rock salt and gypsum deposits. For example, the room-and-square-pillar system of mining is used for the exploitation of the rock salt deposits in the Michigan salt basin (Windsor).

The calculation of the average stress of square pillars was represented in Chapter 12. However, due to an appreciable size of the pillars in the thick salt deposits and mining a high structure (10 m and 20 m high), it is necessary to take into account also the mechanical properties of rock material which influence the pillar stress concentration.

V. D. Slesarev (former USSR)[13] has suggested a pillar formula based on the properties of rock salt, which would come into effect during the action of horizontal stress. He expressed horizontal stress by Rankin's equation for lateral earth pressure:

$$\sigma_H = \sigma_V \left[\tan^2 \left(45° - \phi/2 \right) \right]$$

where:

σ_V = vertical stress;
σ_H = horizontal stress;
ϕ = angle of friction.

Slesarev postulated the following equation for the activated internal resisting forces at the roof and floor boundary of the rock salt pillars (Figure 13.2.1):

$$\sigma_R = \frac{(\sigma_V + c)\, W}{2}$$

where:

σ_R = resisting force (shear stress);
c = cohesion;
W = width of the pillar.

In this equation cohesion is added to the vertical stress by ignoring in many cases the differences between the rock material which transmits the load and rock salt material of the pillar itself.

Slesarev proposed the following formula for calculation of average pillar stress, which was derived from the basic strength of material principles:

$$\sigma_p = \sigma_V \left[1 + \frac{3\, \sigma_H H^2}{(\sigma_V + c)\, W} \right]$$

where:

σ_p = pillar stress;
H = height of pillar.

The calculated stress on square pillars for a height of 10 m and over has been more or less in agreement with in situ instrumentation in some rock salt mines in the former USSR. It is recommended that for pillar design the safety coefficient should be used, so that critical pillar stress does not overcome the overall pillar strength. The safety coefficient should be at least $F_s = 1.2$.

13.2.2 *Design of angular pillars*

High angular pillars of limited length are a common structure for some thick rock salt mining. The stability of this pillar is critical for maintaining large open stopes during and after salt extraction.[8]

M. Stamatiu proposed a formula for the design of large rock salt pillars, on the basis of empirical dependence between strength and relation between width and height of large salt samples, as well as the concept of stress distribution within a large pillar. He proposed that around the room a zone of decreased stress is formed. The destressed zone around a room is represented by the geometrical shape, as illustrated in Figure 13.2.2. By this concept the pillar between two rooms is loaded by vertical stress only in its internal part. Stamitiu,

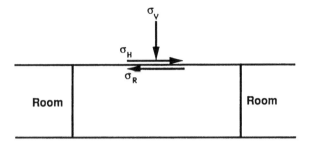

Figure 13.2.1. Sketch of acting and reacting stresses at a pillar roof contact

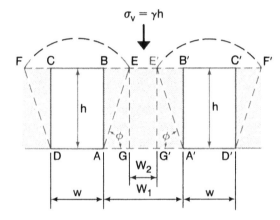

Figure 13.2.2. Loading conditions of an internal part of a pillar

taking into account that maximum shear stress of the high pillar is along the potential slip planes AE and AE' (Figure 13.2.2), suggested the following formula:

$$W_1 = 2H \cot \phi + W_2$$

where:

W_1 = external pillar width;
W_2 = internal pillar width;
H = height of the pillar;
ϕ = friction angle of pillar's rock salt.

He formulated that acting overburden stress on the central part of the pillar has to be less than or equal to the strength of large rock salt cube:

$$\sigma_V = \gamma h \leq \frac{\sigma_c}{n}$$

where:

γ = unit weight of overburden;
h = depth of the pillar;
σ_c = strength of rock salt cube ($W/H = 1$);
n = safety coefficient.

When the dimensional ratio of the pillar is different from the foregoing equation, this is written as follows:

$$\sigma_V = \frac{\sigma_p}{1} \sqrt{\frac{W_2}{H}}$$

where:

    σ_p = strength of rock salt parallelogram ($W/H = 1$);
    W_2 = width of the central part of the pillar;
    H = height of the pillar.

From this relationship the strength of the central part of the pillar has been formulated, which could be rewritten as follows:

$$W_2 = \frac{nH\gamma h}{\sigma_c^2}$$

Including this expression in the equation of the external pillar width, the following equation is obtained:

$$W_1 = H \left(2 \cot \phi + \frac{n^2 \gamma^2 h^2 H}{\sigma_c^2} \right)$$

Some authors consider that room-and-pillars in rock salt mines stay stable for a long period of time. In this case the most suitable assumption is that vertical stress acts on the total length of the pillar, not just on its internal part. The equation can be rewritten as follows:

$$W_1 = \frac{n^2 \gamma^2 h^2 H}{\sigma_c^2}$$

With this equation the average pillar stress distribution is the same as in the case of calculation by the other formula.[14]

Stamatiu's formula for pillar design is primarily used for the layout of pillars with a height which is greater than 12 m. For example, on the basis of this formula pillars with heights between 30 and 50 m have been designed in Romania. Actual mining practice, however, has suggested that the rectangular connection between the roof of the room and the pillar was unstable due to high stress concentration. Stress analysis of the 50 m high pillars suggested that they must be redesigned by changing their upper 20 m to a more inclined shape with a slope of 30° from the vertical axis. This resulted in the formation of the trapeze rooms.

13.2.3 *Multi-level long pillar structure*

The method of design of multi-level room-and-long-pillar mining systems in the salt deposits has been developed by Salustovic and Dunikovski (Figure 13.2.3).[15] The relationship between vertical pillar stress and strength of the pillar can be formulated as follows:

$$\sigma_V = \gamma h \frac{W + w}{W} = \frac{\sigma_c}{n}$$

where:

    W = width of pillar;
    w = width of room;
    σ_c = strength of salt ($W = H$).

Taking into consideration an empirical strength relation of the geometrical forms of salt samples, it can be written as follows:

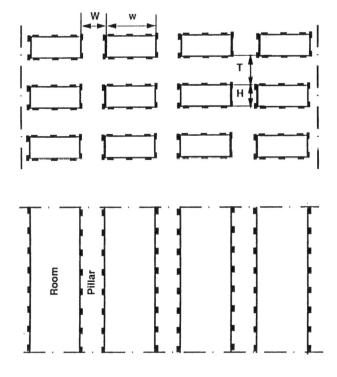

Figure 13.2.3. Layout of multi-level room-and-long-pillar system

$$\sigma_p = \sigma_c \sqrt{\frac{W}{H}}$$

where σ_p is the strength of rock salt parallelogram ($W \neq H$).

Then the equation of the relation between the stress of the pillar and the strength of the pillar could be rewritten:

$$\gamma h \left(w + \frac{w}{W} \right) = \frac{\sigma_c}{n} \sqrt{\frac{W}{H}}$$

or, after transformation, as:

$$\frac{w}{H} = \frac{\sigma_c}{n \gamma h} \left(\frac{W}{H} \right)^{3/2} - \frac{W}{H}$$

For stress analysis of the sill pillars it is necessary to consider a horizontal stress:

$$\sigma_H = \frac{\gamma h}{m-1} \left(\frac{T+H}{T} \right)$$

where:

   T = height or thickness of sill pillar;
   m = Poisson's number.

Also the conditions of strength have to be satisfied:

$$\sigma_H = \frac{\sigma_p}{N}$$

After taking into consideration an empirical strength relation of the geometrical forms of pillars, it is written as:

$$\sigma_p = \sigma_c \sqrt{\frac{T}{H}}$$

Then the equation for the sill pillar is obtained:

$$\frac{\gamma h}{m-1}\left(1 + \frac{H}{T}\right) = \frac{\sigma_c}{n}\sqrt{\frac{T}{w}}$$

After transformation, this can be written as:

$$\frac{w}{H} = \frac{1}{m - 1\dfrac{\sigma_c}{n\gamma h}\left(\dfrac{T}{w}\right)^{3/2} - \dfrac{T}{w}}$$

In addition to horizontal stresses the sill pillar is exposed to bending stresses due to the weight of the rock strata, which mobilize a shear stress at the extension of room walls. Having in mind the appreciable thickness of sill pillars in relation to their span and consequent high rigidity of bending, the bending stresses could be neglected. It is possible to formulate conditions for shearing of the sill pillar in the following equation:

$$\tau = \frac{\gamma_s w}{2} = \frac{\sigma_s}{n}$$

where:

γ_s = unit weight of the sill pillar;
w = width of room;
σ_s = compressive strength of sill pillar;
n = coefficient of safety.

From the foregoing equation it can be seen that the width of the room is limited by the shear strength of the sill pillar:

$$w = \frac{2\sigma_s}{n\gamma_s}$$

It is necessary to determine the optimal dimensions of the rooms, inter-room pillars and sill pillar, which is controlled by the extraction factor:

$$e = \frac{wH}{(1 + W)(T + H)}$$

$$e = \frac{1}{\left(\dfrac{W}{H} + \dfrac{w}{H}\right)\left(\dfrac{T}{w} + \dfrac{H}{w}\right)}$$

The functional dependent of the dimensions of room, seal pillars, inter-room pillars and the factor of rock salt recovery in the case of multi-level room-and-long-pillar mining is diagrammatically illustrated in Figure 13.2.4. The diagram of ore recovery does not give the functional dependence in regard to the mine depth. It is rather based on the geometrical relationship among the dimensions of individual mine structures.[15]

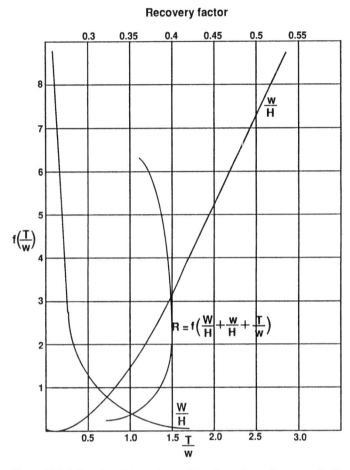

Figure 13.2.4. Functional dependence of the dimensions of room, seal pillars, inter-chamber pillars and the factor of rock salt recovery (multi-level room and long-pillar system)

13.3 STABILITY EVALUATION OF MINE STRUCTURES

From a rock mechanics point of view, instability is defined as a loss of structural integrity, when the rock mass is disturbed by the operation of mining excavation. In the case of thick rock salt mass, with a large excavation of the rooms it is necessary to leave in place large pillars to control mine stability. Strangely enough, new stability analysis indicated that ground control is more effective by yielding pillars, as discussed in Chapter 12. Further consideration on stability evaluation of mine structure is given by the following four topics.

13.3.1 *Deformability characterization of structure*

Deformation and failure of the rooms and pillars are due to the impossibility of maintaining adequate structural stability of the rock salt mass, which was in a state of equilibrium before excavation. The local instabilities of the large rooms could adversely effect mining

operations (Figure 13.3.1). Due to the large size of the opening, the deformation characteristics are more important for the thick salt bed mining, than in the case of mining
moderately thick and inclined salt deposits. As discussed in previous chapters, the
instability of rooms and pillars could be related to two phenomena:

– Roof falls, pillar slabbing and floor heave occurring in association with microscopic
fracturing of the rock salt mine structure. Prevention of rock falls requires internal ground
support.

– Continuous closure occurring in association with the time-dependent deformation of
the rock salt, which is ductile or/and viscous behaviour. To control closure it is necessary to
place mine fill in the room.

At the present time there is extensive research in regard to characterization of deformability of mine structure with an aim of stability prediction. The current model analysis is
still very limited to a reasonably reliable qualitative estimation of the large room stability.
There is an agreement that characterization and control of room stability are best done by
empirical rules ('rules of thumb') which have evolved over decades of underground salt
mining, as discussed below.[16]

1. *The bedded salt deposits*, where the stability of rooms is controlled by stratification
structures, have the following rules for maintaining stability

a) At least 1 m of solid salt should be left in the roof to avoid roof slabbing.

b) For a given room height, larger pillars will suffer less slabbing than smaller pillars
(i.e., pillar slabbing is a function of the pillar width-to-height ratio).

c) In laying out pillar-entry configurations, an entry should not end perpendicular to a
pillar, unless both have been designed for potential roof problems.

d) If roof bolts are going to be installed (as in a permanent roadway), they should be

Figure 13.3.1. Fragmented rock salt on the room floor as a result of falls from the roof

installed following the excavation face as close as possible and no more than one room width back.

e) Floor heave problems can usually be eliminated by cutting slots in the floor at the pillar edges.

2. *The dome salt deposits*, where the stability of rooms is controlled by folded and massive structures, have the following rules for maintaining stability:

a) No opening, including internally drilled core holes, should approach the salt body boundaries less than 100 m; the preferred limit is two times that distance. Distances from the top of the salt should be even greater.

b) Anomalous areas in domes areas salt cause leaks in the shallower zones and gaseous outbursts in the deeper zones. Do not use a continuous miner in anomalous areas. It may trigger blowouts at unforeseen times.

c) Room heights should not exceed 30 m in domed deposits. One major difference between openings in bedded and domed salt bodies is the room height. In bedded salt, rooms are generally up to 20 m high, while in domed salt they are in some cases up to 50 m high (Romania).

The prediction of the creep fractures that are characterized by prolonged periods of slow crack growth before final accelerated fractures and rock falls is a most important factor for ground control.

The monitoring and prediction of brittle fracturing, which occurs rapidly due to unstable crack growth, is difficult. They might be of lesser importance because they are of limited occurrences. However, the prediction of the creep fractures that are characterized by prolonged periods of slow crack growth before final accelerated fracture and rock falls, is of great importance, because it is the predominant mechanism of mine instability. Creep deformation is controlled by monitoring of room closure. In the case of large displacements, their possible cause must be found, in order to solve the problem before it is too late, e.g. before accelerated creep reaches its maximum.

13.3.2 *Evaluation of room deformability*

Evaluation of room deformability is typified by the case of history of convergence

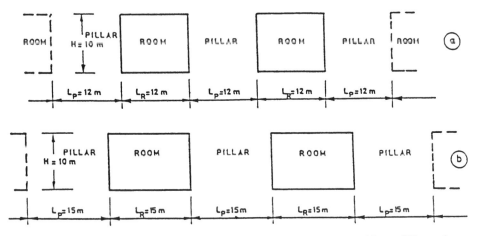

Figure 13.3.2. Bi-level mine section with layout of the room-and-pillar system with two different sizes

monitoring at the Tusanj rock salt mine (former Yugoslavia). The geological characteristics of this mine are discussed in Chapter 4 and in this chapter in Subsection 13.1.6. The monitoring of the room displacement took place at both levels of bi-level mining (Figure 13.3.2). At the level of 190 m, the rooms are excavated with dimensions of 10 m wide and 10 m high; the pillars have the same size. However, at the level of 150 m the experimental rooms are excavated with dimensions of 15 m wide and 10 m high; the pillars have the same size. The measuring displacement stations have been located along the longitudinal axis of the room at 20 m apart. The station has eight measuring points across the room section (Figure 13.3.3). Measuring point consist 1.5 m long anchor, which is placed and grouted in the hole 1.3 m long. The measuring angles of horizontal and vertical displacement have been taken by the instrument Theo 010 Carl Zeis Jena and displacements have been measured by Theo 02 Carl Zeis Jena. The time lag between measurements was approximately six months. The displacements of each point are plotted in a three-dimensional system. The representation of the results is given by three axis vectors and by two displacement vectors (horizontal and vertical).[17]

The convergence of the room (200 m long at 190 m level) has been interpreted by the resultant displacement vector given as the connection of the point from which measurements were taken with the end of the vertical component of the displacement (Figure 13.3.4). Though there seems to be no general regularity in displacements when comparing the displacements of each separate profile or with respect to other room profiles, a better observation allows the conclusion that there is a general increase in displacements as one goes deeper into the room. This is to say that the displacements of the first profile at the beginning of the room appear to be notably smaller than those of the last profile measured.

The convergence of the other room, 1230 m long at 250 m level (500 m depth), showed that displacement components as well as the resultant displacement of the cross-section profiles are greater, compared with those of the room on -190 level depth (Figure 13.3.5). The displacement measurements are carried out in all rooms in the same time period.[10]

The isometric presentation of monitored data for one or more series of measurements makes it possible to construct all vector component displacements. From the projection of resultant displacements it is possible to represent the room convergence, as given in Figure 13.3.6. On this cross-section, the displacement is expected to be nearly perpendicular to the room walls, roof and floor. Figure 13.3.6 offers the concept of the room deformation, namely: heave of the room floor, sag of the room roof, differential vertical displacement of the room walls (left downward and right slightly upwards). Also vector displacement representations indicated that cross-sectional measured profiles have rotation around longitudinal axes.

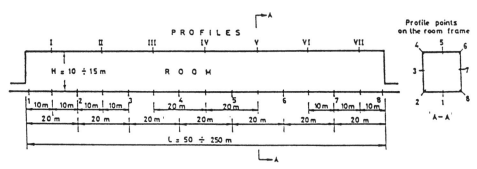

Figure 13.3.3. Longitudinal section of the room and cross-section of measuring points

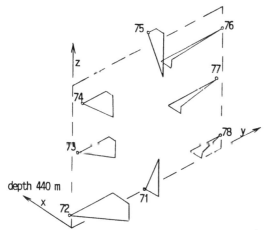

Final mine depth   450 m
Time period of displacement   1 year

Figure 13.3.4. Characteristic section with displacement vectors of one measurement at a level of 440 m

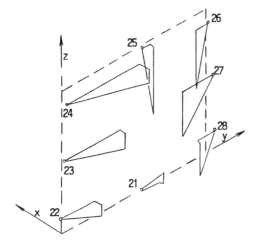

Final mine depth   500 m
Time period of displacement   1 year

Figure 13.3.5. Characteristic section with displacement vectors of one measurement of a level of 600 m

It should be noted that the mechanics of room deformability at the Tusanj rock salt mine is governed to some degree beside time-dependent deformations also by the following natural and geotechnical factors:[18]

1. differential stress state in the rock salt mass prior to excavation;
2. changes of geological-structural features on the short distances along the longitudinal axes of the rooms;
3. different properties and behaviour of individual beds along profiles;
4. size of the rooms and their layout with their spacing;
5. the speed of the rock salt excavation and stoppages during mining;
6. seismicity due to cyclic extraction by drilling and blasting.

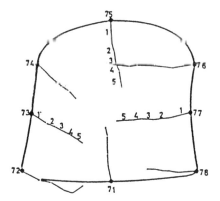

TIME DISPLACEMENTS

1 - 12 months
2 - 6   "
3 - 7   "
4 - 5   "
5 - 11   "

Figure 13.3.6. Projection of the resultant vectors displacement on the cross-sectional plane

However, with the above statement we should not forget the important influence of time on the progress of the room deformability. Figures 13.3.7 and 13.3.8 display the diagrams of displacement of the central part of the room floor, room roof and room walls (four points measuring profile). The measuring in the room at a depth of 440 m. The room size is 10 m high by 10 m wide by 180 m long. All diagrams of vertical and horizontal displacement have a similar trend of deformations, which suggests that a convergence functions does not change as it progresses. The influences of natural and geotechnical factors are important elements of the convergence deformation with time.[19]

It the sill pillars have a large thickness (60 m), the interaction mechanics of multi-level room mining is not in effect. In the case of lesser thickness the interaction mechanics is present, as discussed in Subsection 13.1.6.

13.3.3 *Time-controlled stress relief technique*

This topic has been analyzed in great detail in Chapter 12 for moderately thick bed mining. In this subsection the time-controlled stress relief is represented by the case history of the Goderich salt mine, Southern Ontario. In this mine the thick salt beds were mined by a square pillar system, which had been changed over to the ribbon pillar mining, implementing the time-controlled stress relief technique. The methods of mining at the Goderich salt mine are described in Section 13.1.

The old mining system by square pillars ground stability deterioration was experienced by sloughing and slabbing the room walls and roof. Underground observations showed that the intensity of deterioration, however, was expected to slacken due to the geometry change of the pillars caused by the wall failure itself. This stabilization trend could be undermined by continuing propagation of the roof failure arch. By adoption of Serata's concept of the time-controlled stress relief technique,[2, 20] the problems of stability of the underground mine structures have been largely solved. For example, first investigations in test area No. 1 showed a high degree of roof strata stability in the central of 'protected' room (89W), which

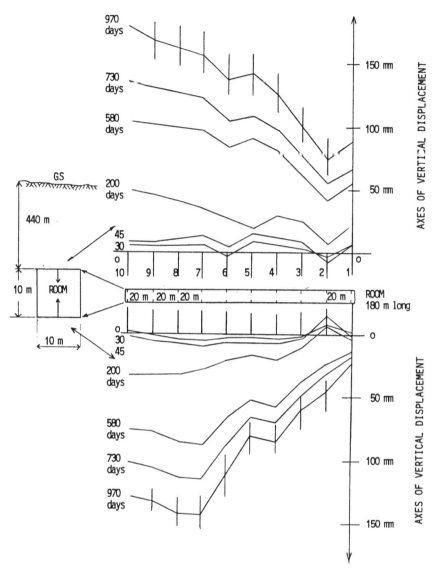

Figure 13.3.7. Vertical displacement of the room along its central axes

has been excavated after the excavation of the two flanking 'control' rooms (99A and 90), as illustrated in Figure 13.3.9. After this three-room experiment, the intersections formed by the meetings of two mutually perpendicular multiple-room entries were investigated. Two such intersections of three-room systems were formed in test area 2, as illustrated in Figure 13.3.10. The long-term results of these experiments demonstrate that for the design geometries used, the protected rooms are even more stable at intersections than away from them, in spite of the much wider roof exposures within the intersection zones.

On the basis of data of the experiments and analysis a new layout of the mining system has been devised, as shown in Figure 13.3.11. The stress analysis of this layout showed that the protective stress envelope is formed around a three-room entry system. The major thrust

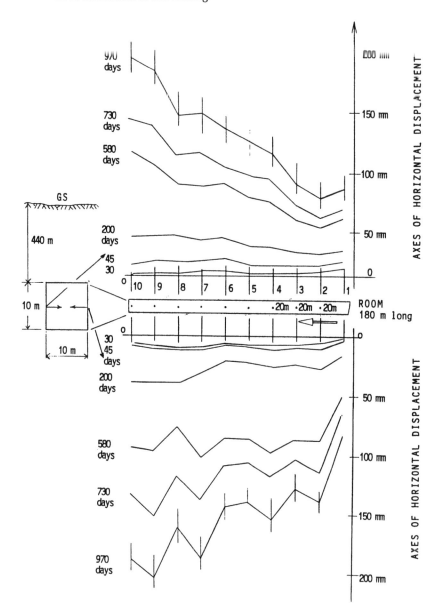

Figure 13.3.8. Horizontal displacement of the room along its central axes

of the stress envelope goes into the hard shale strata, leaving the immediate salt roof beds destressed. Observations confirmed that the centre room (protected room) is very stable, but the two outside rooms suffer large stress concentrations, which results in extensive damage. The centre room is virtually free from creep strain accumulation, even after ten years.

The underground investigations of the centre (protected) room and flanking (control) rooms as well as inter-room pillars and abutment pillars indicated the following:

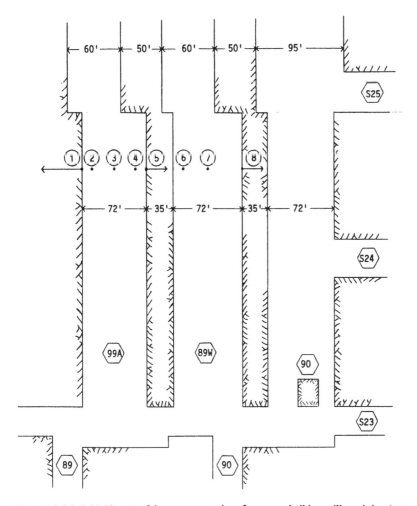

Figure 13.3.9. Initial layout of three-room entries of room-and-ribbon pillar mining (test area No. 1)

1. The convergence of the roof beds (7.5 m thick) of the centre room was less than 3 cm over 500 days. The small convergence of the stope span suggests that the room roof is destressed and that it is a stable structure.

2. The convergence of the roof beds (7.5 m thick) of the outside rooms was 20 cm for a period of 500 days. This is in agreement with the deformation and failure of the observed outside rooms. In all the outside multiple entries, bed separation is exhibited almost immediately from the time of excavation. This occurs in a relatively consistent pattern by separating an immediate roof slab of 3.3 m thick along the bedding planes. After bed separation, no progressive roof failure or rising arch of failure is anticipated.

3. The inter-room, 15 m wide yielding pillars performed satisfactorily, both in creating the desired stress envelope and in providing vertical confinement against destruction of the overlying shaley dolomite strata. The pillars, however, suffered more wall spalling than in the case of the old mining system. In spite of the deteriorating appearance of yield pillar walls, the core of the yield pillars remains ductile without losing its supporting capabilities.

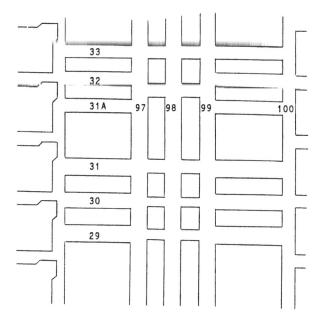

Figure 13.3.10. Advanced layout of three-room entries of room-and-ribbon pillar mining (test area No. 2)

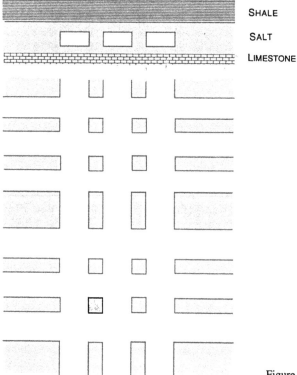

Yield pillar width	15 m
Room width	13 m
Abutment pillar width	46 m

Figure 13.3.11. Design of a three-room entries system utilizing time-control technique for ground control (Goderich salt mine, Ontario)

The maximum scaling of pillar walls is at their mid-height and is not more than 3.5 m (hour glass pillar failure).

4. In the case of the time-controlled stress relief concept, the abutment pillars support the entire overburden load. Their deformation is similar to yield pillars. The wall spalling of the abutment pillars is found to be less frequent than that of the yield pillars. The maximum scaling of pillar walls at their mid-height is in the range between 1.5 m and 3.0 m.

The time-stress controlled technique considered also a mine layout with a four-room system with two central protected rooms. Production panels could be designed by six to nine room panels.

Serata's concept of the room-and-ribbon-pillar mining system is based on the multiplicity of small stress envelopes which have been formed around all of the individual rooms. In spite of the stability of the overburden formations provided by this support, the overburden formation is deforming slowly but steadily, due to the long-term strength within these small stress envelopes. Therefore, a definite amount of subsidence, not exceeding 30 cm at maximum (as indicated by model analysis) over the mined-out area, is now expected on the ground surface.

The surface subsidence has reached a point at which the full weight of the overburden is loading the central portion of the mined-out area. The magnitude of the lake bottom bed subsidence can be calculated from creep deformation of the underground. The calculation uses the basic 'balance equation' which relates surface subsidence to the measured components of the underground deformation, as illustrated in Figure 13.3.12.

With the implementation of the time-controlled stress relief technique, of course the

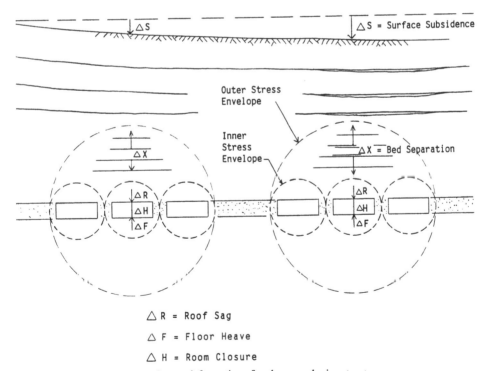

$\triangle$ R = Roof Sag

$\triangle$ F = Floor Heave

$\triangle$ H = Room Closure

Figure 13.3.12. Total balance of strata deformation of underground mine structure

need for ground support is decreased but it is not eliminated, because in some instances local ground support is required. In this case 1.2 m and 4.2 m expansion shell, rock bolts are used. Typically, a 1.5 m by 1.5 m pattern of rock bolt installation is employed.

13.3.4 Interaction bi-level room mining

The interaction mechanics of multiple-level room-and-pillar mining is given by Schmitke in his case study of the Cory potash mine in Saskatchewan.[11]

In Subsection 13.1.6 the bi-level room-and-pillar layout is presented, which is described as follows. The A zone of the mine layout (upper level) employs the time-controlled stress relief technique and allows efficient mining of this zone. As shown in Figure 13.3.13, the stress envelope developed during mining of the A zone encompasses a large area of stress relieved ground below the existing workings. The formation of the stress envelope, sufficiently below the B zone level, allows mining in this zone without requirement of additional stress control methods. Of course this method of strata control stability in the lower level (Zone B) can be achieved when the mining in the first level (Zone A) is completed.

Schmitke analyzed the data of the ground stability monitoring program, with the main objective to assess the stability of the lower level (Zone B) and to provide a comparative analysis with an upper level (Zone A) protected entry. He presented a comparison of horizontal creep rates for B zone and A zone entries (See Figure 13.3.14).

The comparison indicates that for the A zone the horizontal movement occurs immediately after excavation and is approximately four times greater than the initial creep rate experienced in the B zone excavation. Although the long-term closure rates appear to be approaching equivalent values, the stable horizontal creep rate for the B zone excavation is reached in about one-half of the time required to obtain a stable horizontal creep rate in the upper level of the excavation (A zone). Due to the difference in widths of the entries and in order to facilitate a direct comparison of A zone and B zone excavations, the vertical creep rates were compared based on the span of the excavation (Figure 13.3.15).

Similar to the horizontal creep rate relationship, the vertical creep rate comparison illustrates the vertical movement of the A zone immediately after excavation. Also the magnitude of the initial A zone vertical creep rate is significantly larger than the vertical creep rate experienced in the B zone excavation. The vertical creep rates for both excavations approach equivalent values after approximately 100 days; thus indicating an equivalant long-term response within the stress envelope.

In Figure 13.3.16 a comparison of a vertical stress level is presented for various pressure

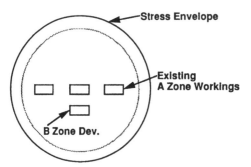

Figure 13.3.13. Design concept of bi-level mining within protective stress envelope

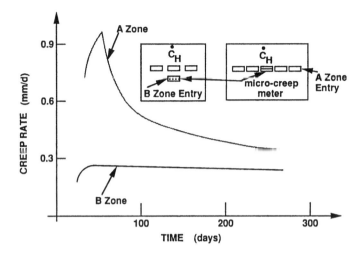

Figure 13.3.14. Comparison of horizontal creep rates for B zone and A zone entries

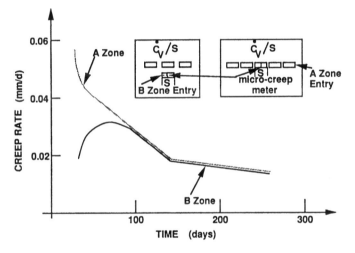

Figure 13.3.15. Comparison of vertical creep rates per unit span for B zone entries

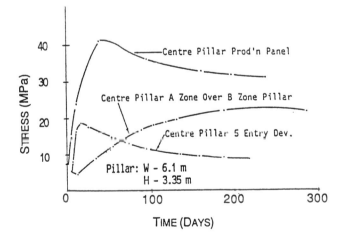

Figure 13.3.16. Relative stress level comparison for vertical stress

cell locations. The comparison indicates that prior to the B zone excavation, the A zone pillar over the eventual B zone pillar retained a minimal stress concentration (1.3 MPa). However, as the B zone excavation commences, the stress level in the pillar, after approximately 220 days since the illustration of the pressure cell, gradually increased to 21 MPa and after 300 days the stress level had decreased to 20 MPa. Thus the stress envelope formed during the A zone mining has undergone a transformation, and a stress redistribution has occurred within the previously stress-relieved ground. The comparison further indicates that the rate of stress redistribution within the stress envelope, as a result of the B zone mining, is significantly lower than the rate of vertical loading experienced by the A zone excavations during the formation of the stress envelope. In addition to the lower rate of loading, the B zone excavations experienced a stress level of approximately one-half of the amount found in the centre pillar of a production panel and the stress redistribution resulting from the B zone excavation was comparable to the vertical stress encountered in the centre pillar of the five-entry developement system.

Table 13.3.1 outlines the vertical creep rates for an A zone entry (age approximately ten years) directly over a B zone entry, and an A zone entry (age approximately ten years) in a stable four-entry system, and a B zone entry (age approximately 100 days) below the location where the A zone creep rate was monitored.

The data presented in Table 13.3.1 indicate that an A zone entry (age ten years) above a B zone entry experiences a vertical creep rate of 0.1444 mm/day and is comparable to the vertical creep rate of 0.126 mm/day experienced in a stable protected entry (age ten years) of a four-entry system. Also, although there is a significant difference in the ages of the entries, the vertical creep rate of 0.099 mm/day measured in the B zone entry is less than the vertical creep rate encountered in both A zone locations.

At the end of his paper Schmitke concluded the following in regard to the stability of bi-level mine structure:

1. The B zone excavations offer increased stability when compared to A zone excavations of similar age. The B zone excavations are not subjected to the high or rapid initial loading as encountered in the A zone and do not experience the high initial creep rates as are prevalent in the A zone excavation.

2. As a result of the B zone mining below existing A zone workings there is a stress envelope transformation. In order to maximize the stability of a B zone excavation below existing A zone workings, the A zone entries must be of sufficient age to ensure that the stress level within the stress envelope, prior to a B zone excavation, has dissipated to a magnitude which can accommodate any stress redistribution induced by the B zone mining. This is particularly important if the A zone protected entries are required for continuing or future mine activities.

3. The A zone entry, located above a B zone excavation, has not to date been adversely affected by the B zone mining. The vertical creep rate of the A zone entry above the B zone

Table 13.3.1. Vertical creep rates effected by undermining of upper level (A zone) by lower level (B zone)

Monitoring location	Creep rate
B zone entry (age approx. 100 days)	0.099 mm/day
A zone entry (age approx. 10 years)	0.144 mm/day
Above the B zone creep rate monitor station A zone entry (age approx. 10 years)	0.126 mm/day

excavation is comparable to the vertical creep rate encountered in an Á zone entry of similar age.

Schmitke's work clearly indicates the character of interaction mechanics of bi-level mining, but also the utilization of this for stability control of mine structures.

13.4 STRESS ANALYSIS OF MINE STRUCTURES

In this section, the consideration of stress analysis of the room-and-pillar salt structure is given only by mathematical modelling. Stress analysis has been chosen for three particular situations. Firstly, the critical stress distribution of the room-and-pillar structure has been analyzed for different mine depths. Secondly, analysis of magnitude of stress distribution around room for three descending sequences of room extraction, and thirdly stress analysis has been implemented for multiple-level mining in the case of the gradual decrease of pillar width and the corresponding increase of the room width. Further discussion of these three particular rock mechanics problems follows.

13.4.1 *Analysis of stress concentration due to mine depth*

The stress analysis has been considered for trapezoidal room-and-pillar mining. The room layout is presented by the trapezoidal cross-section in Figure 13.4.1. Rock salt excavation is performed by drilling and blasting. The analysis of stress concentration of the trapezoidal room structure has been conducted at the Geotechnical Research Centre of Laurentian University, Sudbury (Canada) as a part of the research project on rock salt strata control.

The analysis of the stress distribution and concentration in the final phase of room extraction has been completed by the finite element method. The three levels of rock salt mining for digital stress modelling have been considered as follows:

$h_1 = 250$ m ($\sigma_V = \sigma_H = 5.3$ MPa)
$h_2 = 500$ m ($\sigma_V = \sigma_H = 10.6$ MPa)
$h_2 = 750$ m ($\sigma_V = \sigma_H = 15.9$ MPa)

As indicated above, the primitive stress of the salt deposit is assumed isostatic, with an increase of 5.3 MPa for each 250 m depth. The elastic parameters of the rock salt used for model analysis are listed as follows:
– compressive strength of rock salt, $\sigma_c = 30$ MPa;

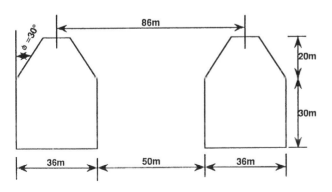

Figure 13.4.1. Cross-section of the trapezoidal rooms, analyzed by finite element analyses

– modulus of elasticity of rock salt, $E = 12$ GPa;
– Poisson's number, $m = 4$ ($v = 0.25$);
– pillar strength with safety factor $\sigma_P = 2/3\ \sigma_c = 20$ MPa.

On the basis of the stress distribution and concentration plots, the following may be concluded:

1. At 250 m depth, the critical stress concentration of the room-and-pillar salt structure is not present, because the induced mining stress does not exceed the pillar strength of 20 MPa. At this level the ground failure is not expected because the surrounding rock salt of the rooms is stable enough to support the superimposed load.

2. At 500 m depth, critical stress concentration around a room-and-pillar structure is present. The stress concentration is mainly located at angular connections of the geometrical elements of the mine structure. It should be noted that the main stress redistribution from the excavation is rather to the surrounding rock salt than within the structure itself. Of particular interest to structural stablity are the critical stress occurrences in the areas of the inclined (trapezoidal) room roof.

3. At 750 m depth, critical stress concentration is extensive and is located both inside the mine structure (inter-room pillar) and outside the mine structure (surrounding salt). The stress redistribution pattern around rooms is a result of interaction stress concentration (overlapping stresses), which causes the intensive and extensive stress concentration zone.

The results of the mathematical model stress analysis are in general agreement with the thick rock salt mining by large rooms as analyzed in this case. In the case of cyclic mining the walls and roof of the room are stress-relieved due to crack formation during drilling and blasting.

The toe of the rooms exhibited stress at failure, at 500 m depth and 750 m depth. The reasoning of this is that if one extends the arches on either side, the arch will cross through the toe. The stress redistribution pattern will be dictated by arch geometry through the toes of the room.

The stress analysis clearly indicated that the dimensions of the rooms and pillars are satisfactory for 250 m mine depth, but not for 500 m and 750 m mine depths. Mining practice suggests that large room mining such as trapezoidal rooms is limited to 300 m depth. The great majority of these rooms are mined between 150 m and 250 m depth. The increase in size of the pillars and decrease in size of the rooms could facilitate some deeper mining. However, it does not make sense because it will require further decrease of the extraction ratio, which is already low at the present mining depth (R = 25%).

The solution for deeper mining by this system might be found eventually by developing a suitable method of internal ground support or by implementation of time stress controlled technique. In mining circles, however, the opinion is that both solutions are not applicable for rooms of 50 m high, and that the solution of mining a deeper portion of salt deposits is by the change-over to the multi-level room-and-pillar mining system with a room height not exceeding 20 m.

13.4.2 *Analysis of sequential distribution*

Here, the stress analysis has been conducted for three separate sequences of trapezoidal room excavation. The sequential stress analysis have been computed at the Geotechnical Research Centre of Laurentian University, Sudbury (Canada) as a part of the research project on the design of salt mining methods.

The stress distribution for three different sequences of trapeze room excavation has been analyzed by a boundary element method utilizing the program 'EXAMINE', which was developed at the University of Toronto. The analysis of stress distribution has been only along the cross-section for all three sequences of stope extraction. The distribution of major principal stress (σ_1) and minor principal stress (σ_3) had been obtained and plotted by computer, as well as the safety factor using Mohr-Coulomb failure criteria. The stress analysis has been conducted for the following geotechnical conditions:

– In situ virgin stress: $\sigma_1 = \sigma_3 = 12$ MPa (isostatic stress).

– Rock salt strength: $\sigma_T = -2.0$ MPa; $\sigma_n = 25.0$ MPa; $\phi = 30°$ (angle of internal friction).

– Elastic constants: $E = 10.0$ GPa (modulus of elasticity); $v = 0.2$ (Poisson's coefficient).

– Symmetry: none.

For all three sequences of the mining there are two diagrams of stress distribution and concentration and one diagram of the safety factor. The results of the model stress analysis are given individually in descending order of sequential mining.

1. *The first sequence of mining* completes the excavation of the 20 m high trapez on which the 2.5 m high rectantular excavation is superimposed, as illustrated in Figure 13.4.2. Each of the three diagrams are separtely described.

a) The major principal stress distribution and concentration are presented in Figure 13.4.2. The diagram clearly indicated that the critical stress concentration is on top of the room, just above the rectangular excavation. This corresponds to the mining situation exhibited by the possibility of roof weakening. In some mines a wooden gallery is installed around rectangular excavation so that roof stability could be monitored and weak rock salt

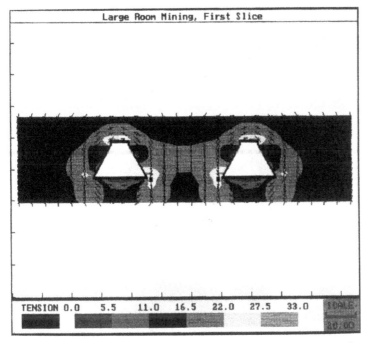

Figure 13.4.2. Major principal stress distribution (σ_1) of the first sequence of trapeze room mining

blocks scaled, as already mentioned in Subsection 13.1.4. The formation of stress concentration, in the pillar between rooms in the vicinity and below the room triangle, should also be noted.

b) The minor principal stress indicates the compressive stress concentration in the pillar in the vicinity of the room triangle, but with lesser extension than in the case of major principal stress (Figure 13.4.3). Tensile stress develops in the lower room walls and room floors in the flanked area of structure. From the present mining practice could be concluded that if tensile stress was in existence, it has been of lower magnitude, e.g. below tensile rock salt strength. Also in this phase of mining the walls and floor of the room are confined by the muck of blasted rock salt.

c) The safety factor distribution indicates the lowest value in the area of concentration of critical principal stresses (Figure 13.4.4). However, the diagram also indicated that the lowest safety factor is not less than 1, which suggests that critical stress concentration is lower than or equal to rock salt strength.

It could be concluded that stability of the room during the first phase of extraction is generally satisfactory, and if some instabilities occurred it could be controlled. However, special attention should be given to control of the roof of the room.

2. *The second sequence of mining* completes the angular room excavation below a trapez portion of 20 m high. The diagrams of stress distribution and safety factor are individually discussed.

a) The major principal stress distribution and concentration are represented in Figure 13.4.5. The maximum stress concentration takes place at the structure with angular geometry, which should be expected. The critical stress at failure could be expected at the corners of the room floor. However, in a mining operation this high stress concentration is

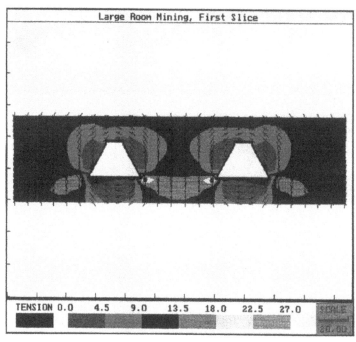

Figure 13.4.3. Minor principal stress distribution (σ_3) of the first sequence of trapeze room mining

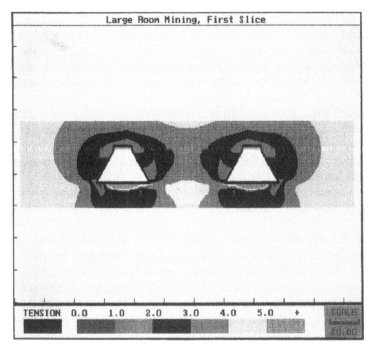

Figure 13.4.4. Safety factor distribution (F_s) of the first sequence of trapeze room mining

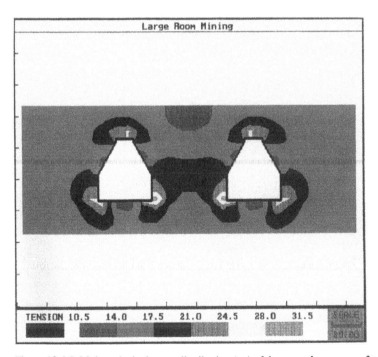

Figure 13.4.5. Major principal stress distribution (σ_1) of the second sequence of trapeze room mining

offset by the reacting force of the rock salt muck which fills up the room.

b) The minor principal stress distribution and concentration are given in Figure 13.4.6. The maximum compressive stress concentration is considerably below the strength of rock salt. Also, the tensile stress is not present in the rock salt mine structure. The stress pattern of the inter room pillar indicates two apexes of the walls of rooms, which overlap. Both apexes might be over 90°.

c) The distribution of the safety factor as shown in Figure 13.4.7 is satisfactory, because its minimum value ($F_s = 1.0$-1.5) is located at the room walls into the pillar In this area pillar failure could be behind the room wall surfaces.

Generally speaking, room stability after completion of the second sequence of mining could be considered satisfactory.

3. *The third sequence of mining* is the final phase of room excavation, which has been extended at 10 m below the previous excavation. Also the three diagrams are discussed as follows:

a) The major principal stress pattern is very similar to the second sequence of mining, as described above. It should be noted that in both sequences of excavation the stress is mostly transferred onto inter-room pillars rather than surrounded rock salt mass (Figure 13.4.8).

b) The minor principal stress does not have much similarity with the stress patterns of previous sequences. After completion of the third sequence of mining two apexes of the walls of the rooms do not overlap and they are located at appreciable distances along the vertical axis of the inter-room pillar. The stress pattern given in Figure 13.4.9 does not have an influence on structural stability, because the minor principal stress is relaxing as room excavation progresses.

c) The diagram of the safety factor is similar as in the case of the previous sequence of mining. From Figure 13.4.10 could be concluded that with the increase in the size of the

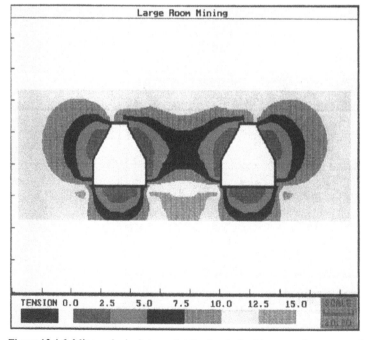

Figure 13.4.6. Minor principal stress distribution (σ_3) of the second sequence of trapeze room mining

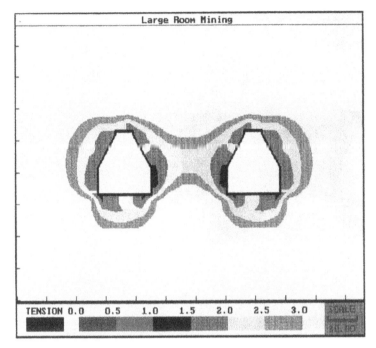

Figure 13.4.7. Safety factor distribution (F_s) of the second sequence of trapeze room mining

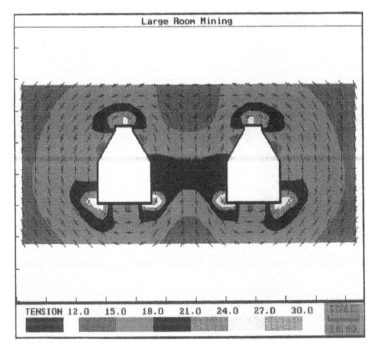

Figure 13.4.8. Major principal stress distribution (σ_1) of the third sequence of trapeze room mining

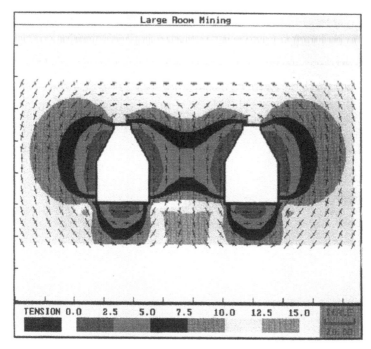

Figure 13.4.9. Minor principal stress distribution (σ_3) of the third sequence of trapeze room mining

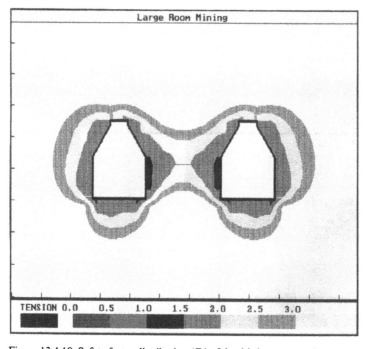

Figure 13.4.10. Safety factor distribution (F_s) of the third sequence of trapeze room mining

room, the safety factor is slightly less but not below $F_s \geq 1.5$, except on the room walls inside of the pillar, where it drops to $F_s \geq 1$. Of course, as mentioned before, in this outer pillar zone failure could be developed.

The third sequence of excavation of the room did not increase the stress concentration in the surrounding rock salt mass – on the contrary, stress has been decreased, as for example in the pillar core.

The results of stress analysis of the sequential mining of trapez rooms are in agreement with the results of stress analysis of the mining of trapez rooms at different mine depths.

13.4.3 *Analysis of stress for room-widening phases*

The stress analysis is based on the research conducted at the Tusanj rock salt mine (Tuzla, former Yugoslavia).[21, 22, 23] The main task of the analysis is to examine the possibility of the room widening and corresponding pillar width shortening with an aim to increase room salt recovery during a room-and-pillar mining operation. However, there is a strong requirement that with maximum economic gain it is necessary also to achieve maximum safety.

The analysis of this problem is based on the mathematical modelling of the pillars and rooms for various relationships with their widths. The stress analyses have been carried out by the finite element method in elasto-plasticity by the program PLANET.[24] In addition, in the Tusanj salt mine an experimental room K-2 has been established on a level of 230 m, to verify the results obtained by the mathematical model. The phases of room widening correspond to the room widths analyzed by the model. The finite element was formed so that computation of the critical stress was in the order of room widening 1 m, then 2 m, 3 m, etc. until the critical room size was achieved. The stress analyses were modelled for seven different widhts of the room.

The finite element mesh includes the complete cross-section of the eastern mining field, where the multi-level room-and-pillar layout is in effect. The room K-2 is on the periphery of the extraction block, the extension of which is in the direction of the solid rock mass (Figure 13.4.11). The mesh around room K-2 is densified, because this structure is the matter of main consideration.

For computation, Druker-Pragerov's conditions of failure are used. Definition of an elasto-plastic material is by non-linear analysis where plastic deformation permits larger displacement of rock mass material without failure. The function of deformation with the time is:

$$t_f(t_t, t_p, t_k) = 0$$

where:

t_t = strain with time;
t_p = total plastic deformation;
t_k = hardening parameter.

For conditions of isotropic hardening as in the case of rock salt material, t_k is a function of t_p, but for conditions of kinematic hardening, t_k could be considered constant (Figure 13.4.12).

The determination of the geotechnical model was on the basis of a certain number of physical-mechanical parameters of rock salt from the mine. The investigations have been carried out by testing numerous rock salt samples. The results could be summarized as follows:

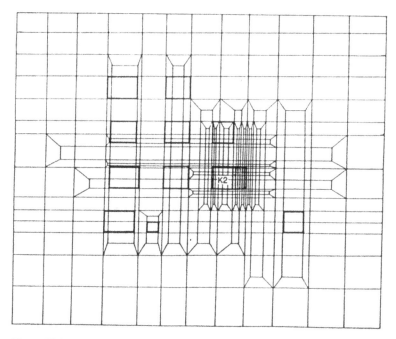

Figure 13.4.11. Final elements mesh of the cross-section of the eastern mining field, with position of room K-2 under consideration (Tusanj salt mine)

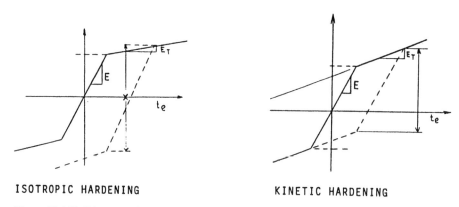

ISOTROPIC HARDENING   KINETIC HARDENING

Figure 13.4.12. Diagrams of elasto-plastic material hardening

- Density of rock salt: $\rho = 21\ kN/m^3$.
- Angle of friction: $\phi = 13°$.
- Cohesion: $c = 0.8\ MPa$.
- Modulus of elasticity: $E = 25\ GPa$.
- Poisson's coefficient: $\nu = 0.35$.
- Coefficient of hardening: $k = 29\ GPa$.

The modulus of elasticity as a coefficient of hardening is calculated from the secondary part of the curve after the boundary of elasticity and upon entering the region of material hardening. Because the model of elasticity in this region is a function of the applied lateral pressure of a trixial test, its determination was at a lateral load of 15 MPa. Due to the hardening properties of the rock salt, cohesion and friction are determined from the stress-strain diagram at 1% deformation.

The digital model computed the magnitudes of shear stress around rooms. The shear stress over 8 MPa is considered critical because it overcomes the shear strength of the rock

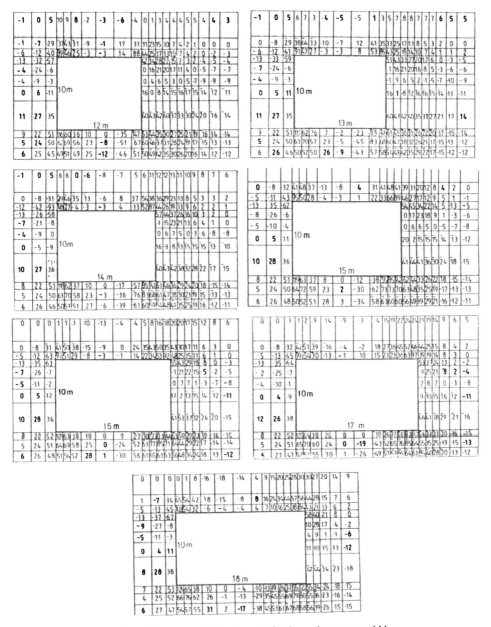

Figure 13.4.13. Magnitude of shear stress in the elements for the various room widths

salt. In Figure 13.4.13 the shear stress distribution is given around room K-2 for its phase increments between 12 m and 18 m width. The shear stress is represented numerically in whole numbers (100 kPa). The shear stress concentration takes place at corners of the rooms of various widths. The concentration of the stress is greater at the room floor corners than at the room span corners. The model analysis indicated that the shear stress distribution as well as shear stress concentration do not particular change in the case of room widening, except its elevation. Also, the principal stress analysis indicated that the changes of the room widths do not influence the stress state around the rooms within the eastern mining field where multi-level room-and-pillar mining is in progress.

The determination of the plastic zone in surrounding rock mass of the room K-2 for each analyzed room width (from 12 m to 18 m) is represented in Figure 13.4.14. The plastic zone delineates the area in which ultimate strength of rock salt material is overcome and the rock failure could be expected. The plastic zones are determined on the basis of possible action of both the critical stresses (compressive and tensile).

The results of the mathematical model analysis are:

1. The stress redistribution from one phase of room widening to another does not show a drastic change, because the stress is altered gradually and continually.

2. The pattern of stress concentration is similar if not approximately the same in all phases of room widening.

3. The critical stress is located in the walls and floor of the room rather than in the sill pillar or inter-chamber pillar.

4. Plastic zones of rock salt material are present in all room corners.

5. Up to a room size of 10 m × 16 m plastic zones are developed only in the corners, but for the room sizes of 10 m × 17 m and 10 m × 18 m plastic zones are present in the room floor and the room span.

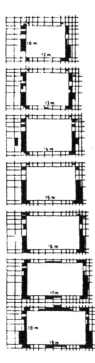

Figure 13.4.14. Distribution of the plastic zones around rooms of various widths

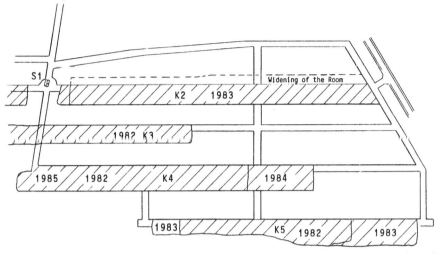

Figure 13.4.15. Layout of the level of 230 m with the room K-2 which was subject to the investigation (eastern mining district)

6. The extension of the plastic zone in roof and floor of a room size of 10 m × 18 m is very serious.

Development of plastic zones in the room span and its propagation in the walls of rooms suggest that widening of the rooms should be limited up to 16 m width. With this widening of the room it is possible to obtain from each room an additional 20 000 tonnes of rock salt (Figure 13.4.15). However, in this case it is necessary to establish a monitoring of the displacement of rock salt mass in the room's roof and room's walls to control stability. The monitoring should be continued after completion of the rock salt excavation from the room extension. To increase stability of the areas with maximum shear stress, it is necessary that all corners of the room are rounded with at least 1 m of radius of curvature.

REFERENCES

1. Muir, G. 1978. Salt mining at Goderich, Ontario.
2. Serata, S. 1982. Stress control methods: quantitative approach to stabilizing mine openings in weak ground, pp. 14-19. In *Proc. of 1st Int. Conf. on Stability of Underground Mining*, August 16-18.
3. Salazar, L. 1987. Methane occurrences in Southern Ontario's salt beds. Manuscript, pp. 1-23. Research Project, Laurentian Univ., Sudbury, Ontario.
4. Geyer, H. 1976. Weitere Fortschritte in der Abbauforderung des Steinsalzbergwerkes Borth. *Kali und Steinsalz*, 48-53.
5. Geyer, H. 1977. Neue Abbaugestaltung auf dem Steinsalzbergwerk Borth. *Kali und Steinsalz*, 205-214.
6. Jeremic, M.L. & J. Moravek 1964. Methods of development and exploitation of salt deposits in Poland. *Min. & Metall. Bulletin* (Belgrade), 2: 32-42 (in Yugoslav).
7. Sonolarski, A. & A. Litonski 1970. Technology of salt exploration in Poland. *Min. J.* 2 (745): 13-19 (in Polish).
8. Jeremic, M.L. 1964. Dry mining and solution mining of rock salt deposits in Romania. *Min. & Metall. Bulletin* (Belgrade) 12: 265-274 (in Yugoslav).

9. Jeremic, M.L. 1966. The mineability of the salt deposit Tusanj. *Geol. Bulletin* (Warsaw) 11: 8-12 (in Polish).
10. Mandzic, E. & M.L. Jeremic 1986. Time dependent deformations of large caverns in rock salt. In *Proc. of a Conf. on Large Rock Caverns,* Helsinki Univ. of Technology, Finland, August 25-28, 1986.
11. Schmitke, B.W. 1983. The application of in situ rock mechanics instrumentation for the comparative analysis of the bi-level mining system at Potash in PCS Mining, Cory division, pp. 291-297. In *Potash technology.* Toronto/New York: Pergamon Press.
12. ... 1980. *Die Kali Industrie in der Bundesrepublik Deutschland* (4th ed.), pp. 8-25. Hanover.
13. Slesarev, V.D. 1958. *Design of various inter-room pillars* (in Russian). Moscow: Mehanika Gornog Dela, Ugletehizdat.
14. Simic, D. & Z. Klecek 1989. Analysis of room stability. In *Principles of rock mechanics*, pp. 151-161 (in Yugoslav). Sarajevo: Posebna Izdanja.
15. Salustovicz, A. 1965. *Rock mechanics of the strata* (in Polish). Katovice: Slask.
16. McClain, W.C. & A.F. Fossum 1981. The evaluation of room stability, pp. 709-715. In *Proc. of 1st Conf. on Mech. Behaviour of Salt*, Penn State Univ., November 9-11, 1981. Clausthal: Trans. Techn. Publ.
17. Baturic, J. & E. Mandzic 1982. Monitoring of rock salt displacement in large rooms. *J. of Faculty of Mining and Geology* (Univ. of Tuzla) 11: 69-75 (in Yugoslav).
18. Mandzic, E. & M. Cvetkovic 1970. In situ determination of dynamic modulus elasticity of rock salt from Tusanj mine by micro-seizmic method. *Bulletin of Min.* (Institute of Mines, Belgrade) 10 (2): 1-10.
19. Mandzic, E. 1981. Time dependent deformations of the large openings in the rock salt mine, pp. 479-483. In *Proc. of Int. Symp. on Weak Rock*, Tokyo, 21-24 September 1981.
20. Serata, S. 1976. Stress control technique – An alternative to rock bolting? *Min. Engng.*, May 1976: 51-56.
21. Mandzic, E., M.L. Jeremic & M. Avdic 1987. Stress analysis of possibility of partial extraction of rock salt pillars in Tusanj salt mine, pp. 144-149 (in Yugoslav). In *Proc. of 16th Yugoslav-Polish Symp. on Techn. of Und. Min.*, Zenica, 23-25 September, 1978.
22. Mandzic, E. 1988. Investigations of exploitation of rock salt from the wall of the room K-2 on level 230 m, Tusanj salt mine, pp. 1-30 (in Yugoslav). Publ. of the Mining Geological Faculty, Tuzla Univ.
23. Mandzic, E. & M.L. Jeremic 1990. Rock salt mass creep around the large horizontal opening, pp. 28-34. In *Proc. of 4th Int. Symp. on Geomech. and Mining Construction Engineering at Great Depth*, Polish Acad. of Sciences, Gliwice, August 28-31, 1990.
24. Owen, D.R.J. & E. Hinton 1980. *Finite element in plasticity*, pp. 157-256 and 511-516. Pineridge Press.

CHAPTER 14

Mining of inclined salt bodies

The inclined salt bodies belong to folded salt strata or salt domes. Some salt bodies have a pitching morphology which is similar to vein deposits and consequently they are mined by the same mining method borrowed from hard rock mining. In this case the consideration of rock mechanics and stability analysis are similar to those in hard rock mining. As in the previous chapters on underground salt mining, the four basic topics will be discussed.

14.1 PRINCIPAL MINING SYSTEMS

The choice of mining method and mine layout depends a great deal on the form of the salt deposit, the angle of inclination, regularity, and continuity. In this regard three groups of salt bodies can be distinguished: regular veins (moderately dipping and steeply dipping), irregular veins (steeply dipping) and lenses (very steeply dipping). The following mining methods are briefly discussed.[1]

14.1.1 *Ascending slice mining*

The inclined room mining by ascending slicing is mostly applied for irregular veins with waste partings. The mining is similar to the cut-and-fill method in hard rock mining. The mine fill is composed by waste rock from underground as well as waste parting obtained during room excavation. The ascending slicing is synchronized with backfilling so that the artificial floor of the stope is formed, from which drilling and blasting are carried on. This mining method has been applied for some parts of the Wieliczka salt mine (Poland). This mine is located 12 km south of Cracov and it has been in operation since the middle of this millennium. The salt deposits in this mine are divided into the upper horizon (salt lenses – 'briles') and the lower horizon (salt beds – 'veins'), as shown in Figure 14.1.1. The ascending mining of inclined rooms is used only within a salt vein.[2]

The inclined room development is by two level drifts (upper – ventilation drifts, and lower – haulage drift) which are driven along a strike of salt beds. In addition, the up raises are excavated at centres of 20 m (Figure 14.1.2). The raises are excavated at the foot wall of the salt vein and they are maintained throughout the mine fill until stope completion. The raise in the mine fill is used as the ore pass and for ventilation.

The principal layout of this method is given in Figure 14.1.3 and can be briefly listed as follows:

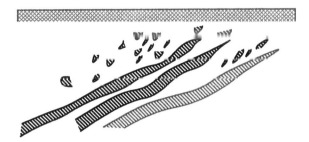

Figure 14.1.1. Longitudinal section of the Wieliczka salt deposits (Poland)

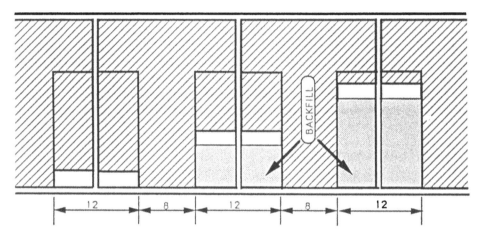

Figure 14.1.2. Layout of inclined ascending room mining with mine fill

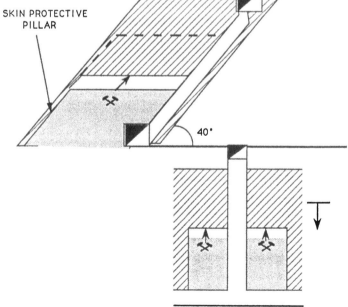

Figure 14.1.3. Plan and cross-section of inclined room with mine fill

Inclined room (stope)
- width: 12 m;
- length: > 25 m;
- height: 7 m.

Rib pillar
- width: 8 m;
- length: > 30 m;
- height: 7 m.

Seal pillar
- width: 4 m;
- length: 60-80 m;
- thickness: 7 m.

The room excavation is sequential. After completion of room over-cut, the salt extraction is from hanging wall to foot wall. A skin pillar of salt is left in the back to prevent roof caving. The length of the pillars and rooms varies considerably due to local geological and structural conditions. For example, the layout of this method for the Wieliczka salt mine under operating mining conditions could be complicated as shown in Figure 14.1.4.

14.1.2 *Descending slice mining*

This mining method is applied for both salt lenses ('briles') and salt beds which have a steep angle of inclination. There are several variations of this mining method, but here only two will be considered; one for regular salt veins and the other for salt lenses and irregular veins.

1. *Mining of regular beds* is mainly for salt beds with an angle of inclination between 30° and 60°. The descending salt excavation is in the direction of the dip of salt vein. This mining method does not permit selective mining as do other methods, e.g. ascending mining with backfill. In addition, descending mining of inclined rooms does not facilitate continuous mining technology. The ore transport from face to haulage drift is through an ore pass by gravity, and haulage by mine cars or trucks.[2]

The descending slice mining method for inclined rooms has been implemented in one

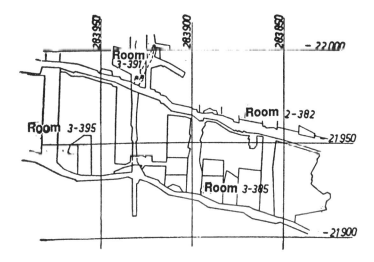

Figure 14.1.4. Layout of inclined room mining at the Wieliczka salt mine

part of the salt deposits at the Tusanj salt mine (former Yugoslavia). The extent of the inclined extraction panel is defined by level drifts driven along strike or ore body (60-80 m) and by raises along dip of ore body (80-100 m). The stope development workings and room extraction occurs between two levels, L-190 and L-250, as shown in Figure 14.1.5. The dimensions of the haulage drift (L-250) are 6.0 × 4.5 m but the dimensions of the ventilation drift (L-190) and all other drifts and crosscuts are 4.5 × 3.0 m. The principal elements of the mine layout are illustrated in Figure 14.1.6. The dimensions of stope structures are as follows:

Inclined room (stope)
  – width: 8.0 m;
  – length: 60.0 m;
  – height: 7.0 m.
Rib pillar
  – width: 8.0 m;
  – length: 80.0 m;
  – height: 7 m.
Seal pillar
  – width: 20.0 m;
  – length: as extraction panel;
  – height: 7 m.

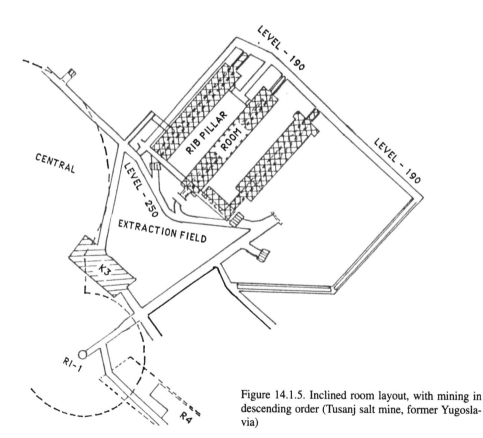

Figure 14.1.5. Inclined room layout, with mining in descending order (Tusanj salt mine, former Yugoslavia)

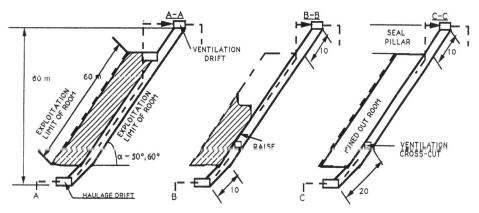

Figure 14.1.6. Cross-section of the inclined rooms for three different phases of excavation

The raise dimension is 2.5 × 2.0 m, and from its very top a first overcut of a height of 2.5 m is excavated. The sequence of excavation of the undercuts is shown in Figures 14.1.5 and 14.1.6. The width of each undercut depends on the bed inclination. For example, a bed inclination of 25° has a room of 12 m wide, and a bed inclination of 65° has a room of 8 m wide. The stope face has a step pattern, which is the result of a sequencing drilling and blasting by horizontal holes toward the hanging wall and vertical holes toward the foot wall (Figure 14.1.7). The descending mining of inclined rooms is applied for salt beds with an angle of inclination between 35° and 60°.

2. *Mining of salt lenses and irregular veins* is discussed from the experiences at the Wieliczka salt mine and the Bochnia salt mine. The stope development is from foot wall level drifts running along strike of mineral deposit. From these the two level cross-cuts are excavated at a vertical distance of 30 m. In the case of irregular salt veins the vertical distance of the level cross-cuts is greater than 30 m, but with the inter-level cross-cut between them (Figure 14.1.8). The level drifts are connected by the raises, which are located in the middle of the ore block. The raises are inclined at 60° and are used as ore passes.

Typically, the lenses have dimensions about 40 m in height and roughly 10-15 m in diameter at the middle, but the dimensions vary from lens to lens. The stoping sequence of the salt begins at the upper-cut of the lens (i.e. The 320 m level) and the salt above the uppercut is drilled manually by stoppers. The upper-cut is used as a free-face into which the salt is blasted with an ANFO explosive. The broken material is then scraped into the ore pass by a three-drum slasher. Care is taken to ensure that a 'skin' of salt remains in place around the outside of the stope. This skin of salt serves a very important ground support role, since it is much stronger than the surrounding clay. The thickness of this salt skin is approximately 1.5 m, and to be effective it should in no case be less than 1 m thick.[3]

The sequence of salt extraction of the salt lens is illustrated in Figure 14.1.9. Each sequence consists of a horizontal slice having dimensions of approximately 2 × 3 × 6 m. Ground breakage is by drilling with a hand-held jackleg and blasting with an ANFO explosive. The fractured rock is mucked by the slasher to the ore pass and then directly loaded into rail cars waiting at the draw point on the haulage level. This process is continued downward, leaving a chamber which is approximately 10 × 12 × 30 m. It should be pointed out that a whole horizontal slice ranging across the width of the lens is not taken,

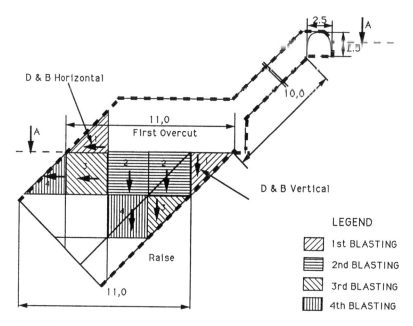

Figure 14.1.7. Sequences of room excavation in descending order

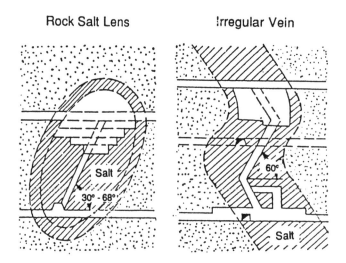

Figure 14.1.8. Development of rooms in a salt lens and in an irregular vein

but rather a 'V-shaped' opening is brought down to direct the stresses away from the salt lens and into the clay strata.

The Wieliczka salt mine is considered to be one of the most beautiful salt mines in the world, so that salt rooms serve an aesthetic purpose. Particularly the rooms which consist of a green salt are impressive. In one of these rooms is a famous underground chapel (Figure 14.1.10).

The mining of irregular salt veins is done in the same way as the mining of the salt lenses. The horizontal slices are also taken progressing downward. The room width and height are

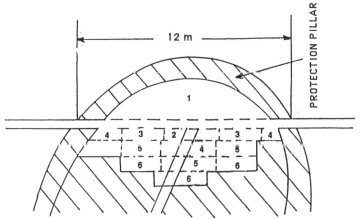

Figure 14.1.9. Slice sequence of room mining (lens)

Figure 14.1.10. Location of the chapel in the old salt room

the same as in the case of the salt lens mining, but the length of the room is delineated artificially – not naturally, because of the salt ore extension along the strike of the vein. In this case, the length of the room is defined as 12 m, with vertical boundaries consisting of the 8 m long salt pillar, as illustrated in Figure 14.1.11.

There are a certain number of layout variations for the mining of inclined and steeply inclined rooms, which are more or less designed to suit local mining conditions. For

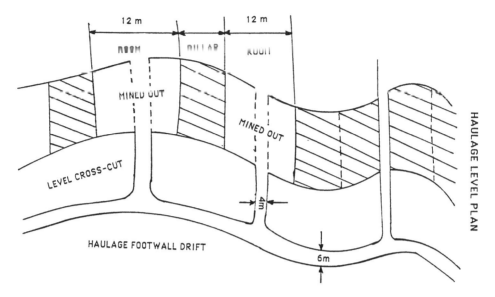

Figure 14.1.11. Layout of rooms and pillar for irregular vein mining (haulage level plan)

example, the Klodawa salt mine implements face extraction by bench mining in the direction of the strike and dip of the salt body, not by horizintal slicing, which is commonly the case in room mining.

Finally, a remark should be made that the structure of the inclined room in salt mining corresponds to the structure of open stope in hard rock vein mining.

14.1.3 *Sub-level stoping*

The sub-level stoping method has been introduced for potash veins located in the salt domes of Germany (Figure 14.1.12). This method of mining was developed by the 'Kali und Salz A.G.' company. Half of the potash production is obtained by this company using sub-level stoping.[4]

The layout and technology of the sub-level stoping in potash mines are fundamentally the same as in hard rock mining, with the one difference that the mined-out area is backfilled. The placing fill in the mined-out area has many advantages: increase of the ore recovery because it is possible to minimize the size of the mine pillars; the mined-out stope back and floor are supported and take some load from the working area; the surface subsidence is appreciably decreased; rock waste is dumped in the mined-out area instead of on the surface. The general outlines of a potash mine using the sub-level stoping method and filling of the mined-out area are shown in Figure 14.1.13.

If the potash beds are sufficiently steep and ore and walls are competent, the sub-levels are blasted in sequence starting at the lowest sub-level. The broken ore is transferred by gravity through the funnel or finger raise to the main level and conveyed to the shaft. The sub-level stoping method consists of three phases, each a highly specialized, independent operation (Figure 14.1.14).

1. *First phase (development phase)*. From the ramp access to the stope panel is gained through the level drifts. The panel is then developed by sub-level drifts, which form

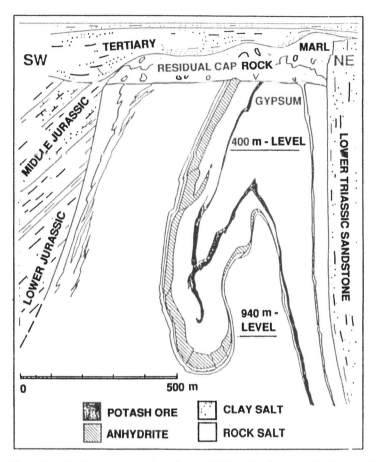

Figure 14.1.12. Cross-section of salt dome which bearing the potash in (Germany)

sub-level extraction blocks of 6-12 m wide, depending on the thickness of the potash vein, and 100 m long and 20 m high. During this phase essential information such as morphology of ore blocks, inclination and strike, tonnage and grade of ore and others are collected.

2. *Second phase (extraction phase)*. After completion of the funnels between the lower main level and sub-level 1, breaking of the benches starts at the very end of sub-level 2. This is done by enlarging the cut 3 m long raise to the full width of the bed. The cut raise, 0.7 to 1.0 m diameter, is drilled upwards from the lower to the upper level. When the bench is broken far enough, the same procedure is repeated on the next sub-level. This method of breaking several benches at the same time permits the remarkable daily output of 5000 to 6000 tonnes per stope. This high production rate also leads to very high utilization of the underground services and to very efficient ventilation.

3. *Third phase (placing mine fill phase)*. This begins when all broken ore is pulled down. Tailings and underground waste are conveyed to the fill holes of the stope at the upper main level, either by belt conveyors or mine trains. This phase does not require any manpower in the mined stope.

Sub-level stoping permits a rate of extraction of 85% and more. The panel layout is either one- or two-winged and in both cases retreat from the panel limit to the central spiral ramp

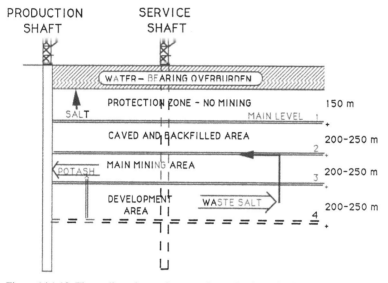

Figure 14.1.13. The outline of an underground potash mine with fill operation

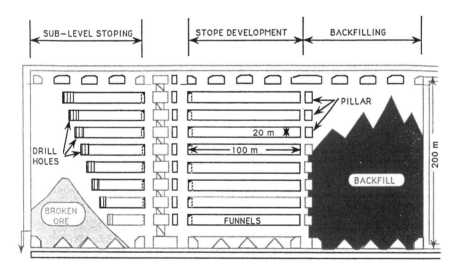

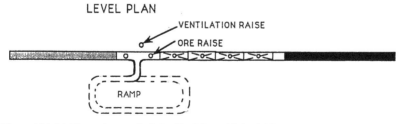

Figure 14.1.14. Three phases of sub-level mining with backfill

is common practice. While one panel is in operation, another one is being developed.

As far as underground transport is concerned, a typical salt dome mine has to haul daily: 10 000 tonnes of raw ore, 1300 tonnes of waste rock, 5000 tonnes of tailings from the preparation plant, 16 300 tonnes total of solid material and 150 to 200 m³ of brine from filled stopes.

The trackless mining equipment required is: drilling jumbos to drill the centre holes (400 mm diameters, 7 m long), drilling jumbos to drill the blast holes (35 to 38 mm diameter, 7 m long), for the burn cut, drilling jumbos to drill the sub-vertical bench holes (about 150 mm diameter), vehicles for transport of explosives (ANFO), and scooptrams with a capacity up to 15 tonnes. Drifts usually have an arched back for self-support and reduction of spallings of the rock. The arch is either blasted with detonation fuse (cushion blasting) or cut. Presently, rock miners are also on trial. A typical sub-level entry is up to 30 m wide, depending on vein thickness, depth and rock conditions. Vertical raises of 1.5 to 2 m diameter, connecting main levels, are mechanically drilled, using one or the other raise boring techniques. Depending on the lateral regularity of the potash veins, belt conveyors or locomotive haulage are used to transport ore and waste on the main levels. Mine cars of the bottom-dump-type with 30 tonnes payload are standard. Up to twenty cars constitute a train with 600 tonnes payload. Ore haulage between main levels is carried out by inclined belt conveyors.

The production rate is 10 000 tonnes of potash ore per day. The mine productivity is very high: OMS 50 tonnes.[5]

14.2 DESIGN CONSIDERATION OF MINE STRUCTURES

The structural concept of inclined room mining is similar to the one in coal mining and hard rock mining of pitching deposits. There are four structural units within inclined vein mining, namely: block of salt delineated by room dimensions (this is a temporary structure because it shrinks as mining progresses), the rib pillars, seal pillars and intersection of these two pillars (these are permanent structures). With regard to mine stability, the type of loading on these pillar structures has a critical influence. In salt mining two types of loadings are exhibited: static loading superimposed on permanent room-pillar structures and dynamic loading which normally affects the salt extraction pillars.

14.2.1 *Static loading of the pillars*

The static loading of the pillars is of primary importance for the inclined room mining of salt beds. During and after mining, all load has to be taken by the rib pillars and the seal pillars. The magnitude of the load superimposed on the pillars depends on the unit weight of overburden rocks, mine depth, extraction ratio and angle of inclination of the salt pillars. The normal stress on the pillar and the shear stresses at the pillar roof and floor are calculated as follows:

1. Calculate vertical and horizontal stress on the unmined salt vein by the following equations:

$$\sigma_V = \gamma h$$

$$\sigma_H = K\sigma_V$$

where:

γ = unit weight;
h = depth to seam;
K = ratio of the horizontal to vertical in situ stress (0.3-3.0).

2. Resolve vertical and horizontal stress of the unmined salt vein into components normal to bedding and in shear parallel with the bedding (Figure 14.2.1). The following equations can be used to calculate the normal and shear stress components:

$$\sigma_N = \sigma_V \cos^2 \alpha + \sigma_H \sin 2\alpha$$

$$\tau = \frac{\sigma_V - \sigma_H}{2} \sin 2\alpha$$

where α is the salt vein dip.

3. Define tributary and pillar areas as follows:

$$A_t = (W + w) \times (L + l)$$

$$A_p = W \times L$$

where:

A_t = tributary area;
A_p = pillar area;
W = least pillar width;
w = width of mined-out stope;
L = length of the pillar;
l = length of the mined-out stope.

4. Solve for the resultant pillar normal and shear stresses on the tributary area:

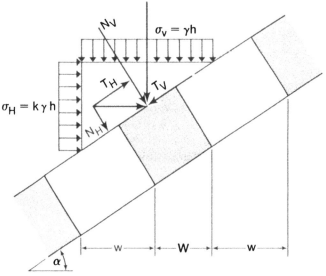

Figure 14.2.1. Average stress of pillar (calculated by the tributary theory)

$$\sigma_{pillar} = \sigma_N \times \frac{A_t}{A_p}$$

$$\tau_{pillar} = \tau \times \frac{A_t}{A_p}$$

For σ_{pillar} positive seam ride (shear) is down dip and for σ_{pillar} negative seam ride is up dip.

These equations determine average pillar load due to overburden and do not include dynamic loading conditions during salt extraction.

It could be argued that calculating the inclined pillar load by the tributary theory may be satisfactory regardless of static stress conditions, because they consider the layout of the pillars after completion of ore extraction. For salt mining this is of great importance, because collapse of overlying strata in the rooms is unacceptable, due to the potential risk of water penetration into the mine. If the inclined pillar load is to be increased in order to increase salt recovery, then artificial support provided by mine fill is required.

14.2.2 *Dynamic loading of the pillars*

The dynamic loading of the pillars is particularly important for sub-level stoping, because of stress transfer on to the solid ore. This phenomenon has been extensively studied in hard rock mining and to some degree in hydraulic coal mining.[6] Of particular importance is dynamic loading of the extraction pillar, which is delineated by the room or stope dimensions, and which is reduced as mining progresses.

The dynamic stress concentration in sub-level blocks is of primary importance during salt extraction. The process of dynamic loading through stress transfer for a sub-level stoping mine layout can be summarized as follows:

1. Stress-transfer during mining extraction is primarily up dip, following the direction of the imaginary triangle (Figure 14.2.2). This stress is the principal factor causing sub-level block-pillar fracturing. The transferred stresses are high and are concentrated in the narrow abutment area of solid ore above. As the salt fractures, the abutment area broadens and the unit stress value is reduced.

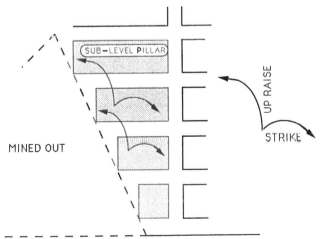

Figure 14.2.2. Dynamic stress transfer on the sub-level pillars

2. Stress transfer along the strike of the sub-level pillar increases as mining progresses because the excavated area is enlarged. However, the magnitude of these dynamic stresses is always lower than the stresses transferred up dip.

3. This is little or no stress transfer down dip because the area below the pillar is mined out.

Dynamic loading of sub-level blocks during ore extraction is a very complex problem because stress changes are difficult to monitor in situ. Consequently, numerical models are commonly used for stress analysis.

In the case of salt lense mining and in some cases of salt vein mining, the slicing sequences in descending order are arranged in such a way that the excavated area of the extraction pillar could be approximated as an inverse pyramid. At the top of the inverse pyramid the vertical stress is zero. Horizontal stresses tend to be transferred outward from the salt lens or salt vein into the surrounding rock strata. Under these circumstances, only the vertical static stress will be in effect, and they will gradually increase as the room size increases.

14.2.3 *The load of the mine pillar skeleton*

The basic supporting element in the inclined bed or vein is skeleton pillar structure, which has been formed by stope or room excavation, leaving in situ rib pillars and seal pillars (Figure 14.2.3).

For the stability of the mine pillar skeleton there are two critical elements, namely: slenderness of the rib or seal pillars which progressively increases as the salt face advances, and the structural connection of the rib and seal pillars at their intersection.

To ensure the satisfactory bearing capacity of the mine pillar skeleton, it is necessary to determine the critical slenderness ratio of the rib pillar and seal pillar. The slenderness ratio of the rib pillars in salt vein mining is limited by the relation:

$$H/W \leq 1.0$$

where:

H = height of pillar;

W = least width of pillar.

In order to maintain the slenderness ratio, the width of the rib pillar must be increased if the height of the inclined room is to be increased.

The intersection of pillars, however, is the key factor in the stability of the mine pillar skeleton structure. The model investigations and in situ phenomenological studies have suggested several stability characteristics of this particular structural element.

1. The stress distribution observed in models indicates that the intersection between the rib pillar and the seal pillar creates a stress concentration which is several times higher than that found in the central part of the seal pillar. This observation is in agreement with other data reported in the literature. The vertical stress magnitude at the middle part of the seal pillar is less than one-half of the overburden load ($> 0.5 \gamma h$), while at the intersection with the rib pillar, it reaches more than three times the overburden load ($< 3.0 \gamma h$), as is illustrated in Figure 14.2.4.

2. Underground observations have also shown great stress concentration at the corners of seal pillars and rib pillars. In some places the peak corner stresses caused failure and the stresses were consequently reduced. The closure measurements in a drift driven through a seal pillar indicated maximum movement at the rib pillar intersection and minimum movement at the centre (Figure 14.2.5).

LONGITUDINAL SECTION

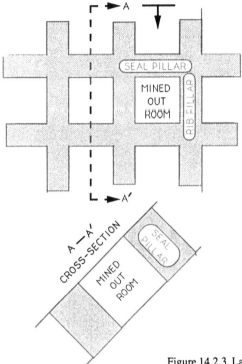

Figure 14.2.3. Layout of inclined rooms, rib pillars and seal pillars

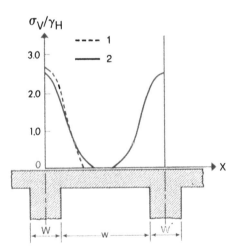

Figure 14.2.4. Distribution of major principal stress at pillar intersections
1. Model analysis; 2. Theoretical analysis

Theoretical analysis based on the model investigations has prompted the derivation of an equation for calculating shear stress at the interaction between the rib and seal pillars

$$\tau = \frac{2Ww\sigma c \tan \phi}{1 - \sin \phi}$$

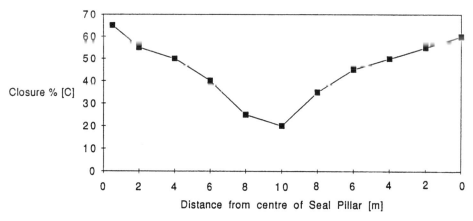

Figure 14.2.5. Distribution of convergence from centre of seal pillar to intersections with rib pillars

where:
    W = width of rib pillar;
    w = width of the room;
    σ = acting stress $(1.6\,\gamma h)$;
    ϕ = angle of internal friction;
    c = cohesion.

The stability of the pillar intersection is controlled by the angle of internal friction of the salt.

At the present time the loading conditions of the pillar intersections as well as the whole mine pillar skeleton structure have been effectively simulated by mathematical models, given the limitation of the occurrences of the material properties of the salt.

14.3 STABILITY EVALUATION OF MINE STRUCTURES

The stability of the mine structure depends on the bearing capacity of the elements which compose a mine layout. The determination of the bearing capacity of the mine structure, however, can only be roughly approximated because of the pronounced creep deformation of the salt and the strain hardening phenomenon of the salt, as previously discussed. For this reason, limited space has been devoted to the calculation of the bearing capacity of the mine structure. The main attention is given to stability evaluation of the salt pillars, as discussed by K. Durr et al.[7]

14.3.1 *Bearing capacity of the pillar structure*

The bearing capacity of the substrata is an important element in the stability of mine pillars. The changes in stability of the pillar structure may be related to the bearing capacity of the footwall, which is influenced by the geological nature of the strata, composed of inter-bedded hard and soft layers. It has been established that an increase in the thickness of the soft layer (potash) overlying a hard layer (marl) will decrease the stability of the mine pillar and vice versa. More rigid layers can accumulated and redistribute stress, but softer layers in the foot wall at higher stress will flow and transfer the excess load to the more rigid layers below. The bearing capacity of the foot wall depends on the thickness and the strength of the hard and soft layers as well as their relative position in the geological sequence.

A.S. Vesic proposed the concept for evaluation of the bearing capacity of the sub-pillar strata by determination of shear strength of the hard and soft layers, as expressed by the equation.[8]

$$q_{ult} = C_1 N_m + q$$

where:

q_{ult} = bearing capacity;
C_1 = shear strength of upper layer;
N_m = modified bearing capacity factor which depends upon the ratio of shear strengths of two layers, $K = C_2 / C_1$;
q = surcharge.

Using this equation, the factor of safety of the bearing capacity of the sub-pillar strata at the Tusanj salt mine has been calculated and is between 1.3 and 1.5. This is in agreement with in situ observations, where sub-floor strata of inclined pillars do not exhibit any stability problems.

The bearing capacity of rock salt pillars for the Wieliczka salt mine has been calculated on the basis of the equation proposed by Brady and Brown:[9]

$$q_b = \frac{1}{2} \gamma W_p N_\gamma + c N_c$$

where:

q_b = bearing capacity of pillar;
W_p = width of pillar;
γ = pillar load unit;
ϕ = internal angle of friction;
c = bearing capacity factor;
$N_c = (N_q - 1)$;
$N_\gamma = 1.5 (N_q - 1) \tan \phi$;
$N_q = e^{\pi \tan \phi} \tan^2 \left(\frac{\pi}{4} + \frac{\phi}{2} \right)$.

In the case of the Wieliczka mine, the calculating parameters were determined as:

$\gamma = 0.03139$ MPa/m;
$W_p = 8.0$ m;
$\phi = 35°$;
$c = 0.15$.

By substituting all the above parameters in the equation, it has been found that the bearing capacity of the rib pillar is $q_b = 31.0$ MPa. Thus, the factor of safety against bearing capacity failure is given by:

$$F_s = \frac{q_b}{\sigma_p} = 1.13$$

where:

σ_p = head on the pillar

This suggests that the bearing capacity of the rib pillars in the Wieliczka salt mine is satisfactory.

It is fair to say that the examples of the calculations for the bearing capacity of mine structures rather represent a simplified stability analysis, which is not often applied in underground salt mining.

14.3.2 *Evaluation of the stability of the mine structure*

This example of the evaluation of the stability of salt rock structures is taken from the Asse potash and rock salt mine in Germany, which is located on an 8 km long ridge about 10 km SE of Wolfenbüttel. The ridge consists of a core composed of Zechstein stratigraphic sequences that were thrust up during the Upper Cretaceous. The mining of the salt vein is along a steeply dipping rock salt bed (Figure 14.3.1). The extraction of the Leine rock salt bed took place in the southern flank from 1916 to 1965 at depths between 490 and 750 m. About 130 rooms were excavated on thirteen levels (Figure 14.3.2). It should be noted that the width of the room and seal pillars in the cross-section changes according to the thickness of the pitching salt bed, but along the longitudinal section, the length and height of rooms and of the seal and rib pillars are maintained according to the mine layout.[7]

Durr et al. analyzed the rib pillar between rooms 4 and 5 at the 535 m level.[7] The pillar had been left after excavations of the rooms in 1956 and had existed for 26 years, e.g. The time when the study had been carried out. Along the strike, it has a width of about 20 m, while perpendicular to the strike it has an average length of 40 m and a height of 15 m. Some stope development workings have been run through the pillar, as shown on its layout (Figure 14.3.3). The rock mechanics investigations were based on the thermomechanical behaviour of rock salt, considering the long-term load effects, exhibiting a highly non-linear function of time, stress and temperature. The viscous behaviour is described in general by a creep law, expanded by a stress term.[10]

$$\dot{\varepsilon}_{eff} = A \cdot e^{-(Q/RT)} \left(\frac{\sigma_{eff}}{\sigma^*} \right)^n$$

where:

Parameters
A = structure factor ($0.18\ d - 1$);
Q = activation energy (54 MJ k mol^{-1});
R = universal gas constant (8.314 kJ mol^{-1} k^{-1});
n = stress exponent (5);
σ^* = normalizing value (1 MPa).
Variables
σ_{eff} = effective deviatoric stress;
$T(K)$ = temperature.

The values given for the parameters are empirical values for steady creep of several types of rock salt.

The parameters for the strength behaviour of rock salt are obtained from triaxial testing. The rock salt failure occurs during higher rates of deformation and small lateral stresses, but with increasing lateral pressure, the visco-plastic deformations take effect. The testing stress at 20% strain is used for determining the maximum load-bearing capacity of the rock salt samples. The extensive uniaxial and triaxial tests have shown that the failure strength, or the bearing capacity, of rock salt increases with a rising strain rate up to a limiting value. This begins at a rate of deformation of about $\dot{\varepsilon} = 10^{-4}$ S^{-1}. Higher rates of deformation do not

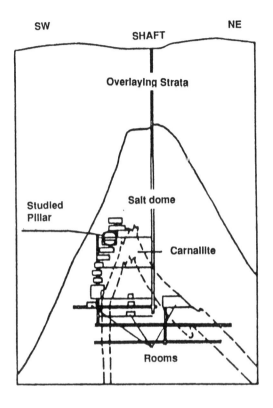

Figure 14.3.1. Mining cross-section of salt dome (Asse salt mine, Germany)

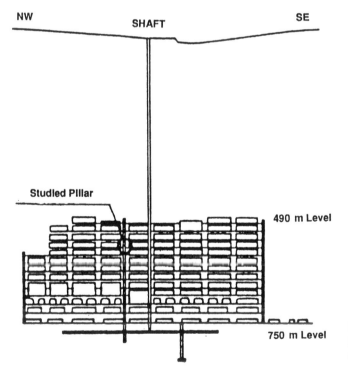

Figure 14.3.2. Longitudinal section of the rooms and rib pillars (strike)

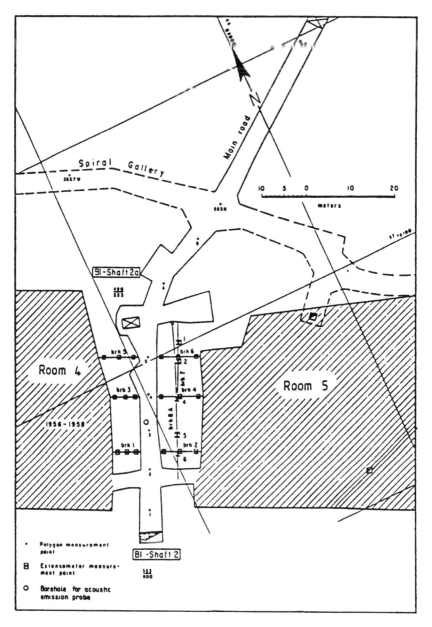

Figure 14.3.3. Layout of rock salt pillar with in situ measurement stations (553 m level)

produce significant improvement of the failure strength or the load-bearing capacity.[7]
    The failure, or load-bearing capacity, criterion is given by the hyperbolic equation

$$\tau_o = \frac{\sigma_o + C_o}{a_o + b_o \cdot (\sigma_o + C_o)}$$

where:

a_o = 0.19, initial increase of the hyperbola;
b_o = 0.012 1/MPa, 1/asymptote of the hyperbola;
c_o = 0.65 MPa, smallest mean main stress;
σ_o = octahedral normal stress;
τ_o = shear stress.
The octahedral normal and shear stresses are defined as follows:

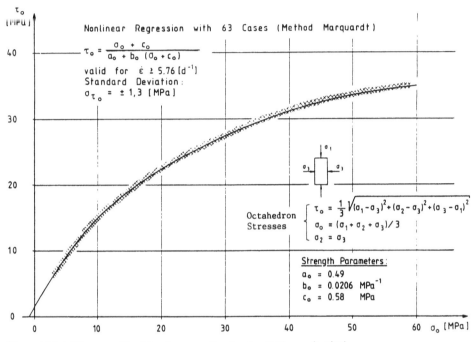

Figure 14.3.4. Diagram of load-bearing capacity of rock salt (Asse salt mine)

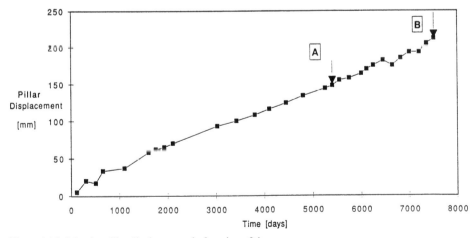

Figure 14.3.5. In situ pillar displacement in function of time

$$\sigma_o = (\sigma_1 + \sigma_2 + \sigma_3)/3$$

$$\tau_o = \frac{1}{3}\sqrt{(\sigma_1 - \sigma_2)^2 + (\sigma_2 - \sigma_3)^2 + (\sigma_3 - \sigma_1)^2}$$

$\sigma_2 = \sigma_1$ (in the case of triaxial tests)

Figure 14.3.4 represents the diagram of the load-bearing capacity of rock salt from the Asse mine. If the computed values of the bearing capacity lie above the curve, then failure or high rates of deformation must be taken into account. A possible safety factor of the mine structure can be calculated as follows:

$$K = \frac{\tau \text{ available (according to load-bearing equation)}}{\tau_o \text{ possible (according to analaytical computations)}}$$

The validity of the results depends upon the correct evaluation of the average stress in the mine due to overburden loading and mining conditions.

14.3.3 *Monitoring stability of the mine structure*

Durr et al. also conducted in situ stability investigations by measuring longitudinal pillar displacement, using a series of extensometers and by recording horizontal pillar deformation in both directions with an acoustic emission monitor.

1. *Displacement instrumentation* in the rib pillar is located between room 4 and room 5 at the 353 m level and has been for a period of over twenty years. Figure 14.3.5 suggests that the displacement curve is linear during the entire period of measurement. In addition, measured and calculated displacement between the polygon points 3 and 4 indicates that a condition of steady-state creep exists in the pillar (Figure 14.3.6).[10] The long-term mean creep rate for this pillar is 3.5×10^{-6} m/day.

The measurements of distribution of displacement showed that the pillar is in state of contraction (compression), which is not evenly distributed along its longitudinal axis. The continued measurements of displacement in the level cross-cut to the no. 2 shaft, however, showed that the adjacent zone is in state of extension (tension). The change-over from compression to tension is at the level drift which intersects level cross-cuts (Figure 14.3.7). The studies of pillar displacement showed that its compression increases linearly with time, and also that the largest transverse deformations occur at the middle of the pillar, where the

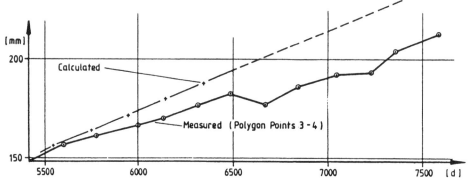

Figure 14.3.6. Measured and calculated displacement of the pillar between the polygon points 3 and 4

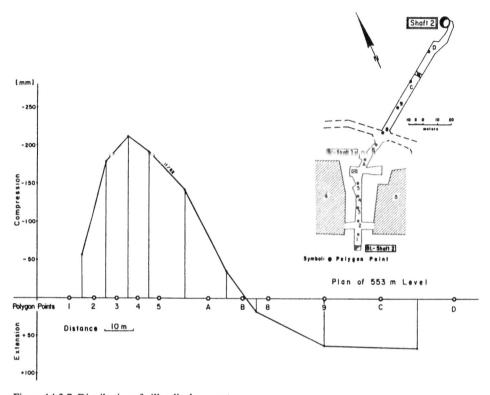

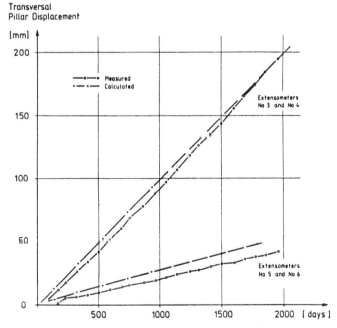

Figure 14.3.7. Distribution of pillar displacement

Figure 14.3.8. Comparison between measured and calculated transversal pillar displacement

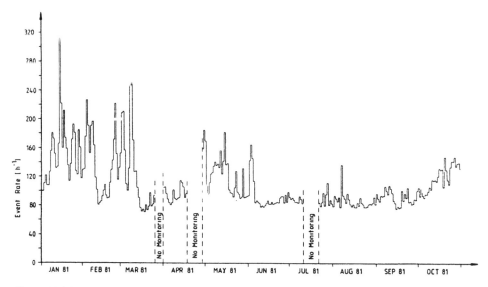

Figure 14.3.9. Acoustic emission rate

creep rate is 5.2×10^{-6} m/day. Creep rates of 2.9×10^{-6} m/day, of 0.1×10^{-6} m/day were determined from the centre transverse displacements. These creep rates related to a measurement period of five to six years. Figure 14.3.8 shows the measured and calculated displacements of the pillar core and the outer zone of the pillar.

2. *Acoustic emission measurements* are illustrated in Figure 14.3.9, where the emission rate varies quite significantly during the period of monitoring. Mining noise has been eliminated in the plot, thus the measured events are thought to be acoustic energy emitted as a result of longitudinal and transverse deformations of the rib pillar. The emission rate curve is not regular and is characterized by many peaks appearing over a period of several days. The reason for this is that the acoustic emission transducer recieves signals from all over the rib pillar, i.e. from the highly stressed core of the pillar as well as from the partly destressed outer zone of the pillar. Consequently, the event rate and pillar deformation do not correspond.

14.4 STRESS ANALYSIS OF MINE STRUCTURES

The stress analysis of the stope structures is mainly accomplished by the use of numerical models. There are several computer programs for stress analysis of the inclined or vertical pillar-room structure of the pitching salt beds. Some of these programs have been applied to hard rock mining as well as to salt mining, particularly in the case of sub-level stoping. Further consideration is given to three examples of stress analysis for each particular method of mining.[10, 11, 12]

14.4.1 *Stress analysis of the vertical rooms structure*

In this case, the stress analysis has been implemented for the Wieliczka salt mine, which has been discussed in Section 14.1. For simplicity, the stopes and rooms are assumed to be vertical.

The stress analysis for the Wieliczka salt mine was conducted at Laurentian University, Sudbury (Canada), as a part of the research project on the design of salt mining methods.

The stress distribution either around the room or within the room-pillar structure was analyzed by utilization of the same boundary element method 'EXAMINE' as noted in Chapter 12. The determination of stress distribution was on plan, longitudinal section and cross-section view for both the individual stope as well as for multiple stopes. The major principal stress (σ_1) and minor principal stress (σ_3) had been obtained and plotted, as was the safety factor using Mohr-Coulomb failure criteria.

1. *The single-room structure* is related to salt lens ('brile') mining. The stress analysis is conducted for the following conditions:
 – In situ virgin stress: $\sigma_1 = \sigma_3 = 11.0$ MPa (isostatic stress).
 – Rock salt strength: $\sigma_T = 3.04$ MPa (Tensile strength); $\sigma_c = 38.0$ MPa (compressive strength); $\phi = 35.0°$ (angle of internal friction).
 – Elastic properties: $E = 15.4$ GPa (modulus of elasticity); $\nu = 0.25$ (Poisson's coefficient).
 – Symmetry: none.

The stress distribution along the longitudinal section of the room, represented by major principal stress σ_1 and minor principal stress σ_3, indicated the stress concentration at room corners and the stress transfer to the surrounding salt strata, respectively. The safety factor is greater than 1, and the zones which are between 1 and 2 are relatively small and are located in the room back and floor. To alleviate instability of the room back, it is obvious that its rectangular shape has to be changed over to an arching shape, which is actually carried on in mining practice (Figure 14.1.10).

The stress distribution along the cross-section of the room indicated a stress concentration similar to the case of the longitudinal section. The safety factor is greater than 2.

The stress distribution in the plan of the room, represented by major principal stress (σ_1), follows the classical principles of the solid mechanics, with stress concentration at the corners and stress relief at the room walls. Minor principal stress (σ_3) is exhibiting some small zone of tensile stress on the hanging wall and foot wall of the room. The safety factor is satisfactory, and is greater than 2.

2. *The multiple-room structure* is related to salt vein mining. The stress analysis was conducted using the same conditions as for the single-room structure. The stress distribution is more complex, as would be expected for a layout of multiple openings.

The major principal stress (σ_1) distribution along the longitudinal section indicated that rib pillars between rooms have a stress overlap and in this regard a higher stress concentration. Due to the strain hardening of salt, the increased stress load does not effect the stability of the rib pillars. The minor principal stress (σ_3) shows a tensile stress concentration within the rib pillars, and may have an adverse effect on the pillars, especially those in the centre of the rooms. The safety factor distribution is unsatisfactory, except at the corners of the rooms ($F_s = 1 - 2$), which may be improved by replacing the rectangular shape of the corners with the oval shape.

The stress distribution along the cross-section of the room indicate the stress concentration in the corners of the rooms. The safety factor is greater than 3, showing adequate stability of the room perpendicular to the salt vein.

The stress distribution in the plan view has a very similar pattern to that of the longitudinal section, where rib pillars are effected by both major principal compressive stress concentrations (σ_1) and minor principal tensile stress concentrations (σ_3).

The utlization of numerical model stress analysis at the Wieliczka salt mine may be considered satisfactory. However, for practical application in mining, the results of the numerical analysis have to be correlated and interpreted in terms of phenomenological investigations of the underground room-and-pillar structures.

14.4.2 *Stress analysis of the sub-level structure*

The sub-level stoping stress analysis was carried out in the Mineral Engineering Department at the University of Alberta, Edmonton (Canada), for application to both soft rock and hard rock mining.[11] This stress analysis of a sub-level layout was accomplished using the finite element method program developed at the Civil Engineering Department of the same University.[6, 11]

1. *Sub-level pillar structure* stress distribution is represented by major principal stress contours (σ_1), minor principal stress contours (σ_3) and the directions and magnitudes of σ_1 and σ_3 at the mesh nodes.

a) The major principal stress contours show a pattern of stress concentrations and relaxations in the hanging wall and footwall, around the sub-level openings. The pillars showed the classic stress redistribution pattern with peak stresses at the abutments and the average pillar stress increasing with depth. Underground studies indicated much greater differences in stress concentrations between upper and lower pillars than the numerical model suggests. This phenomenon could be attributed to the increase of yield of the sub-level pillar as its excavation progresses.

b) The minor principal stress contours show a steady increase with depth; in the vein hanging wall adjacent to the pillars it exhibits a small peak which is mainly due to stress transfer from the individual sub-level drifts. Underground phenomenological studies, however, show greater stress peaks adjacent to the pillars due to some load transfer from the mining faces.

c) The stress vector diagram illustrates the stress trajectories within sub-level pillars, which are influenced by the geometry of mining and geological structures (Figure 14.4.1). An anomaly in the roof and floor of the excavations is caused by the way in which stresses are computed at notes. Above and below the excavationss, vertical stress relief has occurred but the magnitude of the average stress in the vertical direction is greater than the horizontal stress. This contradicts the actual mining conditions.

The most important result of the finite element stress analysis is that increasing the pillar width along strike, the average pillar stress within the pillar decreases. The concentration of abutment stress will be decreased in the vicinity of the sub-level walls, which is the main cause of instability, and is also decreased when the pillar width increases.

2. *Sub-level stope structure* stress distribution is represented also by the principal stress contours and the directions and magnitude of σ_1 and σ_3 at the mesh nodes.

a) The major principal stress contours show the large area of influence of mining on the stress pattern which is coincidental with underground monitoring of strata movement. Also, the diagram of the major principal stress indicates that there is a stress concentration zone in the hanging wall of the mined-out stope. This zone is approximately parallel to the roof strata of the sub-level stope.

b) The minor principal stress contours around the sub-level stopes indicate a stress distribution which is similar to the major principal stress distribution. The diagram of stress

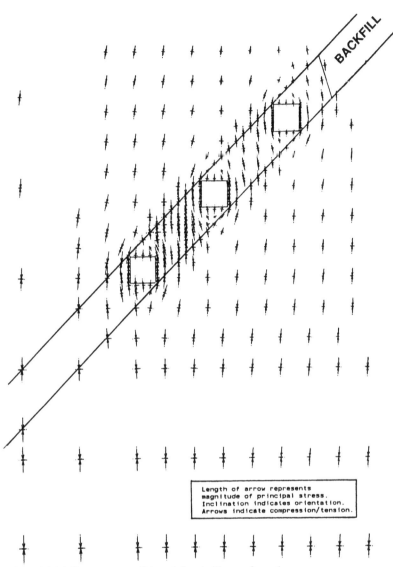

Figure 14.4.1. Stress vectors of the sub-level pillar configuration

concentration suggests that the instability of the hanging wall strata is controlled by very low, or even tensile, stresses above the mined-out area.

c) The stress vector diagram illustrates that the stress field of the sub-level stope is compressive (Figure 14.4.2). This stress state is the main factor controlling roof stability of the mined-out stopes.

The numerical model stress analysis of the sub-level stope structure indicates that unless the hanging wall strata are particularly strong, failure or collapse of the stope is likely. This is an important consideration for sub-level stoping of potash veins in the salt domes of Germany. To ensure that the roof does not cave, the area is backfilled so that further deterioration of the roof is arrested.[6, 11]

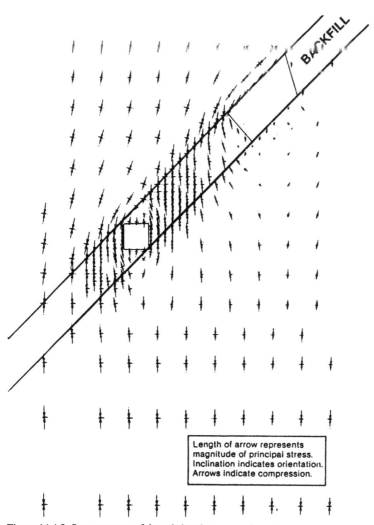

Figure 14.4.2. Stress vectors of the sub-level stope configuration

An analysis of the sub-level structure was also made using the viscoelastic boundary integral method. This is free-dimensional stress and displacement analysis using the boundary integral computation. This method is used for calibrating the model with the results of underground instrumentation, in order to develop a predictive tool with which to compare alternative mine layout.[12] This model is probably more suitable for modelling yielding deformations and associate stress transfer, than elastic model analyses described before.

14.4.3 *Stress analysis of multiple-room structures*

The Asse potash and salt mine, which was analyzed earlier in this volume, is analyzed here using the finite element computations based on the ADINA computer program for non-linear time-dependent solutions.[7]

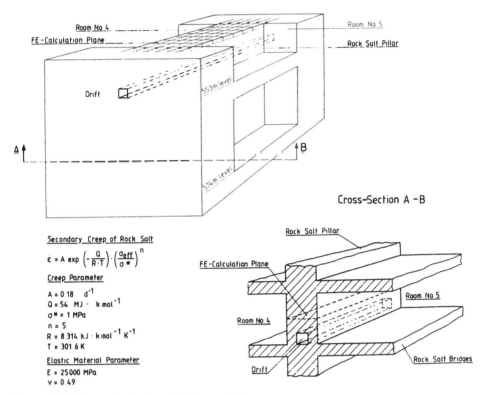

Figure 14.4.3. Geomechanical model of the multiple-room structure

A horizontal section was selected for the computation which roughly characterized the multiple-room structure (Figure 14.4.3).

For the numerical model, a pressure of 10 MPa was assumed as realistic. With the defined geometrical boundary conditions, together with the rock salt constitutive law of creep, the failure criterion and the load distribution, the non-linear deformations were calculated.

Finite element computations for a period of loading of 900 days indicated no significant changes in stress for this period. The larger stresses of $\sigma_{zz} = 10$ MPa on the narrower side of the stope pillar decrease on the wider side to $\sigma_{zz} = 2$-7 MPa. The complete stress distribution is shown in Figure 14.4.4. In the same figure, the creep producing effective stress (σ_{eff}) for two sections (A-A and B-B) are illustrated. In most of the pillar area they were determined to be between 7 and 8.5 MPa. The displacement associated with the stresses is plotted in Figure 14.4.5. In addition to the compression of the pillar in the z-direction, the bulging out in the y-direction is clearly recognizable.

It should be mentioned that the computed results are slightly greater than the monitored ones (Figure 14.3.8). The lines connecting points within an equal safety factor provide an overview of the distribution of the safety factor (Figure 14.4.6). The safety factor isolines are plotted for the last time increment. At the northern and southern parts of the pillar, the local safety factor of $F_s \geq 3$ is greater than in the central area, where it falls to a value of $F_s = 1.9$ which is still satisfactory. This data suggest that the supporting capacity of the pillar is increasing towards the solid salt mass.

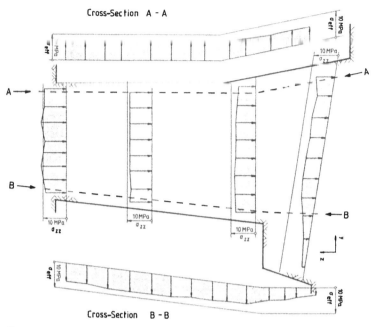

Figure 14.4.4. Distribution of stresses σ_{zz} and σ_{eff}

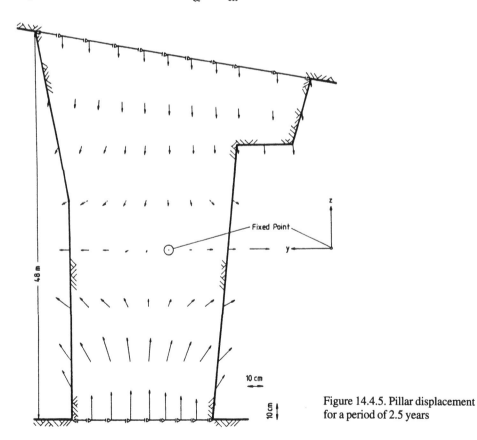

Figure 14.4.5. Pillar displacement for a period of 2.5 years

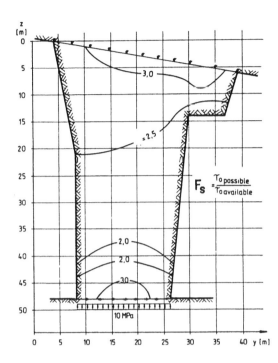

Figure 14.4.6. Isoasphalic lines of distribution of the safety factor (F_s)

$$F_s = \frac{\tau_{o\,possible}}{\tau_{o\,available}}$$

REFERENCES

1. Jeremic, M.L. 1964. Methods of development and mining of salt deposits in Poland. *Min. & Metall. Bulletin* (Belgrade), 32 (2): 264-275 (in Yugoslav).
2. Snolarski, A. & A. Litonski 1960. Technology of salt exploitation in Poland. *Przeglad Gorniczy* 2: 13-14 (in Polish).
3. McClain, W.C. & F.F. Arlo 1984. The evaluation of room stability, pp. 712-720. In *Proc. of 1st Conf. on Mech. Behaviour of Salt.* Clausthal: Trans. Techn. Publ.
4. Hei, W.A. & A.H. Potthoff. Potash mining in steep deposits (salt domes). In *Potash'83. Potash technology*, pp. 79-85. Toronto/New York: Pergamon Press.
5. ... 1980. *Die Kali Industrie in der Bundesrepublik Deutschland* (4th ed.), pp. 8-25. Hanover:
6. Jeremic, M.L. 1985. Stability of stope-pillar structure. In *Strata mechanics in coal mining*, pp. 453-462. Rotterdam: Balkema.
7. Durr, K. et al. 1983. Evaluation of salt rock pillar stability utilizing numerical calculations, mine survey and in situ rock mechanics measurements. In *Proc. of 6th Int. Symp. on Salt*, The Salt Institute, Virginia, Vol. 1, pp. 463-479.
8. Vesic, A.S. 1973. Analysis of ultimate loads of shallow foundations. *J. Soil Mech. and Found.* (Div. ASCE), 99 (SM 1): 37-42.
9. Brady, B.H.G. & E.T. Brown 1985. *Rock mechanics for underground mining*, pp. 331-336. London: Allen & Unwin.
10. Langer, M. 1979. Rheological behaviour of rock masses. In *Proc. of 4th Int. Congr. on Rock Mech.*, Montreux, Switzerland, Vol. 3, pp. 29-96.
11. Jeremic, M.L. 1987. Stress analysis of sub-level caving. In *Ground mechanics in hard rock mining*, pp. 361-365. Rotterdam: Balkema.
12. Dunbar, W.S. & R.D. Hammett 1983. Application of a viscoelastic boundary integral method to the analysis of potash mining scheme. In *Potash'83. Potash technology*, pp. 331-325. Toronto/New York: Pergamon Press.

CHAPTER 15

Solution mining

The abundant existence of natural mineral solutions, such as sea (salt) water, would suggest that solution mining might have a long history. Indeed the records indicate that solution salt mining has gone on for upwards 8000 years. However, the first reported intentional effort to leach on a productive scale was by the Spanish in 1752.

Solution mining has since become an important and significant method of rock salt exploitation, which has been expanded for the recovery of other salt minerals such as potash and trona. Solution mining involves removing a mineral from a deposit without first removing the ore from the ground. As a result of solution mining underground caverns are formed, of which stability is the primary interest.

15.1 PRINCIPLES OF SOLUTION MINING

Principles of solution mining include several aspects; from the mining technology of the caverns solution to the caverns layout.

15.1.1 *Technology and methods of solution mining*

Solution mining started long ago by uncontrolled salt exploitation. This method is still in effect where natural solution of the salt is in progress due to the circulation of aquifers through a deposit, as mentioned in Chapter 8. Actually, under these circumstances, the brines from salt deposits are pumped out directly, without the delivery of fresh water for the process of salt solution. Uncontrolled solution mining occurs in Europe in several countries, such as Poland (Barich), former Yugoslavia (Hukalo & Trnovac) (Figure 15.1.1) and others. This type of solution mining has a catastrophic influence on the surface subsidence leading to collapse of structures, as discussed in the last chapter on subsidence.

Of engineering interest are the technology and method of controlled solution mining. There are two systems of technologies of salt solution as well as two methods of salt production by solution mining.

1. *The technology of salt solution* by borehole is represented by two systems, as briefly discussed below.[1]

a) *System of indirect circulation*, where fresh water is injected between the tubing and the casing. This system is called the 'top injection method', because the water enters the salt deposit at the top of the formation and starts to dissolve the salt near the roof (Figure 15.1.2). The salt brine flows downward to the bottom of the tubing where a sump effect has

Figure 15.1.1. Uncontrolled solution mining wells at Hukalo (Tuzla, former Yugoslavia)

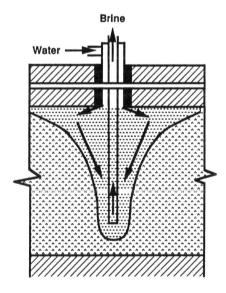

Figure 15.1.2. Method of indirect circulation (top injection method)

been produced by the pump drawing on the tubing. The brine is pumped from the well and then set to the processing area. For roof control of the cavern, a fluid isolant is applied which does not dissolve salt and does not stick to it.

b) *System of direct circulation*, which uses the same type of set-up as an indirect circulation, e.g. the top injection method, except that the flow of water is reversed. The fresh water is pumped down the tubing and dissolves the salt at the bottom of the formation. The brine is then drawn out of the well between the tubing and the casing (Figure 15.1.3). This system is also known as the 'bottom injection method'.

The difference between the two systems is the rate of decay of the well condition. In the top injection method the roof of the cavity is prone to caving since it is exposed more quickly than in the bottom injection method. The caving of the roof soon causes the tubing to plug or break. This causes reduced production or results in the closing of the well.

2. *Methods of salt production by solution mining* are fundamentally represented by the single-well method and by the double-well method, as briefly discussed below.[1]

a) *The single-well method* of solution mining involves the drilling of a single large diameter drill hole into a salt formation. A casing is used in the hole to stop the walls from caving. A second tubing of smaller diameter is then placed in the cased drill hole. The single well is simultaneously assigned to the production well and the injection well. Under this method the solution mining production is related solely to the individual wells. The single-well method is mainly applied for deeper parts of salt deposits.

b) *The two-well method* of solution mining of salt deposits is based on drilling two identical wells into a salt body at a distance from several tenths of meters to several hundreds of meters. One well is assigned as the production well and the other as the injection well. Solution mining starts with the individual operation of each well by one of the single-well methods. After completion of the caverns, and in order to get fluids flowing between them, high pressure water is applied at the injection well in order to cause hydraulic fracturing between the wells. The brine is drawn from the production well and, as the flow between the two wells increases, the opening enlarges by dissolving the salt. As the cavity between the two holes enlarges, the volume of water flow between the holes increases, causing increased production. As the well cavity is enlarged, the danger of roof collapse causing damage to the well is less than in the single-well method, because the area where the salt is being removed is distant from the wells, where collapses cannot cause damage to the well.

The application of the two-well method will be briefly discussed for some countries.

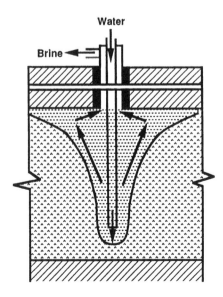

Figure 15.1.3. Method of direct circulation (bottom injection method)

The two-well method of solution mining in Romania has a layout as shown in Figure 15.1.4. The method is known in that country as 'battery solution mining', which occurs where two wells have been connected. Battery wells consist of three types of tubes: casing, tubing for water or brine flow, and tubing for oil flow. The flow of fresh water and the flow of brine are reversed every three days. Through the two-well method, a large lateral cavern is formed due to the connection of the original one-well caverns. The main concern in the stability of the large caverns is roof control. For this type of mining, because of the difficulties of roof control, the depth of salt solution is limited to 300-500 m. At a depth of 1000 m it is very difficult to control roof stability for large spans of caverns and consequently the single-well method must be implemented.

Solution mining of trona deposits in the USA is also carried out by the two-well method. Holes are drilled down to the salt body in pairs, located up to 200 m apart. The connection between a pair of holes is made by hydraulic fracturing and a solvent is circulated through the system until it is saturated with alkali. The brine is then pumped to the plant to be processed into soda ash.

It should be noted that in Poland and some other countries multiple wells are drilled in rows at shorter distances. They are connected by the solution processes of nearby wells, which result in the formation of a large cave. This system essentially forms a system of two wells, where one is the injection well and the other the production well (Figure 15.1.5). This type of solution mining usually results in intensive roof caving and surface subsidence, because the connection between nearby wells creates irregular caverns.

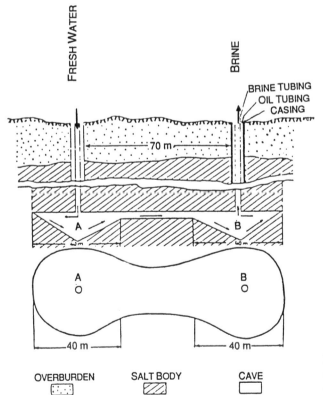

Figure 15.1.4. Two-well method of solution mining (battery wells – Mari shaft, Romania)

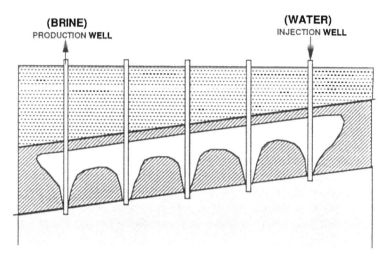

Figure 15.1.5. Multiple-well method of solution mining (Barich salt mine, Poland)

The main advantage of the two-well method is greater salt recovery, up to 15% compared with the one-well method, where salt recovery is up to 12%.

15.1.2 *Techniques of solution processes*

The techniques of the solution process to a great extent depend on the geological-structural features of the deposit and its internal structure. For example, for the salt deposit at Tetima, which is briefly described in Chapter 8, the technique of lateral salt leaching was chosen. The layout has a distance between axes of the drill holes of 120 m and a maximal diameter of the cavern of 60 m.

The solution mining is by a single-well method with indirect circulation during development phases of the cavern and direct circulation during the exploitation phase of the cavern. In the last phase of full production of a single well, the production capacity should be 50 m^3/h of saturated brine. This particular technology of solution mining could be characterized by three phases.[2]

First phase: This phase is represented by the formation of the hydro-undercut at the bottom of the future cavern. The height of the undercut is 15 m. The salt solution flows in a lateral direction, with a protective layer of isolant fluid to control the stability of the cavern back during its formation. The salt solution process ceases when the cavern reaches a diameter of 0.8 D_{max}. During this phase the system of indirect circulation is applied. The capacity of the salt dissolution is 25 m^3/h of unsaturated brine (Figure 15.1.6).

Second phase: After completion of the hydro-undercut a narrow cylindrical cavern is dissolved with a diameter of 5 m and an average height of 115 m, which could be called a hydro-slot. The salt solution of the hydro-slot also has a protective layer of fluid isolant to control roof stability while reaching the final height of the exploitation cavern. During this phase a system of direct circulation is applied with the dissolution capacity of 25 m^3/h of the unsaturated brine (Figure 15.1.7).

Third phase: The two previous phases should be considered the development of the solution mining cavern. The third phase is considered the exploitation phase of the cavern, which has an analogy with dry mining stope extraction. The direction of salt dissolution is

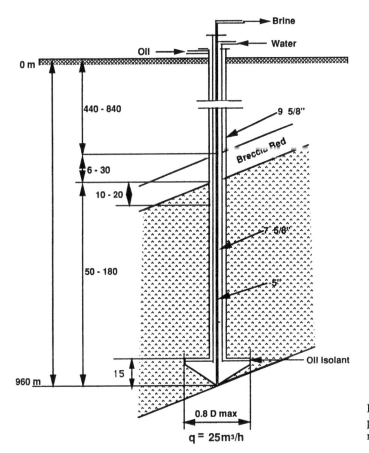

Figure 15.1.6. First phase: cavern development by hydro-undercut

lateral, by widening the cavern and blocking the roof with isolant fluid on the top of the brine. During this phase an indirect circulation is also applied with exploitation tubing installed in the lower part of the cavern. The capacity of salt exploitation in this phase, until cavern completion (D = 60 m) is approximately 50 m^3/h of saturated brine (Figure 15.1.8). During solution mining, it is necessary to follow the development of the cylindrical cavern with the following methods:

    – calculation of the volume increase of the cavern based on the salt mass produced;
    – checking the level of isolant fluid using gamma density logs and the method of differential pressures;
    – analysis of the chemical composition of dissolved salt;
    – surveying the cavern shape by the ultra-sound method, using an echometer.

If the development of the cavern during the solution process is in agreement with the design parameters, then solution exploitation continues. If surveying data record an appreciable deviation to the size and shape of the exploitation cavern, then production has to be stopped. In this case the design parameters must be recalculated and changed, such as the velocity of salt solution, the depth of solution tubing, the level of fluid isolant, and others.

    Without cavern control by calculation and instrumentation it is impossible to execute the correct layout of solution mining. Under these circumstances the inter-cavern pillars, sill pillars and finally protective pillars which envelope the solution mining area might become

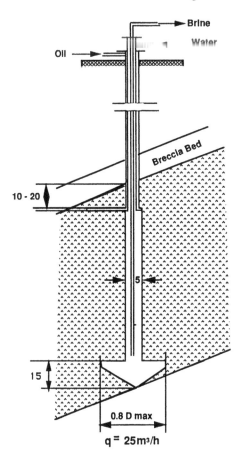

Figure 15.1.7. Second phase: cavern development by a vertical cylindrical slot

unstable, which could jeopardize not only the stability of the solution mine, but also the stability of the whole salt body.

15.1.3 *Multi-component salt solution*

Multi-component salt solution is used for Veendam magnesium-bearing deposits in the Netherlands and is described by Buyze and Lorenzen.[3]

The hexagonal well layout in idealized form is shown in Figure 15.1.9, which corresponds to the formation of cylindrical caverns with inter-cavern pillars, as discussed in subsequent sections of this chapter. However, the layout is unique, particularly because subsequent holes are not drilled from the ground surface, but from the sub-surface through the deviation of the central drill hole, as illustrated in Figure 15.1.10. The spacing between drill holes is 350 m, because the State Mining Inspection Department requires at least 300 m spacing between holes. The drilling of expensive deviated holes for cavern formation is necessary in the Netherlands, which is a densely populated country, and disturbance to the agricultural environment must be kept to a minimum.

The completed well system is shown in Figure 15.1.11. The system consists of two parallel free hanging tubing strings, an 8.5 cm water injection tubing and an 11.25 cm brine production tubing, which are suspended inside a 26.87 cm casing. To avoid salt recrystalli-

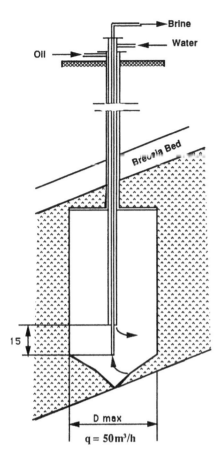

Figure 15.1.8. Third phase: cavern in exploitation stage

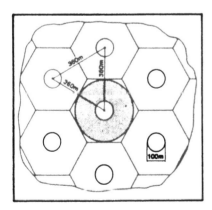

Figure 15.1.9. Idealized well layout

zation in the production tubing due to cooling, dilution is required. This is done through a concentric 7.2 cm dilution string suspended inside the 11.25 cm productin tubing.

The chemistry of dissolution of Veendam salt ore is very complex, because four minerals are mined simultaneously. Their individual solubilities in water may provide an area of the

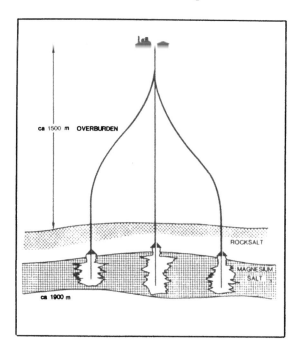

Figure 15.1.10. Cross-section of deviation well geometry

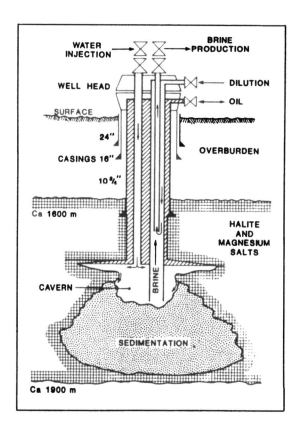

Figure 15.1.11. Schematic brine well completion

wide range of dissolution properties. Table 15.1.1 shows that to dissolve 1 m^3 of pure halite to saturated brine requires 5.8 m^3 of water, while 1 m^3 of pure bischofite requires 0.4 m^3 of water. The same amount of water can dissolve about fourteen times the volume of bischofite than of halite.

The complexity of multi-component salt dissolution, the preferential and interferential relationships of concentrations in salt ore and brine, equilibria, dissolution rates, precipitation, cavern geometry and salt surface conditions cannot analytically be described in a satisfactory way. Therefore, the operational controls of solution mining have instead been developed with a general approach.

Buyze and Lorenzen further discuss a solution mining concept by the method of direct circulation. The injection of water is under a protective layer of oil. The solution mining is in ascending order, where the shape of the cavity will be formed as an inverted cone ('morning glory'). In order to minimize an adverse effect of the 'morning glory' and maximize salt recovery, the total ore section is leached in a series of subsequent hydro-cuts of the height of a few meters from the bottom to the top of the cavern (Figure 15.1.12).

Each time a new hydro-cut is initiated, the injection point is raised to a predetermined level, either by sub-surface tubing cutting procedure or by a snubbing operation, depending on the completion status of each individual level. In each newly initiated hydro-cut a 'non-selective state' of dissolution of all salt components takes place. With carnallite being the source of KCl, the dissolution of all salts will generate concentration of MgCl$_2$, MgSO$_4$,

Table 15.1.1. Salt mineral solubilities in water at 55°C

Salt solution	Halite NaCl	Kieserite MgSO$_4$	Carnallite MgCl$_2$. KCl	Bischofite MgCl$_2$
Salt concentration in brine (% wt)	27	34	34	37
Density of saturated brine (t/m^3)	1.18	1.36	1.28	1.34
Volume ratio m^3 water/m^3 salt	5.8	3.8	1.6	0.4

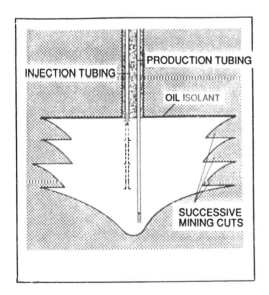

Figure 15.1.12. Hydro-cut mining concept in ascending order

KCl and NaCl increasing toward the bottom of the hydro-cut. When the leaching of a new cut is started, the dissolution process continues in the previous cuts by the 'selective stage' where only the $MgCl_2$ concentration will still increase. An adequate brine quality control has been developed. Brine concentrations are approximately 20% by weight $MgCl_2$ in carnallite and approximately 30% by weight in bischofite.

In order to control the leaching process, the salt available within an imaginary cylinder with the predetermined height of hydro-cut and a designated radius of 50 m is calculated. A sonar survey is used to monitor cavity development in a qualitative sense rather than in a quantitative sense. The surveys suggest that cavity volumes are smaller than the calculated bulk affected volumes as a result of the selective stage of salt dissolution.

Shell Laboratory KSEPL developed an oil/brine interface detector, to continuously monitor the oil protective layer relative to the injection tubing shoe. Due to the constantly applied well head pressure, running in of the tool, and retrieval, requires adaptations to the well heads and special surface mechanical equipment. e.g. flow tubes and a travelling stuffing box spool (Figure 15.1.13). A surface panel provides instant indication as to what level the measuring section is immersed in brine.[3]

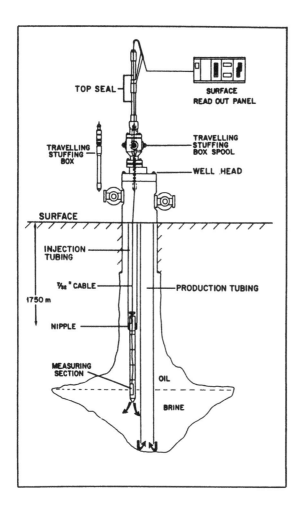

Figure 15.1.13. Continuous isolant/brine interface monitoring system

15.1.4 *Thin bed solution mining*

This topic has been discussed by Yuanxiong and Chengkun for salt deposits of Ziliujing, Szechuan, China.[4] A characteristic of mining of thin salt beds is having the solution advancing in a lateral direction, because the axial direction is limited by the bed thickness. Due to a defficient ratio between the depth and thickness of salt beds, it is necessary to conduct solution mining from one site through deflected drill holes.

The authors described a method of drilling from one drill site but with two holes 1.5 m apart. To achieve the deflection of the drill holes through salt they utilized the existing structural geological characteristic of the strata. The deflection of the drill holes was less than 8° and they were drilled into the rock salt bed according to projected azimuths. One drill hole was directed along the strike and another along the dip of the bed, intersecting the salt between 70 and 100 m apart (Figure 15.1.14). The drilling cost was reduced by 23% and the speed of the borehole completion is increased over two times, because the time-consuming use of conventional borehole deflection tools was not applied. Another improvement of solution mining was obtained by the application of oil isolant, which flowed on top of the brine and protected the thinly stratified roof strata.

The concept of solution mining was by the two-well method, with indirect circulation during the cavity development and direct circulation during exploitation. After the formation of individual caverns, they were connected by hydraulic fracturing or directional jet fracturing. The advantages of solution mining by connecting wells had been known in China since 1892, when local people poured water with rice-husks into the abandoned wells, and the husks were carried up with the brine produced in other wells.

Yuankiong and Chengkun described several case histories of dual-well solution mining,

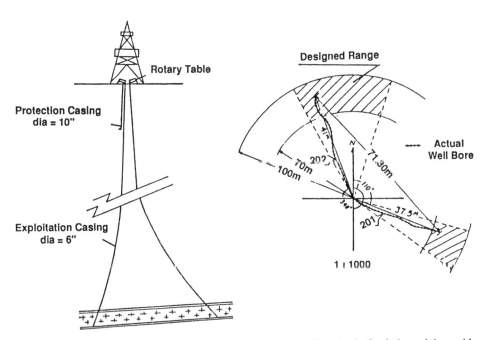

Figure 15.1.14. Layout and horizontal projection of the two-well method of solution mining, with directional drilling

and one of them is briefly described in the following paragraphs.

Two wells were collared for two-well solution mining. The angle of inclination of well no. 213 is 7.4° with an azimuth of 129°, and the angle of inclination of well no. 214 is 3° with an azimuth of 47.5°. The drill hole parameters of the wells are given in Table 15.1.2.

The process of the solution mining by the two-well method is described in two stages.

1. *Cavern development.* This stage is divided into three phases:

a) the phase of low salinity:
   injection rate – 6-10 m³/h;
   injection pressure – 10-20 kg/cm²;
   brine concentration – less than 100 g/l;
   duration – 40-50 days.
b) The phase of salinity increase:
   injection rate – 6-10 m³/h;
   injection pressure – 20-25 kg/cm²;
   brine concentration – 100-298 g/l;
   duration – 100-110 days.
c) The phase of stable salinity:
   injection rate – 6-10 m³/h;
   injection pressure – 25-35 kg/cm²;
   brine concentration – stable at 250 g/l;
   duration – 70-75 days before connection.

Figure 15.1.15 shows the salinity curves before the connection of the cavities in the dual bore controlled by oil isolant. It can be seen that if the height of the cavity is less than 2 m in thinly bedded rock salt, the stability of the salt solution is increased.

2. *Cavern connection.* The connection of the cavern is measured by an echo log along with the calculation of the volume of dissolved salt. The variations of flow rate, pressure and water temperature may influence formation of the connection of the cavities. The connection formation depends also on its length, on the geological features of the rock salt deposit, the location of the connection, the difference between the elevations of the two wells, the well spacing, the injection rate and pressure. It would take 450-500 days before dual bores are put into regular production. Injection rate might be increased because the pressure would gradually drop after the connection of the cavities.

Table 15.1.2. Borehole parameters of the two-well method. Length in meters: radius in centimeters

Item	Well no. 213	Well no. 214
Drillingi depth	1,024.66	1,042.37
Roof depth	1,024.36	1,020.91
Floor depth	1,041.46	1,040.77
Thickness of rock salt	17.1	19.86
Setting depth of brine casing	16.5 × 1,025.69	16.5 × 1,022.81
Length of brine casing in salt	1.33	1.90
Setting depth of annual tube	10 × 1,037.44	10 × 1,037.29
Setting depth of tubing	5 × 1,039.44	5 × 1,039.19
Height of cavity	2.00	1.90

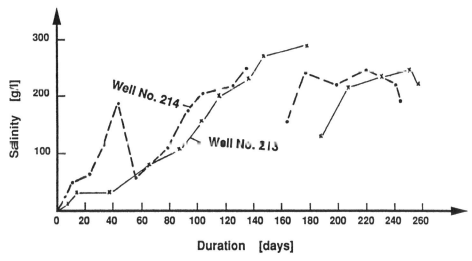

Figure 15.1.15. Salinity diagrams during cavern development (wells no. 213 and 214)

For the control of cavern solution Yuanxiong and Chengxun suggest that particular attention should be paid to the following elements:
– selection of injection flow rate and circulation (reverse or direct);
– thickness of the oil isolant, oil amount injected and the oil injection method;
– measurement of cavity diameter and calculation of lateral dissolvation speed; and
– prediction and observation of the connection of the cavities.

The solution mining of thin salt beds with oil isolant can be effectively controlled by a layer of oil isolant injected in an upward direction. If upward dissolution of salt is not controlled, then the roof strata could be exposed in the cavern back. In this case very thin roof layers could cave in the cavern. The layer forms a thin stable film that prevents fresh water from dissolving salt upwards. According to calculations for the Zilinjing salt beds, the oil injected forms an isolant layer of an average thickness of 1.5-2.0 cm. For example, the cavity formed by solution mining wells no. 201 and no. 202 (Figure 15.1.16) has about 40-50% of the injected oil stored, with the rest having been distributed like thin film with a thickness of 1.5 cm. Injection of oil in the cavern continued until excess oil appears at the surface. Generally, oil injection was undertaken once a day during the first month, once every two days in the second month, and once every three days from the third month on. The actual oil consumption is small, for example the accumulated amount of oil injected in wells no. 213 and no. 214 was 140 m^3.

Solution mining of thinly bedded rock salt deposits indicated that salt dissolution in the vertical direction is two times faster than in the lateral direction, which is 0.16-0.19 m/day.

15.1.5 *Underground solution mining*

Underground solution mining does not have as wide an application as ground surface solution mining. However, in some mines it has found application and profitability. Underground solution mining is implemented primarily by two types of technologies, as briefly discussed below.

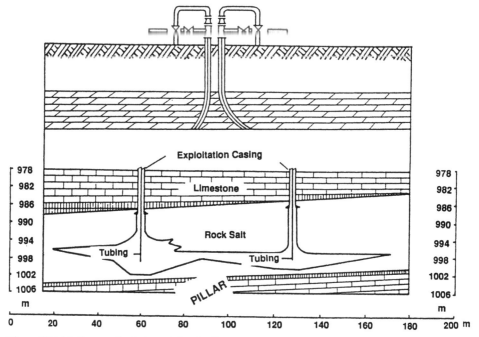

Figure 15.1.16. Cavern of solution mining (wells no. 201 and 202)

1. *Room solution mining* implements a lateral leaching of salt from the initial cut. The length of the room is oriented parallel to the strike of salt deposits. As in the case of dry salt mining, it is necessary to develop stopes by the following development workings:
 – drifts along the strike of deposits to install the pipelines for room leaching;
 – cross-cuts to connect drifts along strike;
 – access drifts for the fresh water pipeline;
 – vertical raises for connection of the drifts at different levels.
The excavation of the development workings is made by fresh water jets. Unsaturated solution is delivered to the room where it will stay until full saturation, then it is drained to the level below flowing to the salt solution collector. This process is repeated until the designed room size is achieved. The principal layout of room solution technology is illustrated in Figure 15.1.17. This type of exploitation technology is beneficial for deposits with a high content of waste partings, which can be left in the rooms as low grade soluble or unsoluble material. Figure 15.1.18 represents a layout of rooms mined out by salt solution which is used in one part of the Solno-Inowroclaw salt mine (Poland). The solution mining in this particular case has some advantages. Firstly, the rock salt mining is selective because the gypsum and anhydrite rock waste is left in the stope as mine fill or sediment. Secondly, if the salt deposit contains a large amount of methane then a great deal of it can be eliminated with gas absorption by water. Thirdly, the cost of underground salt exploitation by solution mining can be as much as four times lower than dry mining.[5]
 2. *Single-well method solution mining* from underground mines is the same as solution mining from ground surface. At the present time there are not many underground operations of solution mining by the single-well method, and none by the double-well method. A successful underground well solution mining is conducted at the Muresh salt mine in

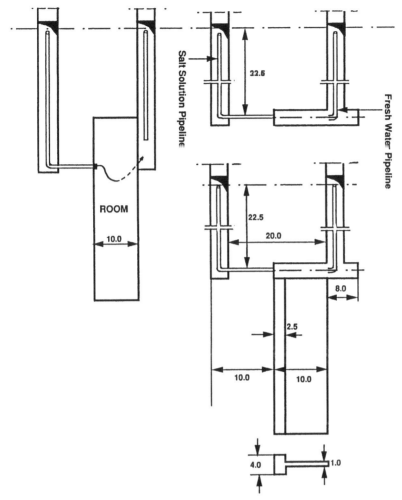

Figure 15.1.17. Layout of the underground solution mining room

Romania. At this mine, single-well solution mining is implemented below a dry room-and-pillar layout at a level of 150 m. At the time of this study, three single wells have been drilled and each of them allows access for the leaching of caverns (Figure 15.1.19). The layout of the boreholes and proposed cavern is outlined as follows:
  – boreholes are spaced at 80 m centre;
  – cavern width of 40 m;
  – inter-cavern pillar width of 40 m;
  – sill pillar thickness of 50 m;
  – depth of borehole 600 m;
  – height of cavern 550 m.
For each exploitation borehole a room of 20 m height from the 150 m level must be excavated. Each room must have sufficient spacing for drilling operations. The depth of drilling is up to 700 m, because underground operating conditions superimpose limitations on deep drilling.

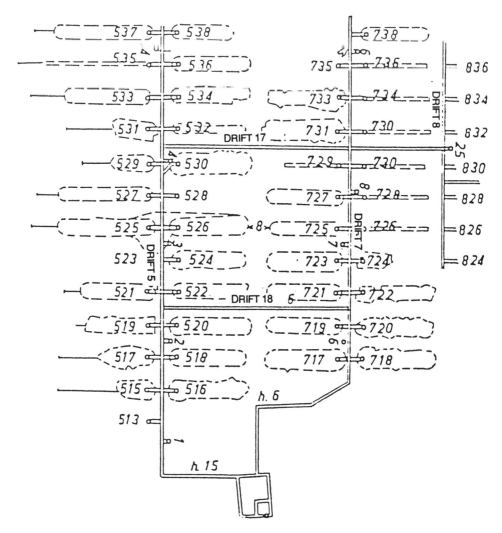

Figure 15.1.18. Layout of the rooms of solution mining (Solno Inowroclaw mine, Poland)

A vertical advance of a cavern of 6.4 m requires 100 days. A higher brine salinity of approximately 310 g/l would result in salt productivity increases.

The pumped brine drains to a collecting station. The brine has to be purified before delivery to the soda plant because it contains some oil used as an isolant to control the cavern back, as discussed in previous sections. The stability of the cavern walls is maintained by a constant presence of water, which exerts an internal pressure on cavern walls and prevents salt slabbing and caving due to external ground pressure.

This method has several disadvantages because it requires access and development of the underground mine which will delay production and tremendously increase capital cost. In addition, the salt recovery factor is low: in the range of 10%.[5]

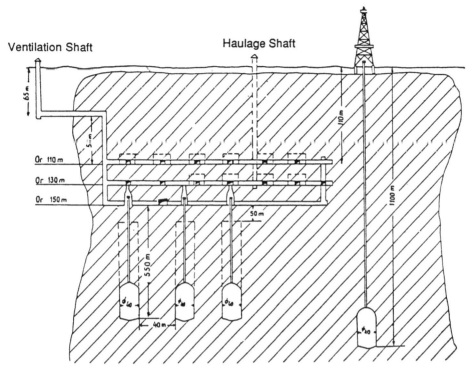

Figure 15.1.19. Layout of dry room mining and single-well solution mining (Muresh salt mine, Romania)

15.2 MECHANICS OF SOLUTION MODELS

At present there are several theories and several physical and mathematical models to evaluate the mechanics of salt stability during solution mining operations. This section is based on the work of Russo, who describes four models.[6]

15.2.1 *Plume model of salt solution*

Because the mixing within a plume is usually rapid, an analysis of plume dynamics based on the assumption of a uniform specific gravity and velocity within the plume (top hat model) is appropriate. Morton presented the result of such an analysis by a set of equations which describe the dynamics of an unconstrained steady plume:[7]

$$\frac{d(b^2 u)}{dz} = 2\alpha b u$$

$$\frac{d(b^2 u^2)}{dz} = 2b^2 g \, (C_o - C)$$

$$\frac{d(b^2 u g \, (C_o - C))}{dz} = 2b^2 u g \, \frac{dC_o}{dz}$$

where:

    b = the effective plume radius;
    C and C_o = fluid specific gravities in and out of the plume;
    u = the plume velocity in the vertical (z) direction;
    g = the acceleration of gravity;
    α = entrainment coefficient.

When the plume is rising through a stably stratified fluid ($\partial C_o/\partial z < 0$), it will rise to a certain level and stop, and its radius will grow indefinitely. This level is denoted by the plume stagnation level in Figure 15.2.1. If the plume is rising in an unstably stratified fluid, it will continue to rise and grow until it interacts with the cavern walls, which then constrain the plume and change its rise rate. The level at which this interaction occurs (the level at which the plume radius equals 0.7 of the cavern radius) will also be denoted as the plume stagnation level, because in either case the entire plume flow is deposited in the fluid cell containing this level.[6]

15.2.2 *Diffusion model of salt solution*

The diffusion coefficient D is a strong function of whether the brine is stably or unstably stratified. For stable concentration gradients, D is just the molecular diffusion coefficient, D_{mol}, which is very small (1.4×10^{-5} cm^2/sec). When the concentration gradient is positive, however (unstable case), a much larger eddy diffusion coefficient, D_e, must be used. An instability analysis indicates that an appropriate mixing length, l, is used in calculating D_e, when all effects are ignored.[8] The diffusion equation is:

$$l = \left(\frac{6\pi}{\alpha}\right)^{3/4} \left(\frac{2v^2 C}{dC/dzg}\right)^{1/4}$$

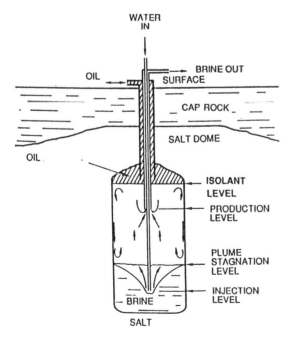

Figure 15.2.1. Cavern geometry and flow regions for direct salt leaching

where:

α = entrainment coefficient;

ν = local kinematic viscosity of the brine;

g = gravitational acceleration.

It is assumed that the eddy diffusion coefficient is proportional to the product of velocity and mixing length, and the mixing length is taken as the minimum of the cavern radius r, and l, as given in the following equation:

$$D_e = D_o \, (dC/dz)^{1/2} \, \text{Min} \, (r^2, l^2)$$

This equation is the final form of the eddy diffusion coefficient to be used in equations where $D = D_{mol}$. The value of D_o used was 31.7 ft$^{1/2}$/sec from the Knapp data, and the value of α which best fit a limited amount of data taken from Bryan Mound well 104 was 0.046. This value of α is not far from other experimentally determined values for buoyant plumes that are typically about 0.08 (Morton).

15.2.3 *Insoluble model of salt solution*

Part of the input data required to run the mathematical model SANSMIC is the specification of the volume percent of insoluble in the salt formation. A sample of salt from the dome at Bryan Mound, Texas, was analyzed to determine the insoluble particulate size distribution. Most of the insoluble were anhydrite particles with an average diameter of between 20 and 400 micrometers, with a peak at 250 micrometers. Assuming Stokes drag on spherical particles, the settling velocity for each particle size over the range of 0 to 400 micrometers was calculated and the integrated fraction that would fall out in an upward velocity field was calculated as a function of fluid velocity. A curve fit to these results, given by the following equation, is included in the code to establish the fractions of insoluble that fill the cavern sump or are discharged:

$$f = 0.5/(1 + 0.00231 \, v) + 0.5e^{-0.002v}$$

where:

f = total fall fraction;

v = upward fluid velocity.

The code keeps account of the insoluble that falls and raises the sump floor accordingly as well as increasing the wall recession rate in proportion to the insoluble freed at each level.

15.2.4 *Numerical model of salt solution*

The numerical modeling is briefly represented from Russo's publication.[6]

The initial radius and concentration for the incremental oil-brine interface level, injection and production levels, and the injection flow rate are defined for each case.

At each time step the concentration in the mesh increment containing the stagnation level is updated by a mass balance between the injection fluid, the remaining brine in the increment volume and the salt which diffused and dissolved during one time step.

The diffusion coefficient is a function of the concentration gradient and is calculated by the equation:

$$D = D_{mol} + D_o \left(\frac{dC}{dz}\right)_+^{1/2} \text{Min} \, (r^2, l^2)$$

where:

D_{mol} = molecular diffusion coefficient;
D_o = empirical determined eddy diffusion parameter;
$(dC/dz)_+$ – specific gravity gradient (when positive);
l = mixing length (determined by diffusion equation).

The solution to any differential equation is determined by its boundary conditions. The boundary condition at the stagnation level is computed at each time step from the values at the previous time step and errors tend to accumulate. The cavern volume and shape are very sensitive to the boundary values used, so it is important to limit the errors on these values. This is accomplished by performing a global mass balance at each time step and computing a correction factor for the concentrations and boundary conditions to be used in the next time step. This forces the mass concentration in the time integration to follow a self-consistent and self-correction path.

The total mass of brine in the cavern, M_T, is computed by the time integral:

$$M_T = m_{co} + \int_o^T \left(\sum_l^N \frac{VSR}{\Delta t} C_{salt} + Q_i C_i - (Q_o + Q_{fill}) C_p \right) dt$$

where:

m_{co} = initial mass of brine in the cavern;
Q_o = outlet volume flow rate for no oil flow;
Q_{fill} = oil volume flow rate;
C_p = brine SG at the production level;
T = time period;
N = number of mesh intervals used;
VSR = volume of salt removed from the increment Δz in the time increment Δt;
m_{co} = initial mass of brine in the cavern.

The total mass of brine in the cavern M_c is computed by:

$$M_c = \sum_{l=1}^N \pi r^2(l) \Delta z C(l)$$

the correction factor for the stagnation level boundary condition is the found by:

$$\text{Corr. Fac} = \frac{M_T}{M_c}$$

This factor is always close to 1 and is printed out with each result. A value of 1 for the correction factor only means, of course, that the calculation is self-consistent, and not that it is modelling any physical situation correctly.

15.2.5 *Comparison between model and field solution data*

Russo presented calculated data and in situ measured data for two solution mining caverns at West Hackberry, Louisiana(USA), which had been leached by the direct mode. The first cavern (WH 101), after 147 days of leaching had a calculated and field measured volume in agreement within 3% however, the calculated radii at the bottom of the cavern are larger than the sonar measurements by up to 16% (Figure 15.2.2). The average measured flow rate of 123 880 barrels/day is most likely accurate, since the volumes agree so well. Because the injection point was, at the end of the flow, buried up to 200 m below the

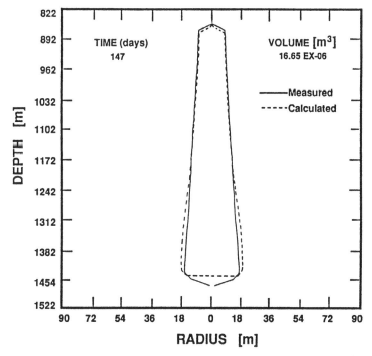

Figure 15.2.2. Comparison of calculated and measured cavern shapes after direct leaching (WH 101)

insoluble level, there were a great deal of entrained insoluble suspended in the lower cavern region, which might reduce the dissolution rate. If that is not the case of this discrepancy, further modifications to the plume mixing model should be considered.

A similar comparison was made for the second cavern (WH 102) which was leached for 156 days with an average measured flow rate of 138 484 barrels/day (Figure 15.2.3). The calculated volume is 12% greater than the measured volume, which indicates that the flow measurements are probably high. The same pattern of over predicting the cavern radii at the lower end is 20%. For smaller diameter caverns where insoluble depth is not so large, this over prediction has not been observed.

For the case of reverse leaching of Bryan Mound cavern 106, the injection level was set at 180 m below the oil blanket and the comparison between the measured and calculated solution mining cavern shape is illustrated in Figure 15.2.4. The total measured cavern volume is less than 1% of the calculated value. Near the bottom of the upper bulb, the radii differs approximately 20%, most likely for the same reason as in the case of the previous two caverns. In addition to errors in the dissolution rate model, it should be remembered that the cavern radii shown are effective values, and that in the majority of cases, the actual cross-section deviates markedly from circularity so that the axisymmetric approximation is poor. In all cases the actual vertical distribution of insoluble was represented by a single number. The estimated error in the meaasured flow rates used in the calculations was 10%.

Russo further described the use of the code to predict cavern growth during oil blanket withdrawal as illustrated in Figure 15.2.5. West Hackberry cavern W.H.11, which was formed during solution mining exploitation of salt, has been filled with oil. The predicted changes in production cavern shape and volume as the oil is withdrawn by displacement

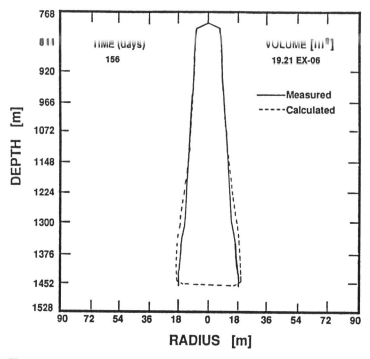

Figure 15.2.3. Comparison of calculated and measured cavern shapes afer direct leaching (WH 104)

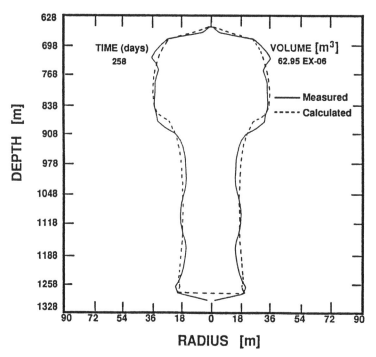

Figure 15.2.4. Comparison of calculated and measured cavern shapes after reverse solution (BM 106)

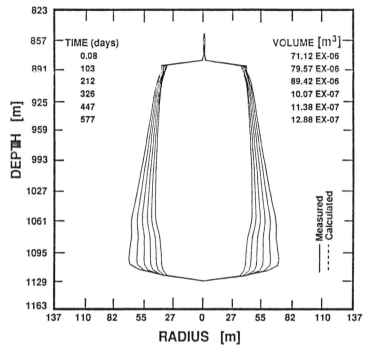

Figure 15.2.5. Calculated production cavern enlargement for five oil withdrawal cycles (WH 11)

with fresh water and refilled for five complete cycles, are indicated in Figure 15.2.5. The lower portion of the cavern which is exposed for the longest time, since the oil blanket rises during withdrawal, enlarges at the fastest rate. This type of withdrawal prediction has been made for each cavern to investigate the possibility of cavern coalescence and to evaluate dome subsidence.[8]

In conclusion it could be pointed out that the mechanics of salt solubility considers a vertically stratified salt mass. It uses a balance model to calculate the bulk brine salinity which is coupled with an unconfined plume model and an empirical dissolution model to determine the wall recession rate.

15.3 STABILITY CONSIDERATIONS OF CAVERNS

The stability investigations of caverns formed by solution mining, are considered for several different aspects in this chapter.

15.3.1 *Cavern convergence*

The assessment of cavern convergence during in situ salt solution is a very simple technique, which has been described by Crotogino.[9]

With the present state-of-the-art of sonar surveying it is not possible to detect local displacements at the cavern wall from $u_r < 0.2$-0.5 m. In addition no reference measuring points can be defined due to the irregularity of the cavern wall. As in the case of a 350 000 m³ cavern, a mean wall displacement of $u_r = 0.2$ m corresponds to a convergence of close to 2%. This method is only applicable for long-term in situ tests. Within reasonable periods only the integral convergence in relation to the brine volume, which is discharged, pumped off or rises within the well hole/cavern, can be measured.

The brine volume is affected by the following factors:

  – cavern closure;
  – thermal expansion of brine due to heat flux from the rock;
  – subsequent leaching;
  – decompression of the fluid due to decreasing pressure (if any);
  – fresh water injection (if any).

Thermal expansion of brine:

$$\Delta V = \alpha_{th} V_o \Delta T$$

$$\Delta T = \frac{dT}{dt} \Delta t$$

where:

  t = time (d);
  T = temperature (K);
  V_o = cavern volume (m³;
  ΔV = volume expansion (m³);
  $\alpha = 4.5\,E-4$ = isothermic expansion coefficient of saturated brine $(1/K)$.

Example of assessment of dimensions:

$$t = 0.03\,K/d\,.\,100\,d = 3\,K$$

$$V = 4.5\,E-4\,1/K\,.\,3.5\,E\,5\,m^3\,.\,3K = 473\,m^3$$

In 100 days the brine in a 350 000 m³ has expanded by 473 m³.

Subsequent leaching:

$$V = -V_o\,0.085\left(\frac{c_2}{\rho_2} - \frac{c_1}{\rho_1}\right)$$

where:

  c = brine concentration (kg/m³) (mass of salt/volume of brine);
  ρ = density of brine (kg/m³).

Example of assessment of dimension:

$$V = -350\,000\,m^3\,0.085\left(\frac{316}{1200} - \frac{310}{1196}\right)$$

$$V = -123\,m^3$$

Subsequent leaching from brine density 1196 to 1200 kg/m³ leads to a volume loss of approximately 123 m³.

Volume expansion upon decompression:

$$\delta V = -k\,V_o\,\Delta p$$

where:

  $k = 2.8\,E-5$ = compressibility of saturated brine at 25°C (1/bar);
  δp = change in pressure (bar).

Example of assessment dimension:

$$\Delta V = 2.8\,E - 5\,(1 \text{ bar}) \cdot 350\,000 \text{ m}^3\,(-1 \text{ bar})$$

$$\Delta V = 9.8 \text{ m}^3$$

A decompression of 1 bar leads to a volume of expansion of 9.8 m³ for a brine volume of 350 000 m³.

The method of convergency surveying described by Crotogino is commonly used in many solution mining operations around the world.

15.3.2 *Cavern creep behaviour*

The prediction of creep behaviour of salt caverns is based on the in-situ temperature, depth of cavern, local virgin stress, and the creep properties of the salt usually determined in the laboratory. The consideration of cavern creep behaviour is particularly important when multi-mineral salt strata exist, and are mined at great depth, as for example magnesium-bearing salts in the Netherlands. In this case, the main element of ground control is to increase the cavern brine pressure which sustains the creep of highly deformable mineral beds of bischofite and carnallite. This should be considered as a preventive measure aimed to maintain an undeformed open cavern of the mining section and to minimize the convergence during salt solution and after.[3]

Buyze and Lorenzen illustrate the differences in creep behaviour of salt minerals of a magnesium-bearing deposit (Figure 15.3.1) as outlined below.

The Veendam lithology comprises only halite layers of substantial thickness, while the creeping minerals, carnallite and bischofite, occur only in association with halite and

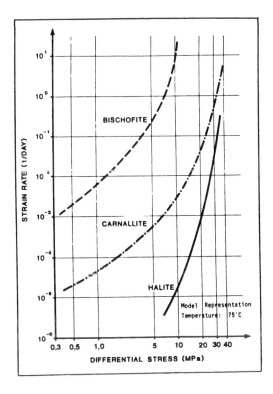

Figure 15.3.1. Generalized creep behaviour of three salt minerals

kieserite. Those conglomerates do not behave as pure minerals. However, to what degree creep flow reduces under local conditions of mixing, structural deformation and sequence of strata is not precisely known yet; assumptions are required for calculations and models. Therefore, field observations must build the bridge between laboratory testing results and in situ reality. Systematic recording of influx and afflux volumes under a variety of operational conditions can provide the basis for real creep behaviour evaluation.

Anticipation response of bischofite creep on even small pressure differences of the order of 1 MPa (equals 10 bar), cavern fluid pressures should be chosen as close to overburden pressure as possible.

At a depth of 1700 m the overburden generates a pressure of about 37 MPa in a layer containing bischofite. To compensate this pressure by a column of injection water, about 20 MPa well head pressure is required, while the brine pressure at the well head may be 16 MPa. So far, the pumping energy to establish these pressure levels is irrecoverable. The expense must be regarded as a tribute to the recovery of a mineral that, due to its flow and dissolution capacity, disappeared from most other evaporite deposits.[3].

15.3.3 *Cavern deformability*

Ore recovery in the case of solution mining of salt deposits at great depth mainly depends on the time-dependent deformations of the cavern structure and surrounding strata. The cavern deformation due to short-time loading may be of interest for shallow mono-mineral solution.

The accelerated convergence of caverns have damaging consequeces such as cavern cave-in, sub-surface and surface subsidence. To prevent such dangerous instabilities, it is necessary to maintain in the cavern near overburden pressure, and bring the convergence to a minimum. For example, a stress difference of only 1 MPa might eliminate the tensile stresses, even in the case of back spans of several hundreds of meters.

Buyze and Lorenzen described for magnesium salt solution in the Netherlands that faster salt influx is, in contrast, anticipated in the carnallite and bischofite zone. As a result, a pressure difference between the cavern fluid and the far field rock is being transmitted at a high rate through the salt layers of high ductility. Consequently, if those layers intersect adjacent caverns, the carnallite and bischofite stress inside the pillars will be affected from all sides and will finally approximate to cavern fluid pressure.

With salt properties established by testing pure samples, pressure equilibrium through caverns and pillars would be reached after about two years, within the well phase of active mining. Necessarily, the overburden would then be solely supported by the salt mass outside the perimeter of the cavern field (Figure 15.3.2).

The described assumptions of this process have led to an unusual model for estimating long-term post-abandonment cavern convergence. The model – 'analytical' and 'finite element method' versions have been worked out – ignores salt pillars and individual cavern size and geometry. Cavern field parameter, layer thickness, pressure deficit and time are the governing parameters.

The volume of salt which flows toward the caverns determines the degree of subsidence which may theoretically occur at the surface. The rate at which underground closure is transmitted upward to the surface and the resulting subsidence geometry depend on local geological conditions. Finite element model analyses for a constant pressure deficit of 1 MPa showed that the potential subsidence might be even a few tens of centimeters after fifty years.[3]

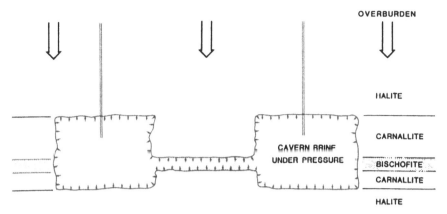

Figure 15.3.2. Theoretical stress equilibrium of cavern pillar due to bischofite viscosity

It should be pointed out that idealization used for model calculations might be to some point in disagreement with the real, natural rock and salt conditions, and hence cavern behaviour, which would be more favourable than finite element modelling indicated.

15.3.4 *Structural stability of caverns*

In regard to the structural stability of caverns two structures are of fundamental importance, namely: salt pillars between caverns (inter-cavern pillars) and salt pillars above caverns (sill pillars). The key factor of the pillars' stability is that their bearing capacity is greater than the superimposed load on them. Further consideration is given separately for each type of pillar.

1 *Inter-cavern pillar* structural stability is determined by three basic parameters.[10, 11]
   a) The long-term loading compressive strength of the salt is approximately 65% of the short-term loading strength (Figure 15.3.3). The decreased strength is:

$$\sigma_{cd} = 0.65\,\sigma_c$$

where the corresponding safety factor is:

$$n = \frac{1}{0.65} = 1.54$$

For the design of the bearing capacity of inter-cavern pillars it is necessary to include a long-term loading strength:

$$\frac{\gamma h}{0.65\,\sigma_c} = \frac{1}{2}\left(1 - \frac{D^2}{S^2} + \ln\frac{S}{D}\right)$$

where:
γ = overburden unit weight;
h = depth of bottom of cavern;
σ_c = short-term loading strength;
D = cavern diameter;
S = distance between cavern centres.

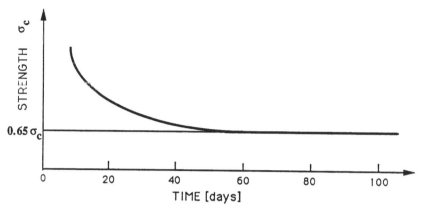

Figure 15.3.3. Compressive strength of salt for a long period of loading

b) The proportionality between the distance of the caverns and the cavern diameter depends on the depth of solution mining (Figure 15.3.4):

$$\frac{S}{D} = f(h)$$

This factor is important for sizing of the inter-cavern pillars which relates to their bearing capacity. In addition the proportionality between pillar diameter (D and pillar height (H), called the slenderness ratio, should be determined. For example, for mining of the Tetima

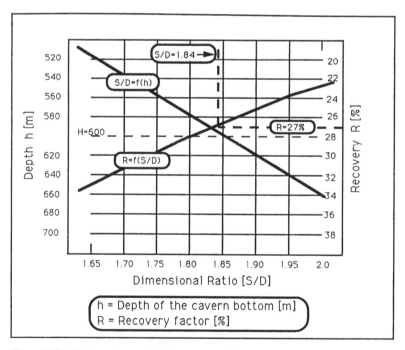

Figure 15.3.4. Design charge of the inter-cavern pillars

salt deposit (former Yugoslavia), the height of the cavern is 80 m and its diameter is 20 m, which gives the following slenderness ratio:

$$\frac{H}{D} = \frac{80}{20} = 4$$

This is considered to be satisfactory.

c) The salt recover factor (R) directly relates to the magnitude of superimposed load on the inter-cavern pillars (tributary concept). The recovery factor can be expressed by the relationship between an equilateral triangle and a circular surface:

$$e = \frac{pD^2/8}{S^2 \sqrt{3/4}} = \frac{\pi}{2\sqrt{3}} \left(\frac{D}{S}\right)^2$$

The extraction factor determined by this equation may be adequate for dry mining, but it is too high for solution mining where salt is recovered from the irregular caverns, not from the perfect cylinders. In the case of solution mining it should be scaled down by up to 50%.

The main stability factor of inter-chamber pillars is that with an increase of depth of solution mining there must be accordingly an increase in their size.

2. *Sill pillars* above caverns must have adequate strength and stability to prevent caving of cavern backs, sub-surface subsidence, water inflow from aquifers and others.[12] The stability parameters of sill pillars (Figure 15.3.5) are listed in the following order:

Q = dead load (salt beam)
    $(\sigma_v + d\sigma_r)\,\pi\,r^2$ = overburden load
    $c \cdot 2\pi r \cdot dh$ = cohesion force
    $\sigma_H \cdot 2\pi r \cdot dz \cdot tgf$ = lateral force.

Considering the boundary equilibrium equation, where all forces are acting on vertical planes

$$(\sigma_v + \sigma_r)\,\pi r^2 + c \cdot 2\pi r^2 \cdot dh + \sigma_H \cdot 2\pi r \cdot tg\phi dh - \sigma_v \pi r^2 - \pi r^2 - \pi r^2 dh\gamma = 0$$

For separated variables the differential equation is given:

$$\frac{dh}{\gamma} = \frac{-d\sigma_V}{2c - r\gamma\, tg\,\phi 2\sigma_V}$$

where:
    dh = element of thickness of sill pillar;
    h = depth of sill pillar;
    σ_V = overburden pressure ($\gamma h = P_V$).

By integration of this equation, the sill pillar thickness is obtained which should correspond to the required structural stability.

$$H = \frac{r}{2\,tg\,\phi}\, \ln \frac{2c - r\gamma + 2P_V}{2c - r\gamma}$$

The calculated thickness of the sill pillar (H) has to be corrected by a safety factor ($F_s = 1.54$).

The same mechanical law applies for sill pillars as for inter-chamber pillars. With the increase in depth of solution mining there must be accordingly an increase in the thickness of the inter-chamber pillars and sill pillars.

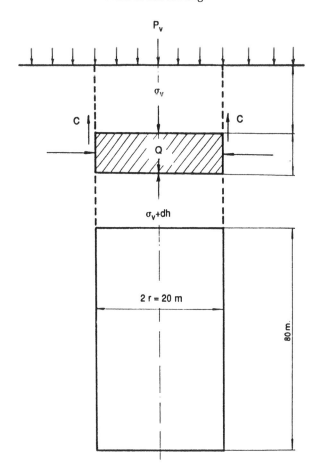

Figure 15.3.5. Loading mechanics of the cavern seal pillar

15.4 STRESS ANALYSIS OF CAVERN LAYOUT

The layout of controlled solution mining might vary in regard to its depth as well as possible utilization of mined-out caverns for storage purposes, as discussed in Chapter 16. The controlled solution mining at greater depth most likely will require utilization of the caverns for storage, to offset the high cost of deep salt exploitation. In this case extensive stress analyses are required by the use of models, as discussed in the next chapter.

Stress analysis of cavern layout is discussed on the basis of Fisher's investigations.[13]

15.4.1 *Stress consideration of cavern layouts*

The general layout of a cavern for deep horizons of solution mining is illustrated in Figure 15.4.1. The layout indicates the lateral extension of mining along the longitudinal section of the salt dome, with a possibility of solution mining also at the second horizon. Consideration of cavern spacing in the lateral and vertical directions is of particular importance for solution mine stability and significantly increases the need for facility design optimization. Case history information on the performance of a cavern in the area of consideration for

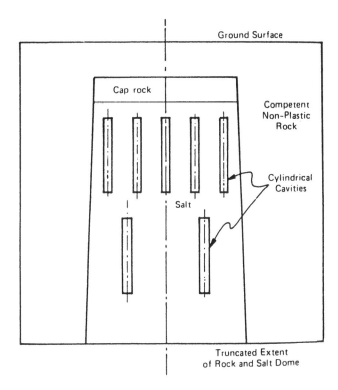

Ground Surface

Cap rock

Competent
Non-Plastic
Rock

Cylindrical
Cavities

Salt

Truncated Extent
of Rock and Salt Dome

Figure 15.4.1. Schematic
idealized layout of caverns

developing new caverns, cannot be used as the sole basis for layout design. To achieve adequate stability of the solution salt mine it is necessary to know the stress distribution and concentration around and between caverns. For stress evaluation it is necessary to know the virgin stress of the salt deposit and the mechanical properties of the salt. Insufficient knowledge of stress distribution and concentration can lead to failure and caving of the caverns.[13]

The main elements for cavern layout analysis are: cavern morphology, cavern pressure, the in situ state of stress, and the mechanical properties of the salt. When the problem is defined, stress can be defined through mathematical models using either boundary element or finite element codes. Figure 15.4.2 illustrates the result of a stress analysis performed by Fisher. Axisymmetry and elastic behaviour of the salt have been assumed. The virgin stress of the salt body was assumed to be isotropic and liberated of shear stress, having completely relaxed over geological time. Yield of the salt has been precluded by the assumption of elastic behaviour. An examination of the maximum shear stress within the salt gives some indication of the disturbance to the salt caused by the introduction of the pressurized brine-filled cavern. The stress pattern around the cavern (Figure 15.4.2) can be used, according to Fisher, for the following purposes.[13]

1. It introduces the approach of axisymmetric idealization.

2. It introduces the concept of examining only a portion of the salt surrounding the cavern.

3. It indicates a large magnitude of shear stress in the immediate vicinity of the cavern and, hence, suggests that plastic yield of the salt may occur.

4. It suggests the extent of induced stress influence due to cavern excavation.

5. It demonstrates the gradual vertical variation of salt stress (away from the ends of the

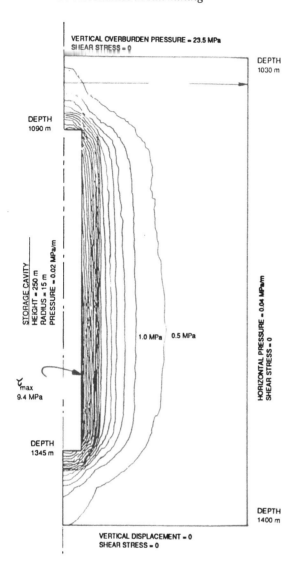

Figure 15.4.2. Pattern of maximum shear stress concentration around a cylindrical cavern filled with brine

cavity) which suggest that, over this region, dimensional stress analysis can be simplified by a one-dimensional radial problem assuming generalized plane strain in the vertical direction.

The following topics further analyze stress mechanics of solution mining cavern layout.

15.4.2 *Stress interaction between caverns*

Stress distribution around a single cavern is rather simpler than for the case of multiple caverns, which are spaced in the domain of stress interaction. The domain of stress interaction in salt caverns is extended, because plasticity solutions indicate that stresses beyond the yield zone surrounding a cavern are greater than would be predicted by an elastic solution.

The problem of stress interaction corresponds to the cavern geometry shown in Figure 15.4.3, where $R_1 = R_2$ and $P_1 = P_2$. The character of stress interaction occurring between two spaced caverns is shown in Figure 15.4.4. The interaction stress consideration is analyzed for a cavern depth of 828 m and where:

$$\sigma_H = \sigma_V + 3\,\tau_o\,/\,\sqrt{2}$$

where τ_o is the octahedral shear strength.

The spacing ratio, a relationship between the distance of caverns (S) and the diameter of the caverns (D), for interaction stress analysis has been varied ($s/D = 7$, $S/D = 4$, $S/D = 3$)

As the caverns become more closely spaced, the regions of yield (shaded parts of stress diagrams) surrounding each cavity coalesce, wherein $\tau_o \geq K_o$. The configuration of stress regions is also observed to change as does the elastic stress beyond the yield region. Under the assumption of vertical plain strain, Serata has found the volume (or area) closure of these caverns[14] to be around 15% for $S/D = 4$ and around 17% for $S/D = 2$. It is obvious that the spacing ratio among caverns is an instrumental factor of interaction stress concentration and cavern deformation. There are some opinions that the spacing between caverns should be large enough to prevent coalescing of the respective yield zones surrounding cavities. In this case, a spacing ratio (S/D) would be around 5-6. Such a spacing of caverns should be excessive in comparison with a spacing of the producing cavern in the field, which exhibits long-term stability with less spacing distance.

The closure of caverns in an idealized infinite layout is illustrated in Figure 15.4.5. Consideration is given to two principal cases.

1. For $S/D > 4$, the radial closure of the infinite layout of caverns is substantially lower than that found in the two-cavity analysis. This appears to be an anomaly until it is recognized that the symmetry planes of the figure above act to 'hold back' the salt from flowing inward into the cavern. This beneficial effect dominates as long as a portion of the salt surrounding the cavity remains in the elastic state.

2. For $S/D < 4$, essentially all of the salt throughout the cross-section yielded. Thus, competent support for the overburden load ceased to exist and large radial closure was in effect. Based on these results, a decision might be made that the spacing factor should be maintained as $S/D > 4$. Fisher pointed out that the problem formulation of Figure 15.4.5 both possesses and lacks conservatism, for example:

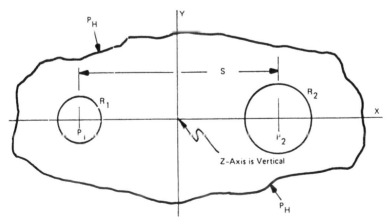

Figure 15.4.3. Layout of two cylindrical caverns in interaction

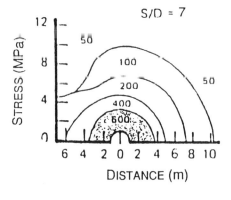

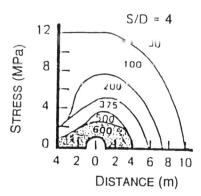

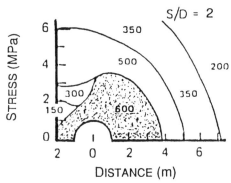

Figure 15.4.4. Octahedral shear stress distribution due to interaction of two identical cylindrical caverns filled with brines for three different spacing ratios (after Fisher)

a) End effects of caverns are ignored (conservative).

b) As applied to large but finite arrays of caverns, no vertical support by unyielded salt or other sedimentary layers has been considered (conservative).

c) The assumption of fixed symmetry boundaries for large but finite arrays of caverns would not be valid (unconservative). There would be a general tendency for all the caverns and salt surrounding the caverns to move inward toward the most central cavern, i.e., the symmetry boundaries would lose some of their effectiveness.

To improve confidence in predicted cavern stability, more sophisticated twodimensional or threedimensional models are needed.[13]

It should be noted that the spacing ratio varies with the depth of caverns, as discussed in a previous sub-chapter, where spacing determined by analytic analysis for a single cavern is lower than in the case where interaction between caverns is considered.

15.4.3 *Stress analysis of the Clovelly dome caverns*

The Clovelly dome is relatively shallow, with its top at a depth of 360 m. At a depth of 450 m, the dome has an irregular circular shape, with a minimum cross-sectional distance of 1000 m. With depth the horizontal area of the dome slightly increases.[13]

The east-west cross-section of the Clovelly dome and the layout of the solution caverns is given in Figure 15.4.6. In this figure a stress envelope around the layout of the caverns is shown. It should be noted that the stress envelope is arching out of the salt dome. Under these circumstances a great deal of vertically induced stress might be transferred outside the

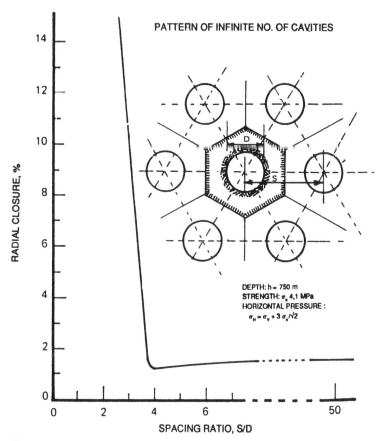

Figure 15.4.5. Radial closure of hexagonally layout of brine-filled caverns in function of spacing ratio (after Fisher)

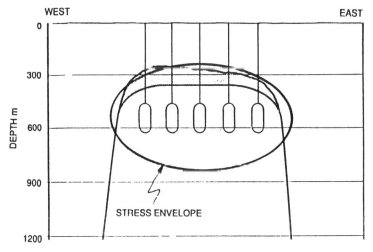

Figure 15.4.6. Clovelly salt dome with caverns layout, which is exposed to stress indicated by stress envelope

salt in the more competent rock which surrounds the dome. It should be expected that the caverns are exposed to stress only by dead load below an arch. This phenomenon is very well known in rock mechanics as a stress diversion on abutment support, as discussed in Chapter 12.

Figure 15.4.7 shows the ring idealization of the Clovelly salt dome. An idealized horizontal cross-section of the dome (circular shape) contains hexagonally arranged cylindrical cavities and an axisymmetrical idealization of a vertical section of the dome caverns. In this case a great deal of symmetry can be introduced, provided the cavities are identical and equally pressurized, as assumed. In particular, radial lines of symmetry exist every 30°. The cavern at the centre of the dome has a ring of six caverns. The T_a radius of this ring is S, which corresponds to a distance between centres of two caverns, e.g. cavern spacing. Surrounding this ring of six cavities is a ring of twelve cavities. The radius of this ring is

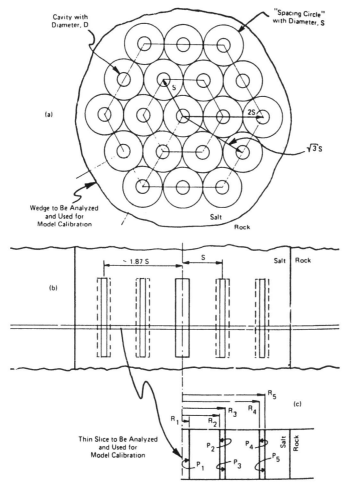

Figure 15.4.7. Idealized horizontal cross-section of a hexagonal cavern layout (a); an asymmetric idealization of the vertical section of cavities (b); and slice element for model calibration (c) of the Clovelly salt dome (after Fisher)

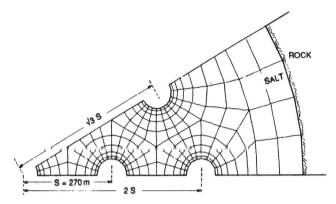

Figure 15.4.8. Wedge of 30° from horizontal cross-section of the Clovelly salt dome. Wedge analyzed assuming generalized plane strain. The radial lines are lines of symmetry (after Serata)

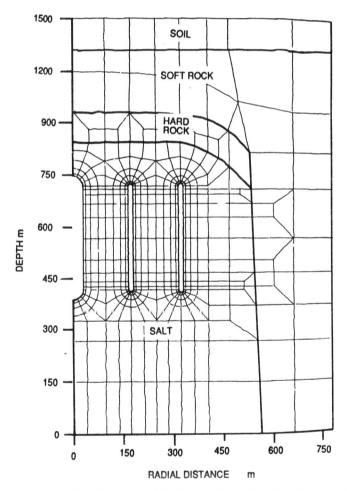

Figure 15.4.9. Axisymmetric idealization of the Clovelly salt dome. Discrete cavities have been replaced by annular rings (after Serata)

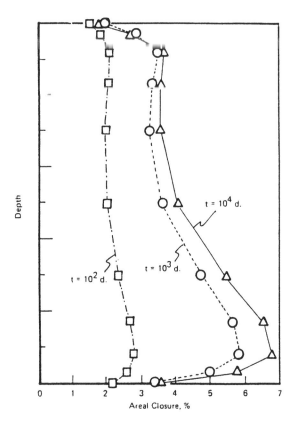

Figure 15.4.10. Sectional closure of centre cavern due to salt creep in function of depth, for the Clovelly salt dome (after Serata).

$$(2 + \sqrt{3})\, S/2 = 1.87\, S.$$

The approximation of axisymmetry is introduced by replacing the discrete cavities surrounding the centre cavity with 'calibrated axisymmetric rings', as shown schematically in Figure 15.4.7b. The mean radius of the inner cavity ring is that of the ring of discrete cavities it replaces, i.e. S. The radial width of this annular ring is computed to make the area of this annulus equal to that of six discrete cavities, i.e., $6(\pi D^2/4)$. Similarly, the outer cavity ring has a mean radius of 1.87 S, while its width yields an annulus with an area equal to $12\,(\pi D^2/4)$.

Fisher gave a generalized plane strain solution, which corresponds to the axisymmetric geometry of Figure 15.4.7c and the 30° wedge shown in Figure 15.4.7a. The proprietary finite element computer program, REM, developed by Serata, was used for these computations. Figure 15.4.8 shows the finite element discretization of the 30° wedge of salt. For both of these analyses, the rock surrounding the salt-cavern system extended radially to 3460 m, and to a depth of 1200 m.

Solutions were sought near the top, middle and bottom of the proposed caverns at a depth of 480 m, 630 m and 780 m respectively.[13]

The calibrated asymmetric rings have been incorporated into the full asymmetric problem shown in Figure 15.4.9. The sectional closure of the centre cavity is shown in Figure 15.4.10 as a function of depth for 10^2, 10^3, and 10^4 days. The maximum closure 7% shown could contract with the volume closure (10-17%).

REFERENCES

1. Jeremic, M.L. 1964. Dry mining and solution mining of rock salt in Romania. *Min. & Metall. Journal* (Belgrade), 12: 265-274 (in Yugoslav).
2. Avdagic, M. 1988. Basic information of salt cavern solution in salt deposits. *Min. & Metall. Journal* (Belgrade), 39 (4): 1-7 (in Yugoslav).
3. Buyze, D. & H. Lorenzen 1986. Solution mining of multi-component magnesium – bearing salt – A realization in the Netherlands. *CIM Bulletin* 79 (889): 52-60.
4. Yuanxiang, L. & N. Chengxun 1983. Technical development of solution mining in thinly bedded rock salt deposits of Zilinjing, Szechuan, China. In *Proc. of 6th Int. Symp. on Salt*, The Salt Institute, Virginia, Vol. 2, pp. 87-90.
5. Jeremic, M.L. 1964. Methods of development and exploitation of salt deposits in Poland. *Min. & Matall Journal* (Belgrade), 32 (2): 32-42 (in Yugoslav).
6. Russo, A.J. 1983. Solution mining calculations for SPR caverns. In *Proc. of 6th Int. Symp. on Salt*, The Salt Institute, Virginia, Vol. 2, pp. 101-109.
7. Morton, B.R. et al. 1956. Turbulent gravitational convection from maintained and instantaneous sources. In *Proc. of the Royal Society*, Series A , 234 (1196): 1-23.
8. Russo, A.J. 1981. A solution mining code for studying axisymmetric salt cavern layout. *Sandia Report* SAN D81-1231, pp. 1-32.
9. Crotogino, F.R. 1981. Salt cavern in situ testing from the constructor's and from the operator's viewpont, pp. 615-628. In *Proc. of 1st Conf. on Mech. Behaviour of Salt*, Penn State Univ. Clausthal: Trans. Techn. Publ.
10. Avdagic, M. et al. 1986. Stability of high cylindrical chambers with a large diameter in salt rock deposit, Tetima. In *ATTI Proc.*, pp. 306-314. Int. Congr. on Large Underground Openings, Firenze, Italy, 8-11 June, 1986.
11. Popovic, R. et al. 1983. Dimension of seal pillars and interchamber pillars in solution mining, pp. 16-27. In *Almanac for Mining and Geology*, Tuzla, former Yugoslavia.
12. Radomski, A. 1981. Exploitation of the salt deposits by borehole solution mining. *Gornictwo* 8: 38-49. (in ...)
13. Fisher, J.F. 1983. An aximetric method for analyzing cavity arrays, pp. 661-680. In *Proc. of 1st Conf. on the Mech. Behaviour of Salt*, Penn State Univ. Clausthal: Trans. Techn. Publ. 9-11.
14. Serata, S. 1978. Geomechanical basis for design of underground salt cavities. *ASME Publ.*, 78-Pet-59.

CHAPTER 16

Storage caverns

Salt deposits are extensively used for underground storage of hydrocarbons and other products as well as the waste disposal and isolation. The storage caverns in salt deposits are made by solution mining technology, which is already discussed in Chapter 15. However, some specifics of the solution technology for construction of caverns are commented.

16.1 UTILIZATION OF STORAGE CAVERNS

Utilization of storage caverns is discussed mainly with four aspects, namely: cavern categorization of salt domes, storage of hydrocarbons, disposal of the radioactive and chemical waste, and storage of compressed air.

16.1.1 *Cavern categorization of salt domes*

The salt deposits, as mentioned before, have a very old and long history of utilization, particularly in regard to solution mining. At modern times the utilization of salt domes has been extended for some other uses, which required an additional knowledge of salt deposits.

Cavern categorization of salt domes have been studied by Martinez, as illustrated in Figure 16.1.1. He suggested an array of current and potential demands of the usage of salt domes:[1]
- dry salt mining;
- solution mining;
- fuel storage caverns;
- gas storage caverns;
- chemical solution storage (brine);
- nuclear waste disposal and isolation;
- chemical waste and disposal;
- air storage.

In addition to these modes of utilization of storage caverns in the salt dome, some consideration has been given to other uses, for example formation of a chemical reaction chamber and others.

According to Martinez, the implementation of some and possibly all of this set of current and potential uses of domes can be best effected by viewing the dome and its surrounding environment as a system.[1]

The investigations of salt domes in the USA have been intensive. Their studies indicated

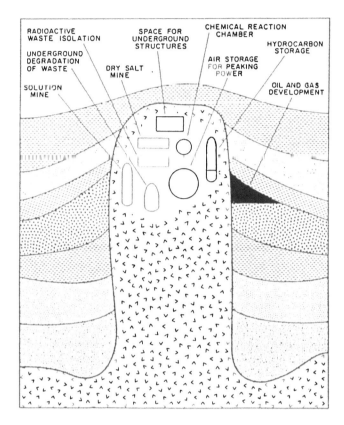

Figure 16.1.1. Cavern structures of salt dome utilization

that the determination of a dome system could be done by three principal geological components:
- the salt stock;
- the cap rock;
- the intruded host rock.

Figure 16.1.2 illustrates a model of these components, along with the various lithological materials that constitute them. In the growth and development of the salt domes a dynamic interaction occurs among the three sub-systems as identified above. In order to resolve problems related to the various subsets of the system, it is important to understand these interactions. The nature of this system and the inter-relations among its components have been recognized by Martinez in his studies. With the introduction of modern highly specialized studies into salt dome research, the value of interdisciplinary efforts focusing on the entire system becomes apparent.

Martinez gives some examples of benefits that may be derived from an understanding of properties of the salt dome system, as illustrated in Figure 16.1.3, and typical critical salt dome characteristics are listed along with operation-design paraameters to which they are related.

16.1.2 *Storage of hydrocarbons*

The storage of hydrocarbons relates to storage of crude oil, petroleum products and gas. It is

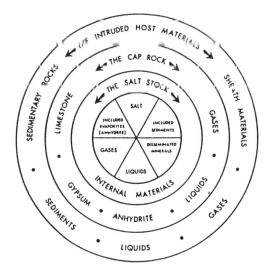

Figure 16.1.2. Model of structural components of a salt dome (after Martinez)

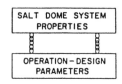

EXAMPLE:

OPERATION–DESIGN PARAMETER	CRITICAL SALT DOME CHARACTERISTIC
• Rate of Cavity Development	Character of Salt and Included Material
• Shape of Cavity	Internal Domal Structure
• Stability of Cavity	Mechanical Properties of Domal Material
• Blowouts	Gas and Stress Conditions
• Contamination	Gas and Liquid Inclusions
• Optimization of Domal Use	Geometric Configuration and System Properties
• Future Adaptation	All Possible
• Environmental	Complete Geomechanical System

Figure 16.1.3. Model of salt dome system properties related to the caverns design (after Martinez)

now possible to store billions of cubic meters of hydrocarbons safely and economically in stable salt caverns. The storage caverns have to meet the criteria necessary to assure stable and tight high-pressure, sub-surface storage vessels.

Certain particularities in regard to construction of storage caverns are discussed by Haddenhorst and Quast, and will be further discussed later.[2]

For construction of storage caverns there should be one drill hole or even in special cases several drill holes. The drilling technique is completely in accordance with the standard of the oil and gas industry. During drilling through salt formation a saturated clay-salt drilling fluid or oversaturated NaCl drilling fluid has to be used. The essential salt solution technology of storage caverns is illustrated in Figure 16.1.4. Fresh water or water with

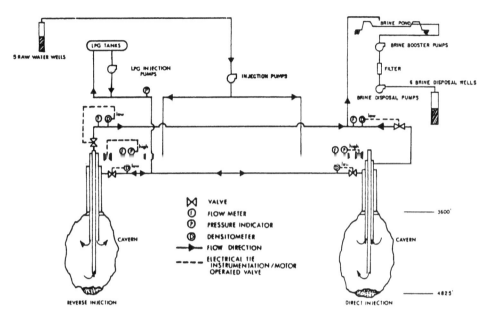

Figure 16.1.4. Construction of salt cavities by reverse and direct salt solution

slight salt content (e.g. sea water) is pumped into the cavern section to be leached. The water dissolves salt at the cavern walls and becomes more or less saturated. The fresh water that has been pumped in displaces the brine to the surface and is drained off. Via the outermost annulus, a protective medium (blanket) lighter than water and non-salt dissolving liquid is pumped into the top section of the cavern in order to limit the leaching process in an upward vertical direction, i.e., at the cavern roof. The depth region in which cavern storage space should be created can be determined by the depth of the brine/blanket interface as well as by the final depths of fresh water and brine strings. The leaching process is planned using a threedimensional finite element computer program. The growth of the cavern, particularly the control of the geometric development of the cavern, is undertaken at certain intervals by the Sonar survey method. A typical cavern shape with the respective intermediate survey as illustrated in Figure 16.1.5. This cavern has a final volume of 430 000 m^3 capacity with a pressure range of 10 MPa. and a working gas volume of 45 million. Such salt caverns in Germany normally are leached with rates of approximately 300 m^3/h. The construction time for such a cavern is approximately two years. In practice it is usual to leach several caverns simultaneously.[2] Haddenhorst and Quast also stated that with this technology the assurance of cavern stability and the minimizing of cavern convergence are of particular importance. For this purpose, extensive laboratory measurements regarding the physical behaviour of the salt are carried out on the appropriate cores. The mechanical stability of the cavern as well as the long-term convergence behaviour are calculated in a finite element program.[2]

The salt deposits are particularly convenient for the underground storage of natural gas. The depth of gas cavern storage varies in Germany between 200 and 300 m. The operational experience has been described by Haddenhorst and Quast, as briefly commented below.

There are about 1000 salt caverns in the Western world with a total geometric volume of

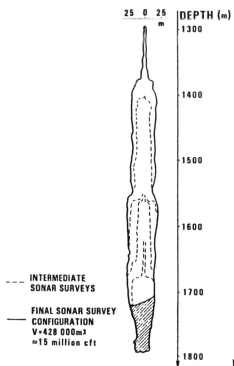

Figure 16.1.5. Phases of progressive cavern solution

approximately 215 million m³ for the storage of liquid and gaseous hydrocarbons. Approximately 10% of this cavern volume is used for the storage of natural gas. The gas volume stored therein is about 2 billion standard m³. The construction of storage caverns in the salt puts certain demands on the geological character of the salt reservoir, such as sufficient thickness and depth of salt deposits (< 200 m), and wherever possible, homogeneous salt composition which should be free of non-soluble or low-soluble layers.[2]

The seasonal usage of storage of natural gas in salt caverns is particularly suitable to meet peak requirements. Unlike aquifer storage, no friction pressure losses occur in the salt cavern. Peak production rates up to 1 million m³/h can be achieved by using a production string diameter of approximately 0.5 m. After overlaying a special fluid on the residual brine at the bottom (sump) of the cavern it is possible to prevent a rise in water vapour content and thereby in the water dewpoint of the gas stored. In this case it is not necessary to install a gas dehydration plant, which will result in the total cost of gas storage.

A schematic diagram of a salt cavern gas storage is shown in Figure 16.1.6, featuring the following:

1. main gas pipeline;
2. compressor station;
3. after-cooler;
4. separator (free water knockout)'
5. dehydration plant;
6. glycol regeneration;
7. methanol container (hydrate protection);
8. sump overlay.

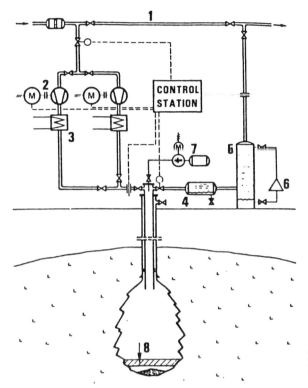

Figure 16.1.6. Schematic diagram of high-pressure gas cavern storage

In the event of a successful sump overlay, the installation of a gas dehydration plant and auxiliary equipment (items no. 4, 5, 6 and 7 as shown in Figure 16.1.6) become unnecessary.[2]

The same number of aspects which relate to the storage of hydrocarbons are directly or peripherly commented in the following subsections.

16.1.3 *Disposal of radioactive waste*

There are presently different concepts for the final disposal of radioactive waste from nuclear reactors, medicine and research. The opinion of the majority of experts is that disposal of radioactive waste in salt formations is a safe method of isolation of materials from the environment and hydrogeological system until the decay of the radionuclides has reached a harmless level. For example, some number of salt deposits formed about 200 million years ago have not been penetrated by water.

Quast and Schmidt proposed the basic concept of how to dispose medium- and low-level radioactive waste in salt caverns as a pellet distributed homogeneously in a cement slurry.[3] Different techniques were investigated to transport this mixture of pellets and cement slurry from the surface into leached caverns through boreholes having a minimal diameter, where it will solidify to a quasi-monolithic block (Figure 16.1.7).

The size of the underground cavern for the storage concept of nuclear waste is a compromise of three diverging options:[3]

a) Long-term use of the surface facilities which are bound to the borehole location.

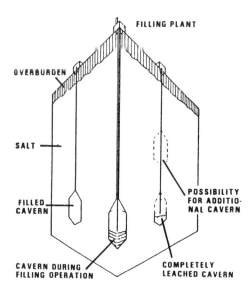

Figure 16.1.7. Cavern layout for disposal of radioactive waste

Despite the possibility of using some components for subsequent facilities, an operational life in the order of ten to fifteen years would be desirable from an economic point of view.

b) Smallest possible number of penetrations through the overburden into the salt dome. This produces greater safety, longer operational life and economical advantages during cavern construction. This requires cavern volumes of about 10^5 m³.

c) A sufficiently accurate statement with regard to stability of large open caverns without inner pressure at depths between 900 and 1000 m. This requires also cavern volumes of 10^5 m³ or several superimposed caverns.

The chosen single cavern volume of 75 000 m³ guarantees the deposition of waste, conditioned according to the storage concept, equivalent to the production of a reprocessing plant for a five-year operation. For this particular purpose, this cavern volume represents an upper limit for the given conditions, according to today's knowledge of rock mechanics and the resulting stability statement. Therefore, it is obvious that the stability analyses ranked very high for this special storage project.

The possibility of nuclear waste disposal in the salt caverns not from surface but from underground mine working has been considered for the Asse salt mine in Germany. The nuclear waste disposal will be in the series of the test caverns with a volume of 10 m³ located in an undisturbed rock region and filled with the waste within cement matrix. The setting behaviour of nuclear waste as well as the temperature development due to hydration heat resulting from the cement suspension, should be measured. These data will be essential for the design of the large 75 000 m³ salt caverns for nuclear waste disposal. The concept of this experimental project is briefly discussed here, as described by Quast and Schmidt.[3]

The underground openings (drifts) have been excavated by conventional technology drilling and blasting in regard to facilitating the solution of five caverns. The mine openings are driven in older halite of the Stassfurt series (Z2) with a length of 123 m and a profile of 4.5 × 3.0 m. The boreholes were drilled from the drift to start the solution of five caverns. The distance between the centre line of the caverns was 20 m to eliminate temperature and stress interactions. The range of depth of the caverns was set to be 15 to 18 m below the drift floor. In addition a pump sump of 1 m below the caverns was made. Parallel to this drift at the depth of 800 m, another drift with a length of 60 m was excavated for deposition of brines produced during the solution of the laid-out five caverns (Figure 16.1.8).

The boreholes for the solution mining of the cavern were 180 mm in diameter and had a depth of 19 m each. A pilot hole having a smaller diameter was drilled and stabilized drill pipes achieved an exact vertical position of the axis. Subsequently, the boreholes were enlarged up to the required diameter of 180 mm. The drilling was carried on by dry drilling methods, meaning that the drill cuttings were entrained by compressed air.

The insoluble or weakly soluble portions in different parts of the cavern ranged between 1.3% and 4%. The insoluble components of the rock salt should sediment in the sump at the bottom of the cavern. For deposition of the brine in the adjacent gallery a brine density of 1.20 was required by both mine management and mining authority. In this way undesirable leaching effects were not possible in this gallery. The total time needed for excavation of the gallery was 50 shifts. All mining work was carried out by the Technical Department of the Institute of Underground Disposal.

The construction of 10 m^3 caverns from the boreholes takes place by solution mining with fresh water. This fresh water had to be transported from the surface to the 800 m level of the mine.

16.1.4 *Disposal of chemical waste*

The Dutch salt industry developed a technology of the chemical waste disposal from the raw brine purification plant.[4] The waste is disposed in the same circuit into the salt caverns of the nearby Hengelo solution mine, as illustrated in Figure 16.1.9.

The waste is composed of the mixture of calcium carbonate, calcium hydroxide, gypsum and magnesium hydroxide. The mixture is slurrified, with waste waters and waste brines from the production plants in a storage tank equipped with a mixer. At slight overpressure the waste slurry is pumped with centrifugal pumps through a pipeline (D = 15 cm) into

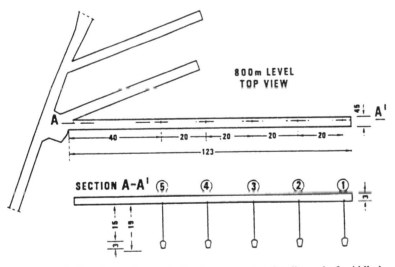

Figure 16.1.8. Experimental array in the Asse salt mine for disposal of middle-level and low-level radioactive waste

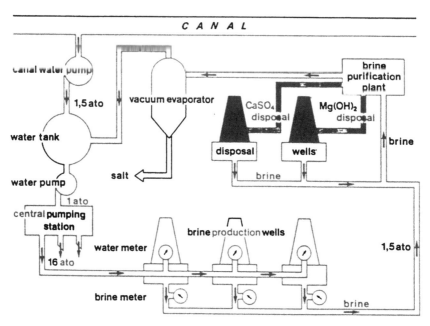

Figure 16.1.9. Brine production and waste disposal system at Hengelo (Netherlands)

cavities of exploited wells. The quality of the displaced brine is controlled and determines the disposal capacity of the cavity. The same cave in roofs of caverns influenced the decision that the caverns should be filled up as far as possible in order to prevent caving and intensive surface subsidence (Figure 16.1.10). To prevent easy settlement of calcium carbonate particles in the pipeline, a ratio of solid – liquid of 30% by volume of solids has been chosen. As Wassmann wrote in his paper 'Cavity utilization in the Netherlands', a total volume of 212 000 m³ of solids were disposed into the solution mining cavity of an original volume of 191 000 m³. The enlargement of this cavity to a volume of about 248 000 m³ was the result of the dissolving action of the unsaturated slurry. 15% of the cavity volume was still filled with brine when the well was closed, which produced an additional 190 000 tonnes of salt on top of the 414 tonnes of salt extracted by solution mining (Figure 16.1.11).[4] Wassmann further described the change-over solution mining system at Hengelo, from a single to double or even triple well system. After closing well 15, disposal activities were directed toward the elongated cavities of a double well. Several of these old hydraulic fracturing wells were also selected for disposal and shut down in an earlier than normal stage of production. This was done to avoid an irresponsible enlargement of the cavity by the unsaturated waste brine during the disposal phase and the problems encountered in keeping the blanket in place during the production stage.[4]

In these elongated cavities the angle of discharge of this waste into a fluid medium is very important to the ultimate filling capacity. In the laboratory tests, an angle of 20° was measured, and this measurement was confirmed in practice by the echo-soundings in several disposal wells. The filling rate can be reckoned to be 70% for the elongated cavities with their relatively unfavourable height-to-diameter ratio. For example, in wells 102-103 about 45 000 m³ of solids were already stored at the moment of the echo-sounding (Figure 16.1.12). It is expected that in total more than 80 000 m³ can be disposed of, exclusive of the

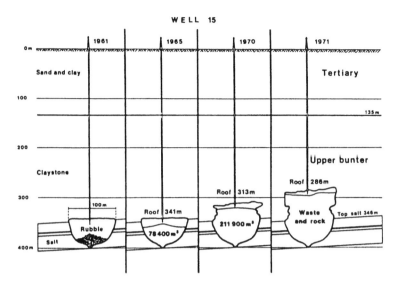

Figure 16.1.10. History of filling cavern by the waste disposal at Hengelo (Netherlands)

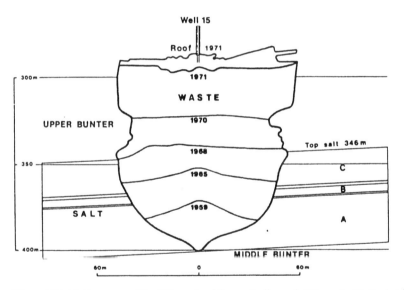

Figure 16.1.11. Summary of the filling rate of the waste disposal at Hengelo (Netherlands)

additional space by the cavity enlargement caused by the dissolving action of the unsaturated waste brine. These wells will be taken into production again after filling their cavities with up to 70% solids from the brine purification plant. This procedure can be followed only if the waste brine is of a similar type and quality as the production brine. There should not be any pollution of the displaced brine by the disposed waste.[4]

Wassmann stated that usage of the cavities produced by solution mining for chemical waste storage by displacement method should not be used because of the chance of pollution of the production brine by waste. A concept of waste storage in the caverns made

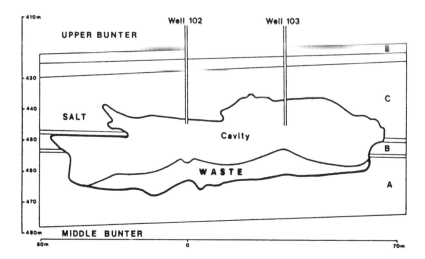

Figure 16.1.12. Waste disposal in a hydraulic fracturing connected well system in the Hengelo area (Netherlands)

for this particular purpose has been developed. The smaller size of storage caverns has been chosen, because each of them can be used for the disposal of a different waste product. However, in normal circumstances a small cavity will be relatively more expensive from the point of drilling costs. For that reason the Akzo Company worked out the idea of a 'string of pearls' being a number of small cavities mined out, one above the other, out of one well (Figure 16.1.13). If a cavity is mined out, it has to be emptied by a deep-well pump. Having been filled up with waste, it has, in turn, to be closed again with an open hole packer, for example, a Halliburton EZ-SV packer and a cement plug. A new cavity thereupon can be mined some 100 m higher up. In that case one group of waste products will require at least two wells, one for the disposal phase and the other for mining a new cavity. Akzo did some experimental work in creating a small cavity at a depth of about 1300 m, and succeeded in making a spherical cavity with a diameter of 42 m and a volume of about 42 000 m^3. This cavity has an ideal shape for emptying and disposing of chemical waste (Figure 16.1.14).[4]

16.1.5 *Storage of compressed air*

Gustin discussed the compressed air energy storage concept, which utilizes off-peak electrical power from a nearby coal of nuclear fueled base load power plant to run a reversible motor/generator which turns a compressor to charge an underground storage cavern with compressed air (Figure 16.1.15). When peak power is required, the compressed air, which is analogous to a turbo charger in an automobile engine, is drawn off, mixed with fuel oil or natural gas, and burned in a combustion turbine.[5]

There are two types of compressed air energy storage, namely: constant pressure and constant volume. The constant pressure concept allows water from a surface reservoir to displace the volume of compressed air as it is withdrawn from the storage cavern, and it is used in hard rock caverns. During the charging mode, the water in the cavern is forced back up into the surface reservoir. The constant volume concept allows the air pressure in the

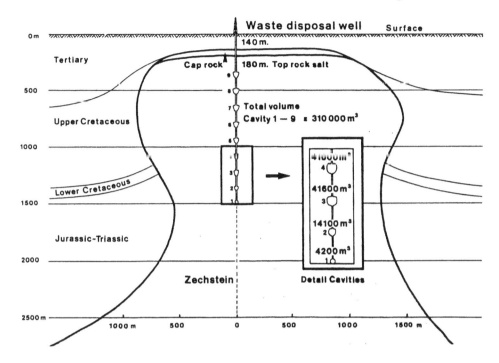

Figure 16.1.13. Proposed layout of the caverns for chemical waste disposal in a salt dome (after Wassman)

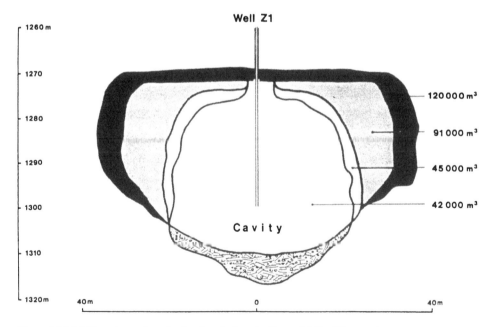

Figure 16.1.14. Experimental cavity for chemical waste disposal (after Wassman)

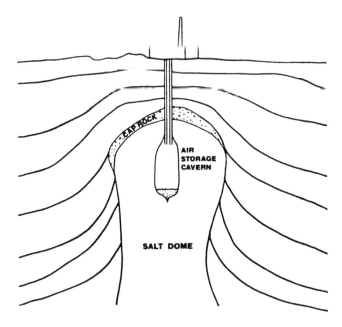

Figure 16.1.15. Compressed air energy storage plant in a salt dome

cavern to decrease during the power generation mode. This concept is favourable for salt storage caverns, since the relatively inexpensive solution mining technology is employed. This storage cavern should most likely use brine instead of fresh water to prevent salt dissolution and enlarge cavern (Figure 16.1.16).[5]

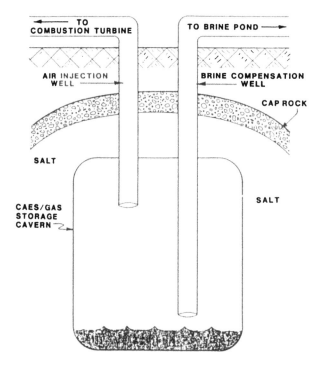

Figure 16.1.16. Brine-compensated compressed air storage

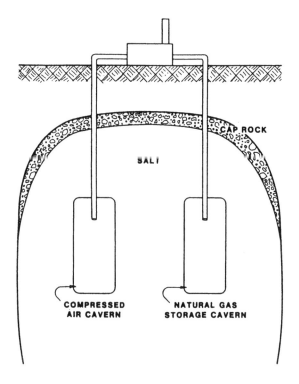

Figure 16.1.17. Co-utilization of a salt dome for compressed air and natural gas storage

Efficient operation of compressed air storage requires that the air is withdrawn from the cavern to power the combustion turbine. Because solution-mined salt caverns are commonly used for hydrocarbons storage, it makes sense to consider co-utilization of the compressed air storage caverns for fuel storage. In this case an adjacent salt cavern, created to store natural gas to fuel the compressed air energy plant, or created to store natural gas for other purposes, could also be leached at the same time that the compressed air cavern is being made (Figure 16.1.17). The fixed cost of mobilizing a salt solution equipment and the cost of installing fresh water wells and brine injection wells would be shared among two or more storage caverns, as commented by Gustin.[5]

16.2 LOADING CONDITIONS OF CAVERNS

The loading capacity of caverns is related to the action of external and internal pressure on their walls. In addition, the transferred load which was supported by salt material before its excavation has to be taken into consideration. The magnitude of transferred stress will depend on the cavern sizes and their spacing. Because the caverns are located at greater depth at which temperature increases, the thermal stresses also might be important within the complex of loading conditions.

16.2.1 *Gravitational stress*

Gravitational load superimposed on the storage caverns is one of the most important factors in determination of external stresses which act on this structure. The magnitude of gravitational load depends on the weight of the overburden, which depends on the depth of caverns (h) and density of overlaying rocks (γ), as given in the following equation:

$$\sigma_V = \gamma h$$

There is an estimate that the state of virgin stress must not be very different from isotropic pressure equalling the overburden load. This can be written as:

$$P = \sigma_V = \sigma_H$$

Its magnitude could be assumed in the general form:

$$P = 0.023 \text{ MPa per m}$$

The assumption of existence of isotropic pressure most likely is valid, which might be confirmed by a mode of cavern deformation observed at a depth between 1300 and 1500 m, where shrinkage in volume could be significant (Figure 16.2.1).

The pressure on cavern walls due to gravity load could be calculated by the assumption that salt behaves like an incompressible viscous fluid and that the density of the brine can be neglected. In addition it is considered a spherical cavity dissolved in an infinite weight medium, where the inertia is neglected. The effect of overburden load is computed by the wall displacement so that boundary conditions will be written on the initial spherical well of cavity, as given by Berest and Minh.[6, 7]

At the infinite pressure is:

$$P = \rho g h$$

where:

    h = depth measured to the centre of the cavity;
    g = gravity;
    ρ = density.

The pressure vanishes to zero in the cavity, so the system of equations is:

$$\operatorname{div} \vec{v} = o$$

$$\eta \, \Delta\vec{v} + \rho\vec{g} = \overrightarrow{\operatorname{grad}} P$$

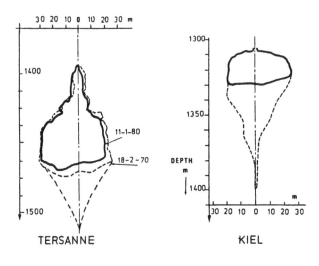

Figure 16.2.1. Decrease of cavities volume due to pressure

And on the wall of the cavity:

$$\sigma\vec{n} = o$$

With spherical coordinates r, θ and φ the following $\vec{v}$ and P offer a convenient solution:

$$v_r = -\frac{1}{3}\frac{\rho g\, R^3}{\eta}\frac{\cos\theta}{r}$$

$$v_\theta = -\frac{1}{6}\frac{\rho g\, R^3}{\eta}\frac{\sin\theta}{r}$$

$$v_\varphi = 0$$

$$P = \rho g\, r\cos\theta - \frac{1}{3}\rho g\, R^3\frac{\cos\theta}{r^2}$$

It is interesting to notice that, if, moreover, the cavity is submitted to an internal traction p δ_{ij}, we must add to the former solution:

$$v'_r = -\frac{p\, R^3}{4\eta\, r^2}$$

For instance, if the fraction is due to the differences between the weight of overburden layers, then:

$$\frac{v}{r} = \frac{-\rho g\, H}{4\eta}\left(1 + \frac{4}{3}\frac{R}{H}\cos\theta\right)$$

If R/H is not too small, the corrective term explains that the load is larger at the bottom than at the top of the cavity.[6, 7]

16.2.2 *Internal pressure*

It should be pointed out that internal pressure of underground storage mainly depends on the nature of the stored products, as discussed by Berest and Minh.[6]

1. For liquid or liquefied products, the central tubing is filled with brine up to the surface, for any movement of products must be balanced by an equivalent movement of brine, so that the cavity and the tubings remain filled up, whatever be the stored quantity of hydrocarbons. The small variations of pressure, due to losses of load during injection or withdrawal, can be neglected, so that the internal pressure is quite close to:

$$P_i = 0.012\, h\ \text{MPa per m}$$

2. For natural gas, the internal pressure can vary to a large extent, only restricted by safety rules. The storage must remain tight and stable; these two restraints lead to select a relevant maximal pressure (for tightness) and minimal pressure (for stability). In France, for cavities of about 1500 m deep, the selected rules were:

$$8\ \text{MPa} \leftarrow P_i \leq 0.0166\, h$$

In both cases for oil and gas cavities, the volume and pressure of the stored products should be kept constant.

Berst and Minh proposed the approximated and simple index of the cavities pressure by the difference between overburden load and internal pressure as given in the following equation:[6]

$$P - P_i = (0.023 - 0.012)\, h = 0.011\, h$$

The index of intensity of cavity loading versus depth for different cavities is illustrated in Figure 16.2.2. The straight line on the diagram represents the relationship for liquids storage (oil, LPG, etc.), but the vertical segments on this line represent the relationship for gas storage, where internal pressure is not constant.

The variations of the internal pressure are kept moderate when liquid products are stored, and can be neglected.[7] At the opposite, these variations are large when natural gas is stored for instance, between 8 MPa and 22 MPa in the case of the Tersanne storage (France). The period of these variations differs from one site to another: it can be (roughly speaking) one year (for a peak shaving storage like Tersanne), one month (for a storage located near a gas field, like the Eminence salt dome) or one day (for a compressed air energy storage like Huntorf).

In situ data do not prove a significant effect of cycling: the experimental cavity of Kiel in Germany has been submitted to monotonous loading; the cavities of the brine field of Vauvert in France exhibit large creep and closure deformations because of their depth (2000 m) in spite of roughly constant internal pressure. At the opposite no special trouble was noticed at the Huntorf site, in spite of daily cycling.

We think, as a conjecture, that cycling does not produce a specific effect (by specific effect, we think of oligocyclic fatigue or ratcheting). Of course, reverse salt creep, after repressurising, is proved by numerous in situ measurements.

Finally it should be noted that if the index of pressure difference is greater than 20 MPa, then cavities experience a large convergence deformation which could significantly decrease the size of their volume.

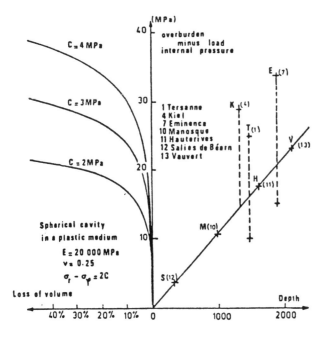

Figure 16.2.2. Index of intensity of pressure differences of cavities (after Berst and Minh)

16.2.3 *Induced pressure by excavation*

Induced pressure relates to the extracted volume of the caverns and their spacing. In the case of multiple-cavern storage, the load on the salt mass could be determined on the principles of a tributary theory as in the case of room-and-pillar mining. In this case the main calculating factors are gravitational load (γh) and extraction factor (e), which is expressed by the very well known equation:

$$\sigma = \gamma \frac{1}{1-e}$$

The cavern spacing on the first place is controlled by the distance between solution wells where two particular factors have to be considered, as Istvan and Querio stated.[8]

1. *The effect of the well deviation.* According to the authors' well control parameters, the last cemented casing shoe must fall within a 7.5 m radius target area around the vertical. The bottom of the cavern must be localized within a 3 m radius target around the new centre line determined by the extremes of the target area for the casing shoe. The most unfavourable situation will arise when both wells are deviated in opposite directions, as shown in position 1 (Figure 16.2.3). At mid-cavern depth, this will result in an 18 m decrease of the pillar thickness between the wells.

2. *The effect of preferential leaching.* Since salt properties vary with location, it is possible that the planned maximum diameter of 45 m is locally exceeded due to preferential leaching.

Istvan and Querio suggested a concept of a multiple-cavity storage structure,[8] which is designed to be operated as a high-low pressure vessel, as illustrated in Figure 16.2.4. The induced cavern pressure for the proposed layout of caverns could be approximated as follows:

$$\sigma_E = \gamma h \frac{1}{1-e} = \gamma h \frac{1}{1-0.2} = \gamma h \times 1.2$$

At the depth of 1060 m the induced stress for the given layout cavern could be assumed:

$$\sigma_E = 0.023 \times 1060 \times 1.2 = 29 \text{ MPa}$$

The maximum cavern pressure could be assumed 75% of the total theoretical external pressure at the depth of the last cemented casing shoe, or:

$$\sigma_c = 0.75 \times 29 = 22 \text{ MPa}$$

The minimum cavern pressure controlled by internal pressure was selected 7 MPa to allow free flow gas withdrawal by means of gas expansion. The suggested minimum pressure should be increased for higher cavern temperature, which deteriorates compressive strength of the salt.

16.2.4 *Temperature stress effects*

Effect of temperature on the property and behaviour of salt is discussed in Chapter 9. This phenomenon is of particular importance for the storage caverns which depth might be in a range between 1200 and 2000 m, where increase in temperature could be significant. For example, the Varangeville salt mine (France), at a depth of 200 m, has a temperature of about 18°C. This is an old mine with some parts mined out a century ago by the

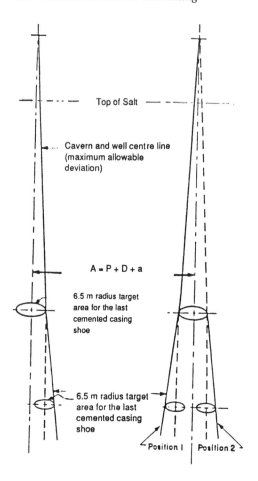

Figure 16.2.3. Distance between drilling wells for salt solution

room-and-pillar mining system. The cavity of Tersanne (France) at a depth of 1500 m (Figure 16.2.1) however, has a temperature of 65°C.

The viscoplastic rate of deformation beside the loading magnitude is strongly dependent on the temperature, what could be expressed by Bingham's viscoplastic model

$$\dot{\varepsilon} = \frac{1}{\eta}\,(\sigma - 2c)$$

where:

$\dot{\varepsilon}$ = rate of deformation;
η = viscosity coefficient;
σ = acting stress.

The temperature-induced stress of deep cavities could be caused by storing hot or cold products. The induced temperature changes have similar stress effects as a geothermal salt temperature. The changes in temperature release very quickly when gas is stored (a few days) and very slowly when liquids are stored (some years). Decrease of temperature decreases the loading stress and vice versa. The influence of the effects of thermal stresses are small when hydrocarbons are stored, but it could be considerable when cold products are stored.

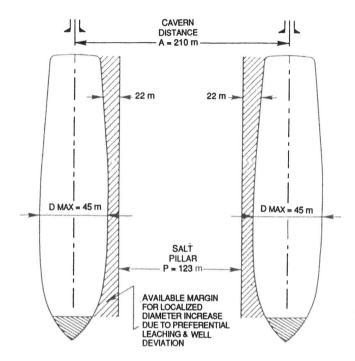

Figure 16.2.4. Layout of a storage cavern (after Istvan and Querio)

16.3 STABILITY ANALYSIS OF CAVERNS

The stability of salt caverns depends mainly on the adequate dimensions of the pillars, namely: seal pillars on the top and bottom of caverns, inter-cavern pillars or rib pillars, and safety pillars enveloping the cavern layout. This section is entirely written on the basis of Dreyer's work.[9, 10]

16.3.1 *Physical model studies*

The estimation of pillar strength for a given load and strength of the salt mass could be carried out by model investigations of equivalent material. The graphical representation of the layout of the cavern-pillar structures is given in Figure 16.3.1. the stress limit, σ_p, is a function of the pillar ratios S/d (seal pillar) and W/d (rib pillar) and it is calculated by assuming only atmospheric internal pressure. The stress calculation of non-cylindrical caverns is based on the maximum cavern diameter. The total volume of salt leached out from the cavern is considered to be minute.

Dreyer in his model studies applied an incremental load on the pillar models to the point of reaching the stress limit, σ_p. At this stress point, fogging and slight coloration of original clear salt grains at the notch of the cavern model is noticed. Rock salt crystals are oriented in the direction of stress indicating very fine micro-cracks, which reflect the top illumination on an increasing scale, thus leading to a foggy effect. This effect might indicate the phase prior to rock salt destruction and it should not yet be critical since the binding strength of the grain boundary remains unaffected. As Dreyer indicated equal degrees of stress, the further deformation of grain binding remained unsheared whereby no new micro-cracks were formed.

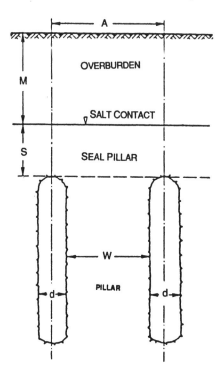

Figure 16.3.1. Section of a storage cavern in study of consideration (after Dreyer)

Visualized failures on the models are tensile and occurred at (or near) the notches of the cavities. In contrast, the model studies showed that, if the top stress is increase still further, the inner crystalline crack formation becomes stronger. The crystallites subjected to the greatest stress become milky white and at this point a structural change takes place which affects the binding conditions on the grain boundaries. With this phase the destruction process is introduced on the pillar skin which, with increasing stress, is transmitted further into the interior of the salt. As a result the system of cracks, no longer bound to the grain boundary, forms joints parallel to the open surface and initiates a spalling process. Storage caverns should be protected from such spalling processes and must be surveyed in such a manner that the stress limited, σ_p, is not exceeded anywhere on the cavern walls.[9]

Also the model studies were carried out by filling caverns with different storage material, to study their stability at various internal pressures. Finally, the stability investigations were carried out in this respect on rock salt models, but should not be accepted without further critical appraisal. With regard to the criteria for evaluating the stability of caverns, the question should be asked whether the beginning of structural changes near the surface around the most heavily stressed portion of the cavern wall would actually characterize that critical maximum load under which the ten-year projected life of storage caverns in salt deposits could be guaranteed. This means that creep tests must be introduced, the purpose of these being, amongst others, to define the safety margin obtainable with these fracture criteria.[9]

16.3.2 *Analytical analysis*

The analytical stability analysis represented here is based on the work of Dreyer.[9, 10]

Considering a crude oil storage cavern from the Ruestringen cavern field, this cavern has a diameter of:

$$d = 50 \text{ m}$$

The cavern top is situated at the depth of:

$$h_{top} = 1293 \text{ m}$$

and the cavern bottom is situated at:

$$h_{bottom} = 1643 \text{ m}$$

so that the cavern height is:

$$H = 350 \text{ m}$$

Figure 16.3.2 illustrates the geometry and associated loading conditions of the cavern.
Using the formula for virgin rock stress (gravity load):

$$\sigma_h = \gamma h$$

and values:

$$\gamma = 0.023 \text{ MPa/m}$$

and:

$$h = h_{top}$$

results in a stress at the top of the cavern of:

$$\sigma_h = 29.74 \text{ MPa (cavern top)}$$

Corresponding to this the virgin rock stress at the bottom edge of the cavern is computed to be:

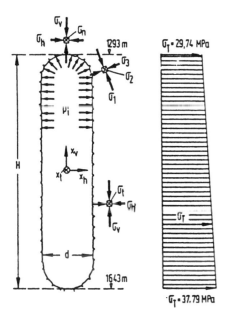

Figure 16.3.2. Storage cavern under isotropic pressure (after Dreyer)

$\sigma_h = 37.79$ MPa (cavern bottom)

It is assumed that virgin rock stress corresponds to isotropic pressure, which is written as:

$$\sigma_V = \sigma_H = p$$

Internal pressure has been calculated to be 7.10 MPa (cavern filled with oil). Investigations of the maximum stress, σ_g, at which opacity appears for the first time in the salt at the roof of the cavern model, the dimensional ratios were found to be:

$S/d = 1.00$ (oil-filled)

$W/d = 5.06$ (oil-filled)

Now relative to the cavern diameter

$d = 50$ m

the minimum seal pillar thickness of salt is computed to be:

$S = 50$ m (oil-filled)

and the minimum rib pillar width is computed to be:

$W = 253$ m (oil-filled)

The distance A between the centre line of the drill holes may be computed using the following equation:

$$A = W + d$$

Neglecting the bore deviation at depth this yields a value of:

$A = 303$ m (oil filled)

It should be noted that the compressive strength of the salt from drill cores in the vicinity of the cavern roof was normal. However, the rock salt quality in the deeper parts of the deposit was somewhat abnormal, especially in the vicinity of the cavern bottom where the drill core exhibited pure, coarse grained, single crystals.

In the case of model tests under internal atmospheric pressure it was found that:

$S = 126$ m (atmospheric pressure)

based on a dimensional ratio S/d of 2.53, and that:

$W = 305$ m (atmospheric pressure)

and:

$A = 355$ m (atmospheric pressure)

With the given network A_{ij} the maximum possible diameter of the cavern is dictated. The linear optimization took into account the oversized measurements of the safety seal pillar and rib pillars and the compound effect with the adjacent caverns. With a trigonal arrangement of caverns this means that, for a central cavern, up to six caverns should be included in the calculation, whereby each cavern functions, step by step, as a central one.

With the condition equation for the seal pillar diameter:

$$d_{i\,\text{permissible}} \geq d_{i\,\text{seal pillar}}$$

the condition equation for the cavern distance:

$$A_{ij} \geq W_{ij} \frac{d_i + d_j}{2}$$

and the condition equation for the rib pillar width:

$$Y_{ij} \leq \frac{2W_{ij}}{d_i + d_j}$$

the necessary arithmetic mean value of the adjacent caverns can be ascertained,

Considering the pillar reserves, the selective criteria of the following target function are introducted:

$$\sum_{i=1}^{n} \frac{d_{i\,\text{permissible}}}{d_{v\,\text{theoretical}}} = \max$$

where d_v refers to the final diameter of the cavern after transferring the content several times. For the linear optimization, a mathematical model analysis is used.

16.3.3 *Mathematical model analysis*

Mathematical model analysis utilized the finite elements method producing a series of computer runs to evaluate stress fields of cavern backs and rib pillars. The stresses computed by mathematical modelling are similar to the stresses obtained by analytical calculations and physical modelling. Mathematical model analysis considered two cases: caverns filled with oil and caverns at atmospheric pressure.

1. *Caverns filled with oil* have been analyzed for induced stresses of the cavern backs and the seal pillars as follows:

a) *The back of the caverns* exhibited a stress distribution and concentration as given in Figure 16.3.3 and 16.3.4. The numerical parameters of these stresses are given in Table 16.3.1.

It should be noted that extension of horizontal stress into the back of the cavity is 53 m before it reaches virgin stress. Vertical stress, however, is of lower magnitude than virgin stress, which is equalized at 28 m into the back of the cavity. The maximum shear stress is at the top of the edge of the cavern. Therefore, the stress state of the cavern back is neglecting pipelines, as illustrated in Figure 16.3.5. The computed data suggest that the thickness of the cavern back should be at least 53 m to avoid adverse effects of nearby cavern excavations. To maintain a back stability it is necessary that the confined compressive strength of the salt should be greater than 40 MPa.

b) *The rib pillars of the caverns* have a stress distribution as illustrated in Figure 16.3.6, which has been computed in function of the distance from oil-filled caverns. The numerical parameters are given in Table 16.3.2.

As a result of equalization of horizontal pillar stress with virgin stress at a distance of 125 m, the minimum pillar width should be $W = 2 \times 125$ m $= 250$ m. This value agrees with the minimum pillar width determined from the physical model analysis ($W = 253$ m). Vertical stress is more or less of equal value throughout the pillar ranging between 32.18 and 32.47 MPa. The tangential stress from its maximum at the outer skin of the pillar decreases the equivalent value of virgin stress at 73 m. The diagrammatic tangential stress distribution (Figure 16.3.7) and the three dimensional state of shear stress at the outer skin of the pillar are graphically represented (Figure 16.3.8). It should be noted that the strength of the salt at

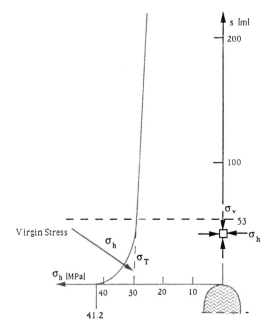

Figure 16.3.3. Diagram of induced horizontal stress at the back of a cavern filled with oil

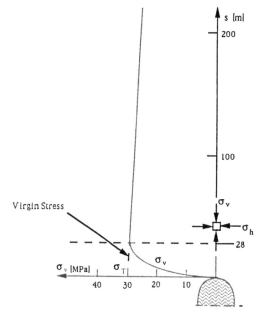

Figure 16.3.4. Diagram of induced vertical stress at the back of a cavern filled with oil

the outer pillar wall could be overcomed by tangential stress, which could cause pillar sloughing and pillar instabilities.

2. *Caverns at atmospheric pressure.* The induced stress of the cavern back and the crown pillars has been analyzed in the same manner as in the case of caverns filled with oil.

Table 16.3.1. Stress distribution of cavern backs

Types of stress	Distance from boundary (m)	Stress magnitude (MPa)
Horizontal (σ_H)	Back wall	41.20
	3.00	38.26
	53.00	29.04
	70.00	28.65
Vertical (σ_V)	Back wall	7.36
	28.00	29.10
Shear (τ_{max})	Back wall	16.90
Virgin stress (σ)	3.00	29.67
	53.00	28.47
	70.00	28.13

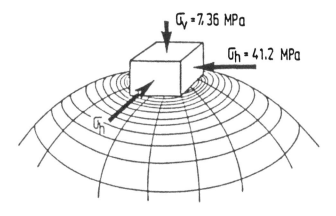

Figure 16.3.5. Stress state at the top edge of a cavern filled with oil (after Dreyer)

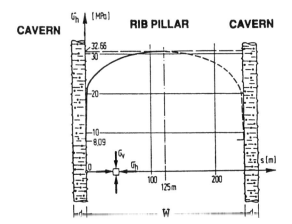

Figure 16.3.6. Diagram of induced horizontal stress of the rib pillar between caverns filled with oil (after Dreyer)

a) *The back of the caverns* exhibited a stress distribution in a similar pattern as oil-filled caverns. The induced horizontal stress is equalized with virgin stress far away from the back surface at 115 m, as can be seen from Figures 16.3.9 and 16.3.10. The magnitudes of the maximum horizontal stress and shear stress are higher as given: $\sigma_H = 52.4$ MPa, $\tau_{max} =$

Table 16.3.2. Stress distribution of cavern wall

Type of stress	Distance from boundary (m)	Stress magnitude (MPa)
Horizontal (σ_H)	Pillar wall	8.09
	125 (centre)	32.66
Vertical (σ_V)	Pillar wall	32.10
	125 (centre)	32.47
Tangential (σ_t)	Pillar wall	50.00
	73.00	32.66
Shear (τ_{max})	Pillar wall	20.96
Virgin stress (σ)	At a depth of 1420 m	32.66

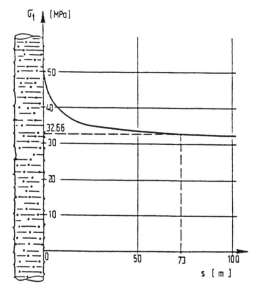

Figure 16.3.7. Variation of distribution of tangential pillar stress of an oil-filled cavern (after Dreyer)

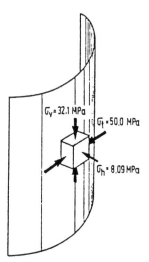

Figure 16.3.8. Three dimensional state of stress at the pillar wall of an oil-filled cavern (after Dreyer)

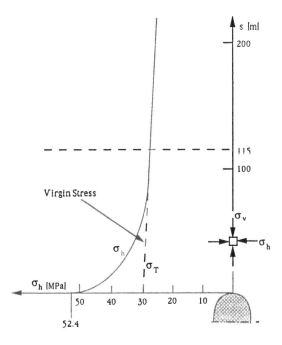

Figure 16.3.9. Diagram of induced horizontal stress at the back of a cavern with atmospheric pressure

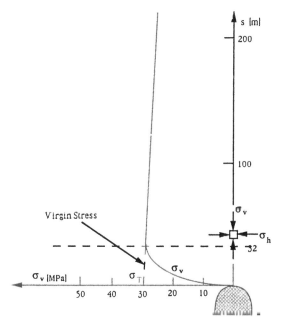

Figure 16.3.10. Diagram of induced vertical stress at the back of a cavern with atmospheric pressure

26.2 MPa. The induced vertical stress extends up to 32 m with a maximum magnitude at the boundary of $\sigma_V = 32.00$ MPa, as also can be seen from Figure 16.3.6. It is obvious that due to absence of internal pressure of the filled gas, the magnitudes and extension of the stresses at atmospheric pressure are higher. In this case, respectively, the seal pillar above the cavern has been thicker, as for example 115 m, to avoid interaction stress effects of adjacent caverns.

b) *The rib pillars of the caverns.* Analysis indicated that induced horizontal stress equalizes with virgin stress on 155 m from its boundary, so that the pillar thickness should be at least 310 m. The horizontal stress at the pillar boundary is 0.00, with a gradual increase to 31.88 MPa at a distance of 155 m from the pillar boundary. The vertical pillar stress, however, is more or less constant, approximately $\sigma_V = 31.9$ MPa, which is very close to the vertical induced pressure of the oil-filled caverns of $\sigma_V = 32.10$ MPa. The maximum tangential stress at 3 m from the pillar boundary is 57.39 MPa and equalizes with virgin stress at a distance of 115 m from the pillar boundary. The low horizontal stress at the pillar boundary and pillar skin will not facilitate sloughing, which could lead to the conclusion that the projected width of a rib pillar could be satisfactory, which is not the case with caverns filled with oil.

Mathematical model analysis could be an important method for the design of bearing capacity of seal pillars (cavern back) and rib pillars, because they give relationship between induced stress magnitude and geometrical dimension of the pillars, for the given depth of excavation.

16.3.4 *Effect of the back shape on the stability of the cavern*

Dreyer made a further contribution to stability investigations of caverns considering three different profiles of the cavern backs (Figure 16.3.11).[9] For each case, horizontal, vertical and maximum shear, the stress magnitude had been computed at the top edge of the cavern back, assuming that the cavern is filled with oil, as shown in Table 16.3.3.

For the case of back profile III the height of the cavern back corresponds to the cavern diameter, and the analytic solution is in agreement with the physical model tests. The back profile is considered to be the most stable form for the cavern back. This parameter should be taken into consideration for the design of the caverns.

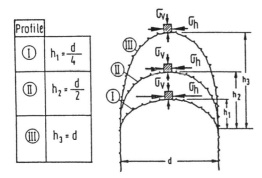

Figure 16.3.11. Three shapes of backs which effect the cavern roof stability

Table 16.3.3. Stress magnitude of oil-filled cavern in function of different shapes of the cavern back

Back profile	Stress at top edge of cavern (MPa)		
	Horizontal (σ_H)	Vertical (σ_V)	Maximum shear (τ_{max})
I	45.2	7.51	18.9
II	41.2	7.36	16.9
III	37.9	7.23	15.3

16.4 GROUND CONTROL OF STORAGE CAVERNS

In this section consideration is given to the storage caverns made by solution mining technology at a depth of 500 to 2000 m. Four topics wil be discussed in regard to strata control of storage caverns.

16.4.1 *The factors of cavern deformations*

There are a certain number of parameters which might have a major influence on the deformational behaviour of deep storage caverns, such as:[6]
- depth of the cavity;
- mean value and variation of internal pressure;
- shape of the cavity;
- geological setting of the overburden;
- vicinity of other caverns;
- temperature;
- properties and behaviour of salt material which surrounds caverns.

In the past, dry mining and solution mining have not developed a sophisticated solution for determining deformational properties of salt. However, extensive investigations for the last few decades, due to the long-term use of caverns made by solution mining for storage of liquid and gaseous hydrocarbons as well as chemical and radioactive waste, resulted in finding a constitutive law that can be used for design and construction of long-term, stable, large-scale caverns.

Several authors proposed a constitutive law of salt behaviour.[6, 7, 10, 11] At this place a brief consideration is given to the salt behaviour suggested by Berest and Minh.[6, 7]

1. *Elastoplasticity*. It appears that the differences of external (P) and internal (P_i) pressures have a major influence on salt cavern behaviour. The in situ measurements indicated a minor convergence of caverns if $P - P_i$ is less than 20 MPa. Also, that decrease of cavern volume is not proportional to the stress intensity. This phenomenon suggests plastic behaviour of salt which could be explained by the perfect plastic Tresca model. The decrease of cavern volume or cavern convergence is given in function of $P - P_i$:

$$\frac{1}{3}\frac{\Delta V}{V} = \frac{4C}{3E}\left\{(1-2v)Q - \frac{3(1-v)}{2}\exp(Q-1)\right\}$$

$$Q = \frac{3}{4C}(P-P_i)$$

where:
  V = volume of cavern;
  ΔV = loss of volume;
  c = cohesion;
  E = Young's modulus;
  v = Poisson's ratio;
  Q = internal pressure constant.

This simple model is in good agreement with in situ measured convergence data if cohesion is taken rather low (less than 3 MPa). Laboratory data for Tersanne rock salt (France) produced a Mohr's Coloumb criterion which cannot be fitted with in situ measured data (Figure 16.4.1). This contradiction has led to a new testing methodology, taking more into

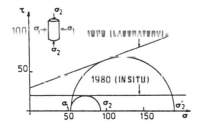

Figure 16.4.1. Comparison between laboratory and in situ determination of plastic criterion

account delayed effects of behaviour. It could be stated that plasticity should explain the order of magnitude of the total loss of volume, e.g. convergence. However, the viscoplasticity could explain the actual history of deformation.

2. *Viscoplasticity*. Plasticity and viscoplasticity give similar results only if the period between initial and final observations is larger than the viscoplastic time constant, so that most of the convergency has been achieved. Experience suggests that this time constant is high (see paragraph 3); then it was useful to dispose of a closed form solution for the problem of the evolution of a spherical cavity submitted to a varying internal pressure. The effects of viscosity on cavern convergence are illustrated in Figure 16.4.2. Berest and Minh analyzed the convergence of $\Delta V/V = 0.3$ for a time of ten years, for the gas cavern storage at Tersanne (France). The cavern depth is about 1500 m, with an external pressure of 34 MPa and an internal pressure of about 8 MPa. The analyses have been done by the triplet of parameters: cohesion (c), viscosity (η) and elasticity (E) in the rather complicated way of equation derivation. These authors finally concluded that if the cohesion is high, then salt has a tendency to perfect plasticity and additional cavern convergence that after ten years would be small in size. However, if the cohesion is low and the viscoplasticity high and constant, the rock salt will behave like a Newton's fluid and convergence will go on during centuries until total cavern closure.

16.4.2 *Closure control of caverns*

The closure is the irreversible displacement of cavern walls in relation to the original cavern volume and generally is defined as convergence. For stability control, in situ tests should be carried out concerning the cavern wall displacement to establish local or overall conver-

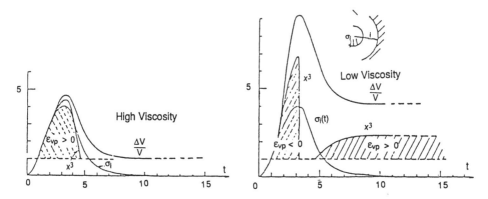

Figure 16.4.2. Effects of viscosity on the intensification of convergence

gence. The measured convergence data could be plotted and drawn, so that the closure deformation could be identified (Figure 16.4.3).

Instrumentation of the closure deformation of caverns is the main factor in regard to strata control. The closure is represented by the volumetric convergence rate ($\Delta V/V$) in function of time. In this regard several possibilities exist, which could affect the stability of the cavern, namely:

- continuous closure until convergence is ceased;
- intensive closure that exceeds economically acceptable convergence rates;
  slow closure that does not exceed economically acceptable convergence rates;
- closure up to failure of the salt back and walls, where caved material beside instability will also shrink the volume of the cavern.

Still there is a lack of knowledge on the closure mechanics for cavern loading over a long period of time. However, present experience is positive in regard to the cavern excavated by solution technology, which volumes are between 100 000 m³ and 500 000 m³ at depths of 400 to 1800 m (Federal Republic of Germany). About a hundred caverns in the FRG and some have been in operation for more than ten years. All are tight and stable under working conditions. Meanwhile, initial findings are available concerning in situ measured convergence. At present there is no evidence that the convergence rate of storage caverns in Northern Germany is more than 1% per year. Recently a method has been developed for correct determination of convergence of storage caverns.[3] Quast and Schmidt described convergence rates of caverns filled with liquid hydrocarbons and they determined them by discharge measurements combined with pressure and temperature measurements. Control of configuration is achieved by sonar measurement methods. However, this procedure makes it necessary to pull up the production string, which is both time-consuming and expensive. For this reason, convergence calculations applying PVT measurements are employed for caverns used for storage of compressed natural gas or compressed air. The rates are measured during a filling or discharge cycle respectively for this purpose, as well as pressure and temperature in the storage at the beginning and at the end of each cycle. This method requires very exact measurement devices. Nevertheless, the error in determining the storage volume is still more than 1%.

Figure 16.4.3. Model of closure deformation of a storage cavern

It has been concluded on the basis of convergence measurements that the caverns for disposal of radioactive waste in the FRG can withstand the external and internal stresses without substantial deformations, if they are excavated by solution technology in suitable salt formations and at adequate depth. The reasonable stability of caverns is also exhibited. They do not show unacceptable convergence rates during the filling process without inner pressure.[3, 12]

In addition, Quast and Schmidt described investigations of closure for the prototype cavern in the Asse II salt mine.[3] The prototype is one of the rare caverns in salt formations that can be explored in detail from a geomechanical point of view because of the availability of inspection and early installed instrumentation (Figure 16.4.4). The phenomenological investigations of the cavern and the shaft area for a period of more than four years could be summarized as follows:

– Deformation around the cavern takes place at a decreasing rate with time.
– Stability of the cavern, as previously mentioned, is guaranteed.
– Deformation behaviour deduced from extensometer measurements is largely homogeneous and can be proved by FEM-calculations.

The data of closure instrumentation are the key factors for a stability control of storage caverns, and a prediction of their stability in the following years.

Finally, it should be noted that rock salt deformational parameters could be determined in situ during drilling, e.g. before the cavern starts to be leached The advantages of the drill hole tests are:

– Due to the small diameter it is in principle possible to record depth-dependent displacements by caliper logging (Figure 16.4.5).
– The results are available even prior to the construction phase of the cavern.

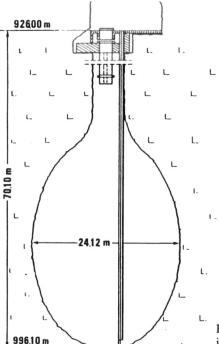

Figure 16.4.4. Prototype cavern for storage of radioactive waste at the Asse mine (Germany)

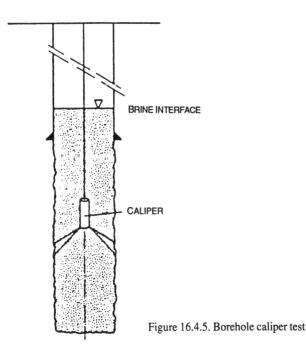

Figure 16.4.5. Borehole caliper test

BRINE INTERFACE

CALIPER

– In the event that drilling of the well and start of its leaching are not directly consecutive, longer measurement spans are possibly available.

The disadvantages, owing to a certain number of limitations, are:

– A cavern geometry cannot be evaluated after the leaching phase and its influence on the cavern stability.

– Inhomogeneities which occur in the cavern vicinity at a greater distance from the well.

– The leaching phase with the respective stress rearrangement.

If the costs for borehole tests or cavern tests are identical, one will probably opt for the in situ test on the cavern due to the greater assurance of the test. For this reason the borehole test can only have a chance if the required expenditure is considerably lower.

16.4.3 Configuration control of caverns

Whereas measurements of the cavern shape in liquid-filled caverns have been state-of-the-art for years, these measurements were not possible in gas-filled caverns until recently. Also, previously used methods of measurements have been relatively inaccurate in regard to configuration control (thermometric method). At present, configuration control of gas-filled caverns is with sufficient accuracy using a modified sonar measurement method and laser method, as further discussed.

1. *Sonar surveying* is carried out by echologing. The results of sonar surveying are represented in Figure 16.4.6. The mean depth of the cavern is 730 m and is operated as a seasonal storage. As Crotogino described, there is an interval of five years between both measurements. Within the bounds of the anticipated measurement accuracy no quantitative statement regarding convergence can in fact be made; on the other hand, it is clear that the differences cannot be very large. Even if the resolution is the same in brine as in gas, a

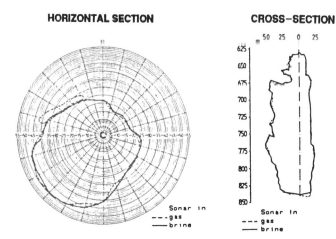

HORIZONTAL SECTION   **CROSS-SECTION**

Figure 16.4.6. Sonar survey of an HP-gas cavern

greater accuracy can be expected with gas, as the density distribution in gas is horizontally considerably more homogeneous than that of a brine-filled cavern at the end of the leaching process, a time when total saturation cannot yet be assumed. Sonar surveying in gas caverns represents a substantial step forward in the long-term in situ measurement of storage caverns, since it makes it possible to obtain essentially more exact results.[13]

Quast and Schmidt briefly commented on the configurations of a 400 000 m³ cavern after completion of the leaching process and after four years of gas storage operation. During this period the cavern was maintained with inner pressures between 25 and 160 MPa. The shape of the cavern has not undergone any substantial changes within the limits of measurement accuracy, which means that neither peeling nor caving could be detected, and convergence rates remain lower than a quantifiable magnitude of several percent (Figure 16.4.7).[3]

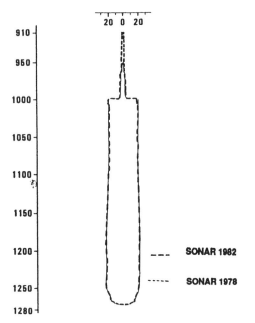

Figure 16.4.7. Convergence section of a gas cavern

2. *Laser surveying* first time has been used for laser probes compressred air storage cavern (1980).[13] The main advantage of laser surveying compared with the sonar survey, is the narrow focusing of the light and the interrelated considerably larger resolution. The measuring accuracy of the distance probe – wall is within the same dimension range as that of the sonar survey. The measurement results and by way of comparison the sonar survey of 1976 can be seen in Figure 16.4.8. This figure shows a horizontal section of the cavern at a depth of 760 m, as measured at the end of the leaching operation 10/1976 and the result of the laser survey 9/1980 at the same depth, projected into this section.[13]

A laser method for instrumentation of caverns is developed for compressed air storage, because of defficiency of the sonar surveying of these particular storages. Also the sonar method of surveying of the caverns for natural gas storage did not produce satisfactory results and in this case a laser method comes in effect.

16.4.4 *Stability control of filled caverns*

Stability control of filled caverns at this place has been considered for liquid-filled caverns and gas-filled caverns. The representation of this topic is entirely based on the Crotogino work.[13]

1. *Control of liquid-filled caverns* is related to particular measurements and tests. They are individually described here.

a) *Efflux measurements* to establish a normal internal pressure are easily done by measuring the flow of brine which is discharged at the well head. The brine will be allowed to discharge pressureless, so that the convergence does not diminish by a pressure build-up. Normally this test is used for caverns filled with liquid products in order to check the convergence trend over the operation period. In this case the pressure is allowed to rise to the permitted value, so as to keep the convergence as low as possible. Once the pressure has reached the maximum level, a certain quantity of brine is discharged. If it is deducted from the volume flow, the measured volume differences due to heating, leaching and decompression, then the convergence rate could be found related to internal pressure corresponding to the brine column plus well head pressure.

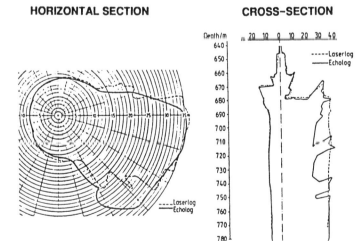

Figure 16.4.8. Laser survey of a compressed air cavern

b) The *pumpout test* is used to measure reduced internal stress, with the aim of obtaining results for verification of applied creep law and adoption of parameters as soon as possible prior to cavity commissioning. Lowering of the brine-air-interface offers two advantages. Firstly: due to reduced internal pressure, convergence increases and this leads to greater accuracy in determination, because the interfering factors (thermal expansion, etc.) remain constant. Secondly: for a short time the cave back has a rapid increase of pressure, what gives better adaptation of the calculation and verification of the material flow. As illustrated in Figure 16.4.9, a submersible pump is inserted in the casing so that the brine column corresponds to the selected internal pressure ends above the pump. First of all the brine is pumped off until the interface reaches the intended depth. As the brine interface would have risen to the well head again within a matter of hours to days due to convergence etc., a pressure device governs intermittent pumping off of the brine. In this way the internal cavern pressure can be kept quasi constant. The advantages of this method are a simple stabilization of interval pressure as well as easy evaluation of the measured data, as the volume flow can be measured above ground.

c) *Additional investigations* should be taken in order to eliminate the influence of interferring factors on the change of the brine volume, such as: thermal expansion of brine, subsequent leaching, decompression, fresh water injection. The most important investigation is the recording temperature logs prior to and after measurements respectively. If subsequent leaching is in effect, then brine samples must be taken from several levels in order to determine their NaCl concentration.

In regard to stability control of liquid-filled caverns, duration of the tests should be determined with the aspect of rock mechanics, e.g. that time intervals facilitate the

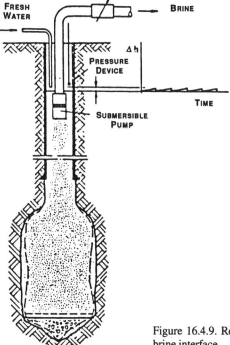

Figure 16.4.9. Reduced internal pressure at constant height of the brine interface.

ascertainment of a close curve for time-dependent convergence (could expect 100 days or more).

2. *Control of gas-filled caverns* at the present time is possibly by thermometric method. Due to its accuracy it permits determination of only large convergence. The general equation is:

$$V = \frac{mZRT}{p}$$

Where;

V = volume;
m = gas mas;
Z = super compressibility of gas;
R = gas constant;
T = gas temperature;
p = gas pressure.

As the errors in the determination of the gas mass increases more and more with the growing number of fill/withdrawal cycles, the actual gas mass must be determined in another way. There are two possible methods:

a) *Difference method*. The cavern is filled or emptied over a period of some days and the differential gas mass m is recorded as accurately as possible. From the foregoing equation follows the evaluation equation:

$$V = \Delta m\, R \left/ \left(\frac{p_2}{Z_2 T_2} - \frac{p_1}{Z_1 T_1} \right) \right.$$

Due to the varying internal pressure and the resultant cavern wall displacement, the volume V is not a real constant volume. Measurements undertaken by KBB on a cylindrical cavern in the salt dome have shown that this percentage is negligible within the bounds of the accuracy obtainable. Mean values in relation to cavern depth must be inserted for gas pressure, temperature and with this compressibility for the state at the beginning and the state at the end of the test.

b) *The hydrogen tracing method* is based that after injection of a measured quantity of hydrogen into a gas storage cavern, the gas content is ascertained on the basis of the percentages of hydrogen measured by chromatography in the volumes of gas subsequently withdrawn. It is thereby assumed that within a short time the quantity of hydrogen injected forms a homogenic mixture with the natural gas found in the cavern. When the gas mass within the cavern has been determined in this way, the geometrical volume can be calculated using the method described above.

Due to mathematical simplifications and measuring errors the determination of the cavern volume in regard to closure is much more inaccurate, compared with the procedures described for brine-filled caverns.

REFERENCES

1. Martinez, J. D. 1983. Energy programs – A contribution of salt dome knowledge. In *Proc. of 6th Int. Symp. on Salt*, The Salt Institute, Virginia, Vol. 2, pp. 235-245.
2. Haddenhorst, H.G. & P. Quast 1983. Underground storage of natural gas in West Germany. In *Proc. of 6th Int. Symp. on Salt*, The Salt Institute, Virginia, Vol. 2, pp. 203-209.

3. Quast, P. & M.W. Schmidt 1983. Disposal of medium and low level radioactive waste in leached caverns. In *Proc. of 6th Int. Symp. on Salt*, The Salt Institute, Virginia, Vol. 2, pp. 217-235.

4. Wassman, T.H. 1983. Cavity utilization in the Netherlands. In *Proc. of 6th Int. Symp. on Salt*, The Salt Institute, Virginia, Vol. 2, pp. 191-201.

5. Gustin, J.D. 1983. Energy storage in salt. In *Proc. of 6th Int. Symp. on Salt*, The Salt Institute, Virginia, Vol. 2, pp. 177-182.

6. Berst, P. & D.N. Minh 1981. Stability of cavities in rock salt, pp. 473-478. In *Proc. of Int. Symp. on Weak Rock*, 21-24 September, 1981, Tokyo.

7. Berst, P. & D.N. Minh 1981. Deep underground storage cavities in rock salt, pp. 555-572. In *Proc. of 1st Conf. on Mech. Behaviour of Salt*, Penstate Univ. Clausthal: Trans. Techn. Publ.

8. Istvan, J.A. & C.W. Quero 1983. Storage of natural gas in salt caverns. In *Proc. of 6th Int. Symp. on Salt*, The Salt Institute, Virginia, Vol. 2, pp. 183-190.

9. Dreyer, W.E. 1981. Crude oil storage of salt caverns, pp. 629-660. In *Proc. of 1st Conf. on Mech. Behaviour of Salt*, Penn. State Univ. Clausthal: Trans. Techn. Publ.

10. Dreyer, W.E. 1969. Geomechanische Untersuchungen an Kavernen im Steinsalz und Schlussfolgerungen für die unterirdische Gasspeicherung. *Bergakademie* 21: 404-412.

11. Hardy, H.R. et al. 1983. Laboratory and theoretical studies relative to the design of salt caverns for the storage of natural gas. In *Proc. of 6th Int. Symp. on Salt*, The Salt Institute, Virginia, Vol. 1, pp. 385-409.

12. Quast, P. & S. Backel. Derzeitiger Stand der soltechnischen Planung von Speicherkavernen im Salz und die damit erzielten praktischen Ergebnisse. *Erdöl-Erdgas*, Heft 6, pp. 213-218.

13. Crotogino, F.R. 1981. Salt cavern in situ testing from the constructor's and from the operator's viewpoint, pp. 614-627. In *Proc. of 1st Conf. on Mech. Behaviour of Salt*, Penn. State Univ. Clausthal: Trans. Techn. Publ.

CHAPTER 17

Salt mining subsidence

The influence of underground mining on ground surface damage was first studied in the nineteenth century. The mechanics of displacement of overlying strata above a mined-out area have been determined for the case of maximum subsidence. The mechanisms of the interaction between salt exploitation and surface deformation have been studied extensively throughout this century. This chapter is composed of four topics, namely: underground mining subsidence, solution mining subsidence, sub-surface ground arching, and ground surface damage.

17.1 DRY MINING SUBSIDENCE

This section discussed salt exploitation by dry underground mining, primarily flat salt deposits or flat mine workings. The first part considers the fundamentals of subsidence as elements and types of subsidence. The second part considers the practical application of subsidence with regard to prediction and to comparison with actual surveyed data. The third part considers the dependence of room deformation on surface subsidence.

17.1.1 *Principles of ground subsidence*

The mined-out area of salt bodies experiences closure deformation and possible fissuring and caving of the overlying strata. As a result of mining activities causing overlying strata deformation and disturbance, the ground surface experiences continuous displacement, characterized by surface bending and the formation of a depression zone. The subsided area has a greater lateral extent than the mined area below.

The characterization of basic subsidence elements is given in Figure 17.1.1.

1. *The zone of influence of subsidence* is exhibited by different configurations of ground surface deformation. These zones of influence are categorized as follows:

a) The central zone trough (AA') is of a flat configuration, because horizontal displacement is zero ($\Delta U - 0$). The vertical movement is maximal ΔS_{max}) and is evenly distributed.

b) The internal zone compression (AB, AB') has both vertical and horizontal displacement. The vertical displacement is downwards.

c) The external zone extension (BC, BC') also has both vertical and horizontal displacement, but with vertical displacement upwards.

It should be noted that at the point of change of the boundaries between the internal zone

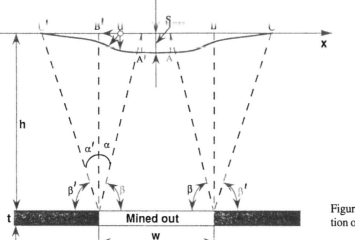

Figure 17.1.1. Characterization of mining subsidence

and the external zone, the horizontal displacement is maximal (U_{max}).

2. *The characterization of displacements* is briefly given here.

a) The vertical component of displacement, which is generally called 'subsidence', has an influence on the inclination of the ground surface and its curvature of deformation. This type of deformation has a large influence on the stability of engineered structures.

b) The horizontal component of displacement is directly related to the strain distribution which influences the strength and failure of engineered structures.

The stability of surface engineered structures is dependent on the position and distribution of the components of displacement. Position is defined by the horizontal distance from the mining limits.

3. *The angles of influence of subsidence* are important elements of subsidence consideration. Those angles are:

a) The angle of caving (β) defines the limit above a mined area, which has a maximum of sub-surface subsidence and possible strata failure, fracturing or caving.

b) The angle of draw (α) determines the same angle as above, but is measured between the vertical axis and the exploitation limit.

c) The angle of a subsidence limit (β') defines the boundary between the intact ground surface and the deformed ground surface.

The angles of influence of subsidence depend on the mechanical properties and behaviour of the rock strata. For example, the potash mines in New Mexico have a maximum angle of caving $\beta = 51.5°$.

4. *A classification of types of subsidence* has been made for flat salt deposits or flat mine workings which, in the majority of cases, belong to the controlled mining category. The classification of subsidence types, based on the relationship between the depth (h) and witdh of a mined-out area (w), is given here.

a) *Supercritical subsidence* occurs where the mine depth is less than the width of the mined-out area ($h < w$). This type of subsidence is characteristic of shallow mining and affects the ground surface substantially (Figure 17.1.2). A low ratio of depth/width for an excavated area results in full subsidence, commonly in the shape of flat-bottomed troughs at the surface. An important aspect arising from this is the high probability of appreciable surface fissuring occurring mainly along the longitudinal flanks of the subsidence trough. Such fissuring is usually time-dependent where pseudoplastic rocks occur at the surface.

b) *Critical subsidence* occurs where the mine depth is equal to the width of a mined-out area ($h = w$). This type of subsidence is also characteristic of shallow mining, and it has a similar characteristic to subcritical subsidence (Figure 17.1.3). The geological structural features of the rock strata are also important with respect to surface ground deformation, as discussed previously. The rock lithologies do not appear to have a marked influence on subsidence in shallow mining.

c) *Subcritical subsidence occurs where the mine depth is greater than the width of a* mined-out area ($h > w$). Under this condition, the sub-surface area is prone to deformation and caving above the mined-out zone but below the ground surface and the draw planes do not intersect the ground surface (Figure 17.1.4). This type of subsidence does not cause damaging deformations or fissuring of the ground surface. The subsidence of deep mines continues after mining ceases. For example, in the case of very deep salt mines, say over 1000 m, the subsidence could continue for up to ten years after mining ceases.

It is not difficult to predict subsidence, its extension and intensity with the time if mining is carried out in a controlled layout, without collapse of the mined-out area, as discussed in the next sub-section.

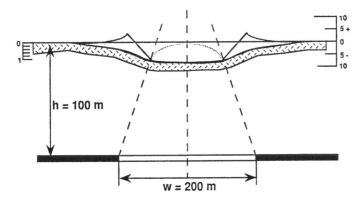

Figure 17.1.2. Supercritical subsidence

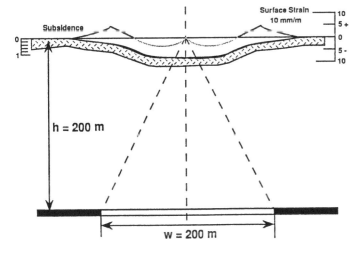

Figure 17.1.3. Critical subsidence

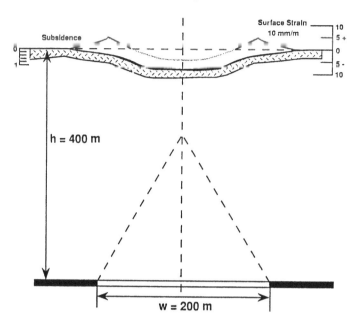

Figure 17.1.4. Subcritical subsidence

17.1.2 *Prediction of subsidence*

Prediction methods may be either empirical or analytical. Empirical approaches are best suited for the prediction of subsidence in salt mining, particularly based on the concept of influence functions and complementary influence functions. The zone area method, used for subsidence prediction of Saskatchewan potash mines, is a simplified version of the influence function method which lends itself to computerization, as discussed by Steel et al.[3] The mining area underlying a point which will influence the subsidence of the point, is represented by a cone of revolution. At any particular depth, the zone of influence is given by a circle of critical radius. The influence function method uses a continuous function to represent the effect of the extracted element on the subsidence of a point and integrates this influence over all extracted elements in the zone of influence to determine subsidence. The zone area method divides the zone into a series of concentric rings, each assigned a particular weight value, as illustrated in Figure 17.1.5 and as described in the following paragraphs.

$$S = f(\sigma_1 A_1 + \sigma_2 A_2 + \sigma_3 A_3 + ... \sigma_n A_n)$$

where:
S = subsidence;
$\sigma_1 - \sigma_n$ = weight of ring $1 \to n$;
$A_1 - A_n$ = proportion of ring $1 \to n$ extracted.
The extraction within each ring is determined and multiplied by the weight of the ring. The sum of the products from each ring is used to determine subsidence.[3]

Steel et al. analyzed five mines, three of them with a chevron layout and two with a pillar panel layout. Observations of the angle between the horizontal line and the line connecting the edge of the workings to the point of zero subsidence (limit of subsidence) ranged between 46 and 68°. The subsidence in all cases is subcritical due to the appreciable depth of mining. Central Canadian Potash surveying consisted of 21 stations overlying a chevron

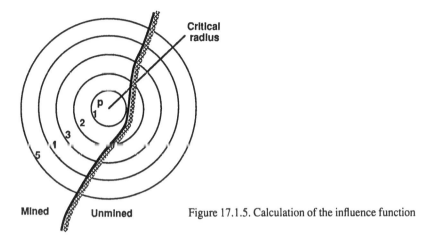

Figure 17.1.5. Calculation of the influence function

panel mining area, achieving a panel extraction of 60% with an overall mining extraction of 26.5%. Mining retreats toward the shaft. The survey is aligned in the direction of mining advance. The general layout of the mining panels and the location of survey stations is illustrated in Figure 17.1.6. The surveying data suggest that once mining advances beyond the zone of influence of a surface station, subsidence continues at a remarkably constant rate. Comparison between the constant subsidence rates above Central Canadian potash workings and the extraction within the range of influence of the monuments indicates a nearly linear relationship between extraction and subsidence (Figure 17.1.7). The analyses considered the extraction of the whole panel, including both rooms and yielding wing pillars. The mining subsidence assumes a salt creep in response to mining-induced stress and removal of strain. The salt deformation has been observed in two stages:

a) initial room convergence;
b) flow of the salt in the rooms.

The pillars, floor and back of the rooms react to the change in the stress field by creep into the opening.

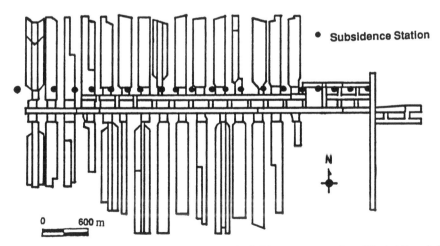

Figure 17.1.6. Pillar panel workings – Block II and subsidence survey stations (Central Canada Potash)

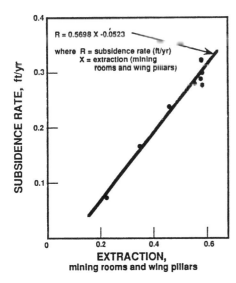

Figure 17.1.7. Comparison of subsidence and extraction rate – Block II (Central Canadian potash)

The prediction of subsidence, based on the above mentioned observations, requires the following input for the influence function:

a) the weighting of the contribution of extraction within the zones surrounding the point to the subsidence of the point;

b) the function relating the influence of extraction to the long-term constant rate of subsidence of a point;

c) Steel et al. wrote a computer program to predict subsidence based on extraction and the related subsidence functions. The required input includes the geometry of mine workings, location of the survey stations or location of desired subsidence points, and knowledge of the three functions previously listed.[3]

Mine plans are modified to a block model form, in which the plans are reduced to a grid of blocks of 15 m square. Each block in the grid coinciding with an excavation area is coded by a mining code and the year of excavation. The mining code refers to the particular method of mining of the room, which is assigned an extraction ratio in subsequent analysis.

A file containing the coordinates of the subsidence survey bench marks is used to generate the circular zones of influence surrounding each point. These zones are superimposed on the block model, and the proportion of mining in each zone, by year, is determined. The cumulative mined proportion of each ring, by year, is determined and the weighting function associated with each ring applied. The sum of the contribution of all the rings for a year produces an extraction factor for which the functions relating extraction to subsidence rate and initial subsidence are applied to determine yearly subsidence.[3]

Steel et al. represented three cases of subsidence for the Saskatchewan potash mines, which have been compared with the predicted subsidence modelled by the computer program described in the previous paragraphs. The mining height for all three mines was approximately 3.4 m.[3]

Central Canadian potash. With existing data, subsidence rate was found to be proportional to the panel extraction ratio within the area of influence. The computer-modelled subsidence value was adjusted until a visual best fit of the data resulted and took on the following value:

$$S = 1.5\,e$$

where:

S = subsidence attributed to the influence of extraction;

e = extraction factor including both rooms and wing pillars.

A log value of 1 year illustrates the subsidence prediction to the actual plot in Figure 17.1.8. The reason for discrepancy between actual subsidence and modelled subsidence is not understood.

Lanigan mine. The mining method is a pillar panel method with an extraction ratio of between 34 and 48%. Prediction of subsidence using the function applicable to the Central Canadian potash mines indicated very shallow subsidence compared to the actual surveyed subsidence. When a function relating the influence of extraction derived from German coal mines was used, the predicted subsidence at the Lanigan mine more closely matched the observed case, as illustrated in Figure 17.1.9.

Cory mine. This mine uses a chevron panel mining method, with a panel extraction ratio of approximately 64% and an overall extraction in the range of 38-42%. Mining below the survey line began in 1976, with subsidence surveying beginning in 1977. Zone weighting, using the method of Central Canadian potash mines, again resulted in a prediction of shallow subsidence as compared to the actual case. However, using the method of the Lanigan mine, the predicted subsidence resulted in a fairly good correlation with actual surveyed data.

The surveyed subsidence at the Saskatchewan potash mines clearly indicates two stages of deformation, first initial subsidence, followed by long-term steady subsidence until complete closure of the rooms occurred. The magnitude of initial subsidence and the rate of long-term subsidence are site specific and depend on the mining method and the overburden strata rheology. The subsidence mechanism is also influenced by depth of mining, sequencing in mining excavation, seismic events and others.

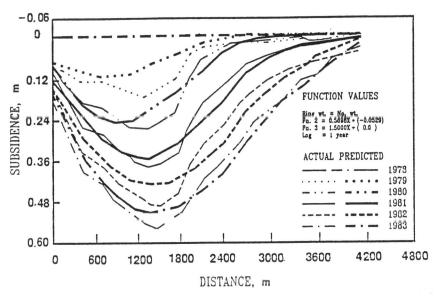

Figure 17.1.8. Comparison of subsidence prediction and surveyed subsidence (Central Canadian potash)

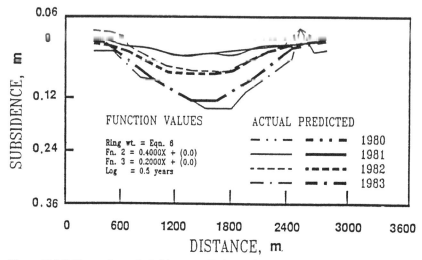

Figure 17.1.9. Comparison of subsidence prediction and surveyed subsidence (Lanigan mine)

17.1.3 *Dependence between room deformation and surface subsidence*

The dependence between room deformation and surface subsidence has been studied for Tusanj salt mine (former Yugoslavia), mainly with two aspects.

1. *Dependence between change of room volume and surface subsidence* has been studied in regard to bi-level room-and pillar mining (level 190 and level 250). The dimensions of the room-and-pillar layout are:

– width of room $W = 10 - 15$ m;
– height of room $H = 10$ m;
– length of room $L \leq 150$ m;
– width of sill pillar $S = 10$ m;
– width of room pillar $w = 10 - 15$ m.

The excavation of the first rooms started in 1967 on both levels, Level 190 (depth 466 m) and Level 250 (depth 526 m), and continued until 1975, as illustrated in Figure 17.1.10. The surface area, where subsidence had been monitored in relation to room excavation, was about 0.8 km^2, and had 44 instrumentation stations. The measurements in this area were taken during 1968/69, 1969/70, 1970/71 and continued for some time after room completion.

On the basis of surveyed data the equation for dependence of surface subsidence on the volume of the excavated room was formulated as follows:

$$S = a \cdot e^{bV}$$

and:

$$S = a \cdot V^b$$

where:

S = surface subsidence;
a, b = coefficients;
V = volume of the room.

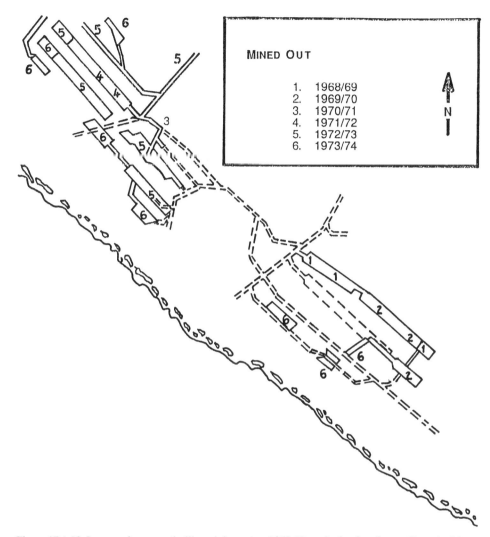

MINED OUT

1. 1968/69
2. 1969/70
3. 1970/71
4. 1971/72
5. 1972/73
6. 1973/74

N

Figure 17.1.10. Layout of room-and-pillar mining at level 250 (Tusanj salt mine, former Yugoslavia)

In Figure 17.1.11 the coefficients of correlation for each form written above are given considering an average mine depth of 480 m. In the same figure the graphs are representative of each form computed by the method of least squares, as well as the data obtained by in situ instrumentation.

The computed data indicate the dependence between the volume of room excavation and surface subsidence for the approximated mine depth of 480 m. The surveying data indicated that the maximum subsidence at ground surface was recorded mainly after one year of room completion. The proposed equation of subsidence as a function of excavated volume is valid, because it expresses the maximum subsidence. Unfortunately, these equations do not take into account the rheological effects which have a definite influence on the deformation.

2. *Dependence between vertical room movement and surface subsidence* has also been

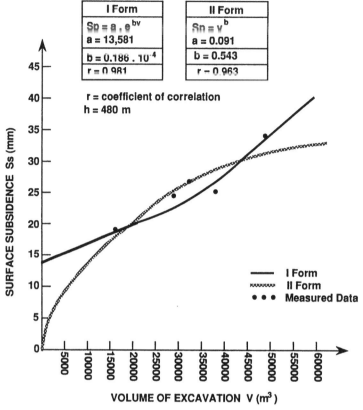

Figure 17.1.11. Dependence of surface subsidence on the volume of room excavation

studied in the same area discussed previously (Table 17.1.1). The time-dependent deformations have been measured for two chosen rooms on level 190 (R 191 and R 192). These rooms are adjacent to each other and divided by an inter-chamber pillar. In the larger room instrumentation was installed along three sections and in the smaller room instrumentation was installed along two sections. Each section consisted of eight instrumentation stations, stabilized by rock bolts. The instrumentation was carried out by measuring the length and angle of ground movement, so that displacement coordinates for each point could be monitored. The average error of point coordinate displacement was ± 1.5 mm. The readings

Table 17.1.1. Measured data of room volume, room movement and surface subsidence

Year of excava- tion	Tusanj salt mine	Depth (m)	Volume V (m³)	Vertical room move- ment S_R (mm)	Surface subsi- dence S_S (mm)
1968/69	East	526	15 700	70	−18
1969/70	West	466	35 000	45	−24
1969/70	East	526	28 700	100	−24
1969/70	East	466	26 700	45	−23
1969/70	East	526	42 400	130	−30
1970/71	West	466	49 760	62	−35

of point movements were taken at regular time intervals. On the basis of analysis of the measured data the equation for dependence of surface subsidence on the vertical room closure is proposed in the following form:

$$S_S = a_o \cdot S_R\, a_1$$

where:

S_S = surface subsidence;
S_R = vertical room closure;
a_o, a_1 = coefficients.

The relationship between vertical closure and subsidence was computed by this equation, separately for each mine level, as illustrated in Figure 17.1.12. It should be noted that computed curves in the diagram correspond to the data for synchronized monitoring in situ of both vertical room movement and surface subsidence. The same statement made previously also applied to this relationship, because the equation does not take into account the rheological effects which have a definite influence on the room and ground deformations.

Finally, it should be noted that the existence of nearby solution salt mining may have some limited, indirect influence on the deformation of sub-surface and surface structures, which is very difficult to determine.

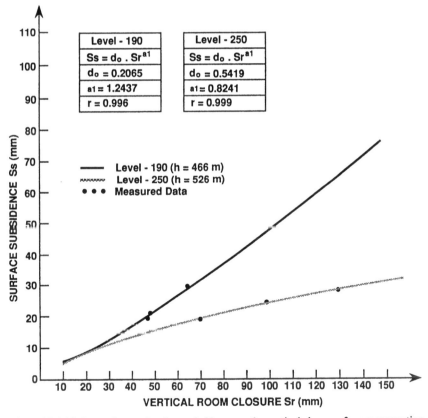

Figure 17.1.12. Dependence of surface subsidence on the vertical closure of room excavation

17.2 SOLUTION MINING SUBSIDENCE

Solution mining subsidence is a complex problem of rock mechanics, particularly in the case of uncontrolled solution mining. Three types of subsidence in regard to solution mining methods are discussed and a fourth type considers subsidence development as a function of time.

17.2.1 *Subsidence from single-well mining*

Subsidence from single-well mining can be designed and controlled. This type of subsidence is similar to that in room-and-pillar mining, because the caverns are irregular rooms which are supported by pillars. To control subsidence, it is of paramount importance to design a proper thickness of sill pillar (salt pillar above caverns) and the required thickness of the inter-cavern pillars as discussed in Chapter 15. (Subsection 15.3.4). The skeleton of the pillars is uniformly deformed, so that subsidence consideration is given as for a block of salt mass in which caverns are located (Figure 17.2.1). Under this assumption, further consideration of subsidence is given as in the case of room-and-pillar mining, e.g. the case

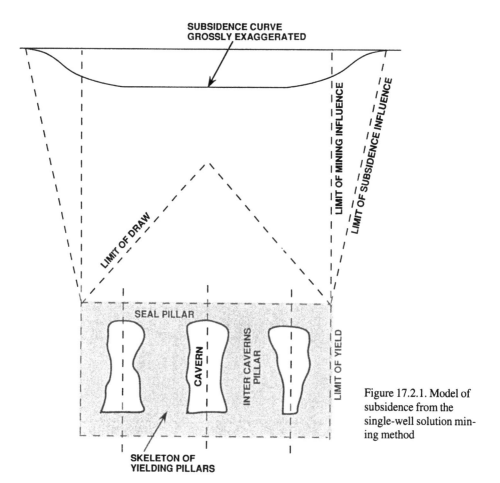

Figure 17.2.1. Model of subsidence from the single-well solution mining method

of uniform subsidence of the entire caverns field area. In order to evaluate surface subsidence, the roof subsidence of the caverns is estimated over the volume of convergence. The best control of cavern subsidence is achieved by a trigonal layout of the single wells, where the contact points are preferred. Subsidence distortion analyses and subsidence surveys suggest that the amount of surface deformation is small. Even in the most unfavourable situations, damage to buildings by subsidence is not possible, because the amount of ground tipping is negligible.

The cave-in of caverns or collapse of a cavern system leads to damage of the ground surface only in the case of shallow solution mining. Under these circumstances, damages at the ground surface could be devastating due to the formation of sink holes. At greater depth (over 1000 m), regardless of the collapse of the cavern system, destruction at the ground surface is impossible.[4] This is of paramount importance for storage caverns, which must maintain their stability for a long period of time. The important requirements for storage caverns are that deformation of the surrounding rock does not fracture to the surface and that the flow of salt maintains small enough deformation that subsidence will also be small.

Several mathematical models have been developed for the analysis of the conditions created in the rock mass surrounding a cavern system with the aim of predicting subsidence, particularly in the case of storage caverns. All of them have some deficiencies because they do not incorporate the discontinuous nature of the salt rock mass.

17.2.2 Subsidence from two-well mining

Solution mining by two or more wells requires a close location of the caverns, so that they can be connected without difficulties. Under these circumstances large lateral caverns with irregular shapes are form, that are called 'galleries' by some. It is obvious that large caverns/galeries experience a higher stress concentration, surrounded by rocks of deteriorated strength due to high moisture contents induced by the dissolution process. These structures are susceptible to closure and cave-in, and could result in the catastrophic collapse of overlying sediments, and significant surface damage through sink hole subsidence, as discussed later in this chapter.

Further consideration of subsidence from two-well solution mining is given by the case history of the Michigan salt basin, as described by Coates et al.[5]

1. *The Grosse Ile solution mine (USA)* started brine exploitation in 1941 by the conventional single-well method of injecting water down a tube and forcing brine up through the annulus between the tubing and well-casing (direct circulation system). The single-well caverns formed by this system later coalesced into two major galleries, labelled 'north' and 'central' (Figure 17.2.2). In 1954, elevations were recorded. Down-warping up to 0.6 cm per year was considered acceptable. Total subsidence of around 1 m over the entire area was also acceptable as long as the down-warping did not cause tension breaks in brine pipelines or other structural damages. It was estimated that the subsidence in the brine field would be limited to only 1 m or little more. The maximum subsidence was manifested at two sink holes, through which the centre of the subsidence profile is represented (Figure 17.2.3). It should be noted that it took over ten years before sink holes were formed. The sink hole in the north gallery was formed four months before the sink hole in the central gallery.

In the first phase of solution mining, with a single-well method, a large amount of roof support of the single cavern was removed relative to the volume of salt that was extracted.

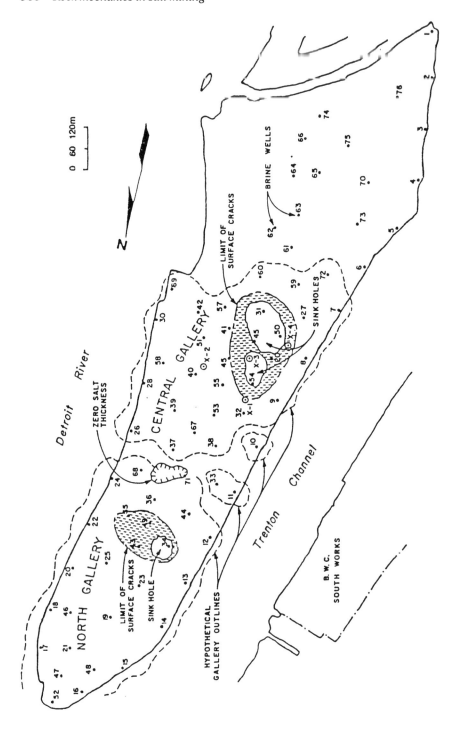

Figure 17.2.2. Location of brine wells, galleries and maximum subsidence with sink hole (after Landes and Piper)

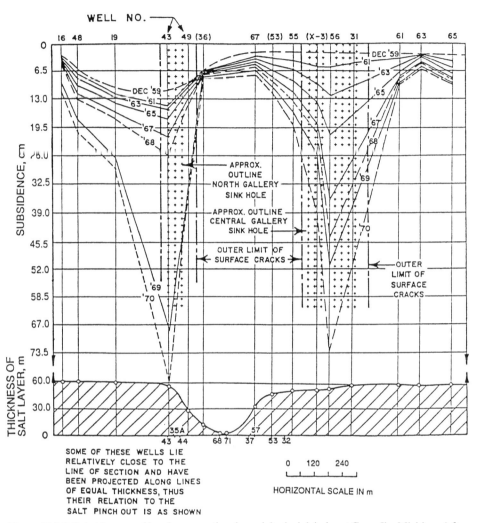

Figure 17.2.3. Subsidence profiles along a section through both sink holes at Gross Ile, Michigan (after Nieto-Posatto and Hendron)

This occurred because of the high rate of salt solution at the top of the cavity, immediately under the roof. Connecting the wells by undercutting the salt layer was meant to alleviate this problem by forcing lateral dissolution, but by that time some roof support had already been lost. A salt pinch-out is believed to have contributed to the formation of large cavities by causing concentrated solution exploitation on either side of this axis. The extensive use of inlet wells, which accounted for almost 45% of the total production in both zones of maximum subsidence (north and central galleries), was responsible for the concentrated extraction in the area adjacent to the pinch-out. The maximum subsidence occurred on both sides of the salt pinch-out, and when maximum subsidence reached 0.65 cm, sink holes were formed. The relocation of inlet wells in 1961 and 1968, in the central and north galleries, respectively, reduced the subsidence rate, but did not prevent the eventual formation of sink holes.

The postulation of the mechanism of subsidence at the Gross Ile solution salt mine has been given by Nieto-Pescetto and Hendron (1977). The removal of roof support by salt solution induced roof sagging and the concentration of vertical stresses around galleries. This, in turn, accelerated the creep of the salt forming the walls of the gallery and caused further subsidence of the ground surface. As sagging progressed the roof beds either developed concentric tensile cracks or existing cracks opened up. The maximum crack concentration and propagation were at the centre gallery, where the tensile stress reached a maximum. With additional salt removal by production, the vertical tensile fractures propagated to ground surface and facilitated shear displacement along them, which resulted in a progression of the subsidence. Further, by rock strata loosening and caving, ground surface subsidence progressed, and sink holes formed.

2. *Windsor solution mine (Canada).* Here, solution mining was by conventional gallery mining, which involves pumping fresh water into the salt beds through input wells and removing the brine through production wells. The gallery is formed by a merging of the individual salt caverns. Inadequate roof support above the gallery is the controlling factor of localized subsidence. Coates et al.[5] summarized the events of subsidence as follows.

The existence of subsidence had been recognized in 1948 by cracks in a number of plant buildings. Investigations reported subsidence up to 3.7 cm between October 1948 and October 1950, occurring in an irregular pattern. Between October 1950 and October 1951 the subsidence increased slightly to over 6.2 cm per year north of the liquid chlorine plant (Figure 17.2.4). An increase in subsidence of 7.5 cm was noted east of the hydrogen tank between October 1951 and October 1952. However, subsidence at the liquid chlorine plant was 22.5 cm from October 1952 to October 1953. A bowl shape depression with a radius greater than 300 m could be discerned. Differential settlements between holdings on the plant site showed that the maximum subsidence in the plant area did not exceed 1 m before the major subsidence. On February 19, 1954, a rapid subsidence occurred. Sounds and vibrations resembling the bumping of rail cars and the rumbling of small earthquakes were heard and felt, apparently originating below the offices of the Canadian Salt Company Ltd. The noises became louder, and the vibrations were severe shortly after 9 a.m. At 9.30 a.m., the first physical signs of subsidence could be seen (Figure 17.2.4). A depression about 0.5 m deep was filled by water. Subsidence caused the rupturing of brine lines and water lines, which contributed to the flooding. Close to the north rim, a rapid subsidence resulted in a bowl-shaped depression with a radius of about 300 m and a depth of about 1 m. An elliptical lake was formed in the centre of the depression.[5]

The mechanics of subsidence, discussed by Karl Terzaghi, was based on the theory of localized rock subsidence in which the surface of the bedrock area within the sink hole settles.[6] Ruth Terzaghi's theory is based on the ground surface subsidence being approximately equal to the subsidence of the rock surface.

17.2.3 *Subsidence from uncontrolled solution mining*

An estimation of subsidence from uncontrolled solution mining is difficult because the intake of fresh water into the salt deposit is not known since it is intersected by aquifers. The outlines and the extension of the solution galleries are not controlled and generally they are not accurately known. It should be noted that controlled subsidence of two-well solution mining could become uncontrolled at the point of large gallery formation through the connection of a number of individual caverns, as for example at the Windsor salt mine, Canada.

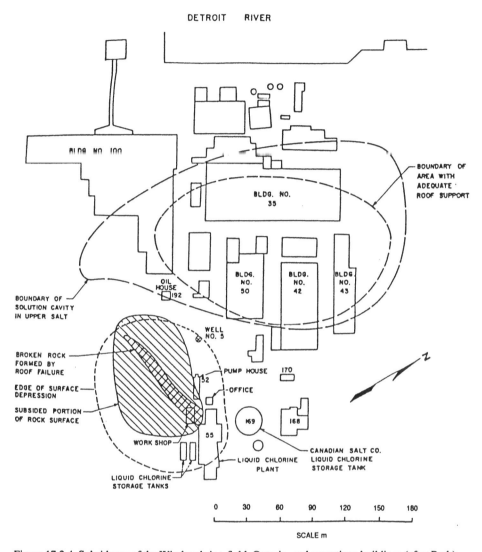

Figure 17.2.4. Subsidence of the Windsor brine-field, Ontario, and operations buildings (after Peck)

The most significant subsidence recorded from uncontrolled solution mining, either had been initiated originally or came by conversion from controlled solution mining. To better understand solution mining subsidence, it is necessary to analyze the mechanics of interaction between sub-surface deformation and ground surface damages, as further discussed.[7]

1. *Sub-surface subsidence* caused by solution mining has no satisfactory mathematical or empirical theories to explain its mechanism due to the fact that it cannot be observed. The quantity and geometry of the rock salt remnants left by solution mining are unknown. It is almost impossible to relate the configuration of one or more caverns in underlying salt beds to the collapse of the rock salt and overlying strata. The general hypothetical model of sub-surface subsidence is based on the assumption that profiles of the caverns in most rock

salt deposits exhibit a form similar to a loaded thick beam with clamped edges. The events of deformation might be in the following order

a) The postulated thick beam comprised of the roofing strata is subjected to gravitational body forces, whose magnitude depends on the strata thickness (Figure 17.2.5a). As a result of solution mining, the thick beam will deform. The deformation slowly increases.

b) Slow lateral movement at an approximately constant rate permits gradual bending of the postulated thick beam. During the extraction period, several caverns in the salt formation might be formed, controlled by individual wells. It is assumed that more or less similar sagging takes place in all caverns. As a function of time, creep deformation will be prolonged before rupture occurs (Figure 17.2.5b).

c) Dr Bays suggested that sooner or later most salt wells become connected through solution channels to each other and finally to a large gallery. At that moment, the strength of the overlying sediments might be overcome and rupturing initiated. The rupturing will be progressive, and the arch of the overlying rock will be formed (Figure 17.2.5c).[6]

d) When the rock salt remnants can no longer carry their own weight and the weight of unconsolidated overlying rock in the area of a large gallery, they will collapse. At this moment a large gallery will be totally filled in by fragmented rock and the arch will be extended upwards at the accelerated rate. Probably, when the parabolic slope of the arch of draw intersects the ground surface, a sink hole will be formed, with a vertical ground displacement of 1 to 10 m (Figure 17.2.5d).

2. *Surface subsidence* obviously is a product of sub-surface deformations. From many published data of surface subsidence caused by solution mining, several common features of this mechanism can be inferred as suggested here:[7]

a) The surface effects are associated with vertical and lateral ground movements. Probably all lateral surface displacements are toward the centre of the cavern. Vertical surface displacement starts well beyond the lateral limits of the extracted area, with its maximum at the centre of the cavern.

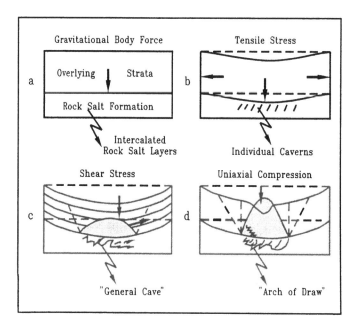

Figure 17.2.5. Hypothetical model of subsidence development, with stress field delineation

b) As a result of this displacement, a curvature of the ground surface results. The curvature in some cases represents Gauss' function, when the subsidence reaches a final settlement. The deformations of the ground surface can be described by three zones (Figure 17.2.6).[10]

c) First, the zone of maximal vertical displacement (simple uniaxial compression) which should be above the centre of the large gallery in the rock salt formation. This zone of deformation is represented by a sink hole filled in by water. It is located at the middle of the bowl of surface subsidence and it is filled by the broken rock material. The profile of subsidence in this zone caused by solution mining has a convex curvature.

d) The second zone of ground surface deformation surrounds the sink hole. The slope is represented usually by a concave curvature (Hutchinson, Kansas;[8] Barich, Poland[9]) and it is formed by vertical and horizontal ground movements (shear stresses). It is interesting to note that while the radius of curvature varies, the geometrical shape is always the same.

e) The third zone represents an area of minimal vertical movement, but horizontal extensions occur (tensile stress). The slope of this peripheral part of subsidence could be very gentle, with a larger lateral extension.[10]

Theoretical analysis of the deformation on the ground surface caused by mining operations in bedded deposits is very well explained by the trough subsidence.

Unfortunately, there is no similar analysis for subsidence caused by the uncontrolled solution mining. From an engineering point of view, it is very difficult to represent the relation between surface ground deformations and the geometry of the extraction area. However, in a general sense, subsidence curvature of deformation might sometimes be used as a parameter for an approximate estimation of the extension of the large cavern and the tip of the parabolic slope of the arch of drow.

17.2.4 *Subsidence as a function of time*

Surface and sub-surface subsidences are time-dependent deformations. For example, the

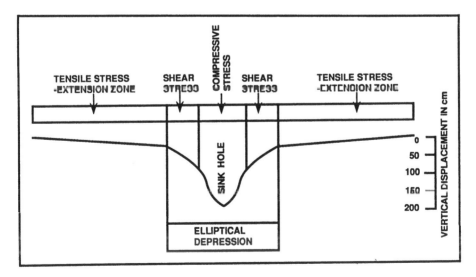

Figure 17.2.6. Subsidence of the Tuzla brine-field (former Yugoslavia) and the stress fields delineation

surface subsidence of most rock salt deposits excavated by the room-and-pillar method is very shallow and regular, without rupture of the ground surface. In this cases, subsidence is not manifested immediately after excavation, and it can only be measured after several years.

Usually a couple of years after underground excavation ceases, the surface comes to rest. However, on the contrary, subsidence caused by solution mining may cause an eratic and unpredictable surface deformation. An instructive example of the subsidence progress is published by R.D. Terzaghi in a paper entitled 'Brine-field subsidence at Windsor, Ontario'.[6] The rate of subsidence at Windsor had many features in common with that of other surface deformations caused by solution mining in many parts of the world (Figure 17.2.7). At Windsor, the salt formation is about 180 m thick and consits of shale, dolomite, gypsum, anhydrite and rock salt. The thickness of the overlying sediments is about 270 m. The subsidence deformation as a function of time can be summarized as follows:

a) The period interval before subsidence is evident, is much longer for solution mining that for an excavation formed by underground mining methods. Usually subsidence above an excavation due to mining is noticeable after two to three years. However, this period is much greater for solution mining of rock salt deposits, and subsidence may perhaps be only noticeable after decades and in some cases after centuries, depending on the intensity of solution mining. For example, at Windsor, solution mining began in 1902, but the first noticeable surface deformation was in the late 1940's.

b) The first phase settlement of subsidence for solution mining progresses at a very slow rate, and it can last for several decades. At Windsor, the first phase of surface subsidence was from 1940 to 1952, forming a convex shaped depression.

c) However, the phase of accelerated subsidence takes place within a period of a couple

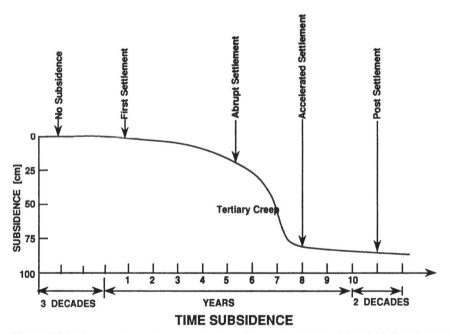

Figure 17.2.7. Diagram of subsidence as a function of time at the Windsor brine-field, Ontario (modified after Terzaghi)

of years. During this time, the ground surface deformation forms a bowl-shaped depression, which might subside up to 40 cm. Accelerated subsidence at Windsor occurred during the years 1953-1954.

d) Abrupt subsidence can occur within several days or hours. For example, at Windsor (February 19, 1954) the catastrophic subsidence happened with an interval of only several hours. The bowl-shaped depression began to fill with water and rapid subsidence formed a sink hole. The sink hole appeared along one side of the rim of the bowl-shaped depression and it was up to 75 cm deep. By mid-afternoon, movement had virtually ceased. In a similar manner, the sink hole at Tuzla (former Yugoslavia) was formed, but the process of abrupt subsidence lasted through several days.

e) Post subsidence settlement was immediately followed by irregular movements of small magnitude. At Windsor, the average annual rate of settlement over the fourteen-year period following the sink hole was up to 1.25 cm per year.

If the curve of the subsidence deformation, as a function of time, is analyzed, many similarities with a generalized creep curve can be noticed. A typical three-stage creep is developed in which the deformation rate accelerates with time. In time (decades and years), a rock salt formation with some ductile properties may be deformed by the creep mechanism with three distinct component processes. In accordance with this suggestion, the ground surface subsidence is rather the product of the creep deformation than of elastic yielding deformation.

17.3 MODELS OF SUBSIDENCE DEVELOPMENT

Three case histories are considered as models of subsidence development in solution mining of salt. Each model has certain structural characteristics with regard to natural factors and also the form of maximum subsidence.

17.3.1 *Subsidence developed at the Tuzla brine-field*

The industrial exploitation of the rock salt deposit by solution mining started in 1886 and has been increasingly continued ever since. Dipping to the NW, the rock salt deposit has a strike in the SE-NW direction, covering an area of about 1.5 km^2 (Figure 17.3.1). With more than 230 wells drilled for exploration and production, the deposit is well defined except at its NE boundary. The rock salt occurs in four beds of a total thickness of 200 m. The salt layers are interbedded by marl, consuming about 30% of the total volume in the upper part, and about 50% of the total volume in the lower part of the deposit. The overlying strata of salt deposits consist of banded and clayey marls with some sandstone intercalations. The uppermost salt series, about 15 m thick at a depth between 100 and 300 m, is almost mined out. The second salt series which is in the exploitation process is about 60 m thick.

Banded marl beds and the upper salt series from a uniform hydrogeological recipient. On the outcrops of the salt series the recipient is fed with water from the Solina River on the eastern side of the deposit and the Jala River on the southern side. The average water inflow through porous sandstone of the overlying strata is about 29 l/sec and runoff water is about 8.45 l/sec. The total area of runoff water flow is about 9.7 km^2. Solution mining is uncontrolled because it allows the inflow of fresh and brackish water from the upper salt series. The brine exploitation is achieved by pumping through holes with a tubing of 52 cm

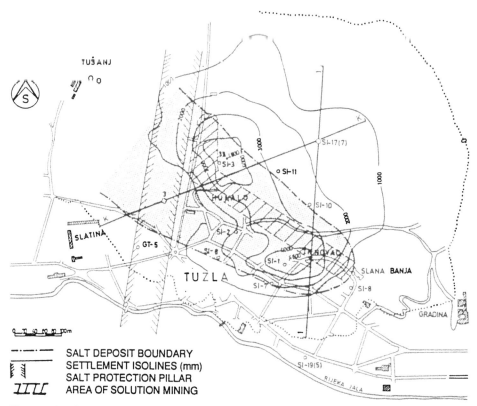

SALT DEPOSIT BOUNDARY
SETTLEMENT ISOLINES (mm)
SALT PROTECTION PILLAR
AREA OF SOLUTION MINING

Figure 17.3.1. Salt deposit divided by a protection pillar on a dry salt mining area and a solution mining area (Tuzla, former Yugoslavia)

diameter. Because the natural water inflow into the deposit is not equivalent to the pumped brine-outflow, the underground water level is decreasing steadily. For example, for the period from 1886 to 1972 it has been lowered by 107 m. With increased brine pumping, surface subsidence and damage to structures in the town of Tuzla have progressed. The subsidence recorded between 1972 and 1978 is as much as 7.5 m. The maximum subsidence in this brine-field is called the 'pot-hole', due to its large lateral extension, which is controlled by the formation of a large gallery in the salt deposit.

Subsidence development has been analyzed by mathematical modelling, for four distinct phases of exploitation. The finite elements have been considered to be subjected to gravity loading and, below the observed underground water level, to vertical uplift components. Seepage forces due to hydraulic gradients have been neglected. However, for the third phase and fourth phase of solution mining the hydraulic potential field was assumed to be unchanged, because to maintain satisfactory water levels it is necessary to discharge fresh water from the ground surface.[11, 12]

The results of mathematical model analysis are supplemented by actual surveying data, as illustrated in Figures 17.3.2 and 17.3.3, and as briefly described here.

a) The modelled stress distribution along cross-sectional planes is represented by the horizontal (σ_y) and vertical (σ_z) normal stress components and in addition the corresponding σ_y/σ_z plots are given.

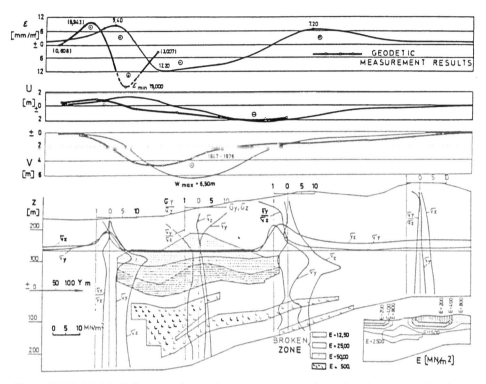

Figure 17.3.2. Modelled diagrams of normal stress components (σ_z, σ_y) and stress ratio (σ_y / σ_z) compared to surveyed surface subsidence displacement (v, u) and horizontal strain (ε) for the second phase of solution mining (1976)

b) The surveyed surface subsidence is given by settlement (v) and horizontal displacement (u) of the surface points as well as the corresponding strain diagrams.

The stress ellipses are represented in Figure 17.3.4 and the second phase of solution mining (1976) and in Figure 17.3.5 for the fourth phase of solution mining (2036). In addition, Figure 17.3.6 represents the isolines of the p_c/p_f ratio, p_c being the radius of the stress circle and p_f the distance from the centre of the stress circle to the failure envelope, measured in the direction perpendicular to the failure envelope [11, 12]

From all represented data it can be concluded that the numerical model analysis and the long-term surveyed subsidence data are in reasonable agreement. When comparing the computed and measured values, a correlation between settlements and mined-out salt volume exists.[11, 12]

The diagrams representing the stress ellipses as well as the vertical normal stress versus depth plots, show very pronounced arching effects of the immediate overlying strata. According to the analysis performed so far, the arching remains effective even in the completion phase of solution mining.[11, 12]

The intolerable ground deformation at the beginning of the 1980's resulted in an intersection of the cave-in of overlying strata above a large gallery with the ground surface. This phenomenon produced a large pot hole with a maximum depth of up to 10 m, in which some part of the core of the downtown collapsed. The large gallery below the subsided part of the city had been filled by caved rock, and after a couple of years, settled.

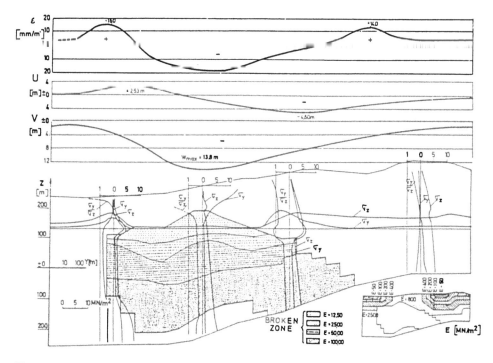

Figure 17.3.3. Modelled diagrams of normal stress components (σ_z, σ_y) and stress ratio (σ_y/σ_z) and modelled surface subsidence (v, u) and horizontal strain (ε) for the fourth phase of solution mining (2036)

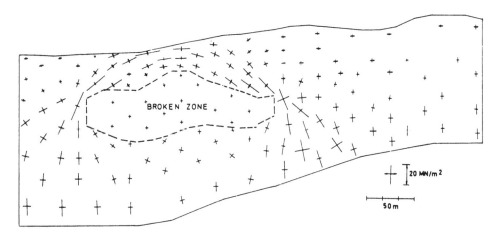

Figure 17.3.4. Modelled cross-section of stress ellipses for the second phase of solution mining (1976)

17.3.2 *Subsidence developed in the Windsor-Detroit area*

The history of solution mining in the Windsor-Detroit area has been previously mentioned in subheading 17.2.2. The following is a discussion of the mechanical model for sink hole development as described by Nieto et al.[13]

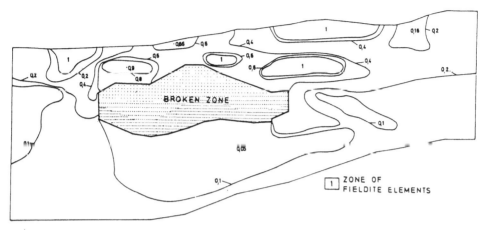

Figure 17.3.5. Modelled cross-section of stress ellipses for the fourth phase of solution mining (2036)

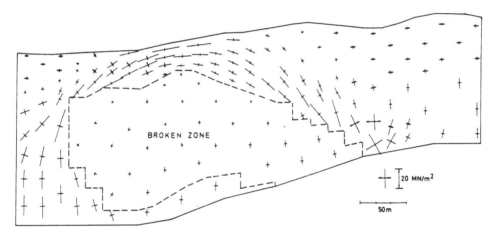

Figure 17.3.6. Cross-section of the isolines of the p_c/p_f for the second phase of solution mining (1976)

A bowl of subsidence was formed due to concentrated solution mining by the single-well method of the salt horizons (Figure 17.3.7). This subsidence was first manifested at the surface as a bowl of gentle gradient with limited extension, but then grew as much as 2 mm/m with a diameter up to 500 m. The increase in subsidence resulted from an increased flexural deformation of the underlying strata. Any layer or group of layers between the surface and the top underwent similar deformations. However, the intensity of flexure increased with depth.

Nieto describes the following mechanisms for sink hole development.

a) As the underlying layer sags and/or collapses, the layer in question deflects elastically and develops horizontal compressive stresses in its upper half. No tensile stresses develop in the lower half because of existing joints.

b) Using the elastic theory, it can be demonstrated that if the layer in question has a lower modulus than the underlying layer, there is virtually no separation at the contact between the two layers as subsidence continues (provided that the two layers being disucssed are of

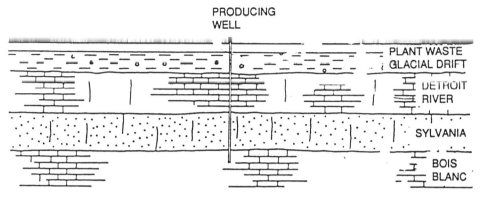

Figure 17.3.7. Pre-solution mining conditions (lateral stress in sylvania beds: 12-14 MPa)

equal thickness). If the layer has a modulus equal to or greater than the underlying layer, the former separates along a bedding plane or other horizontal discontinuity and 'hangs' unsupported.

c) As subsidence continues (wider bowls and steeper gradients), the stresses in the upper half of the layer, either supported or unsupported, continue to increase until the layer either fails in shear, because the horizontal compressive stresses along the upper half exceed the available strength, or simply collapses ('turns inside out', 'snaps') without shear failure.

d) As subsidence continues, the horizontal compressive stresses in the failed or collapsed layer decreases and eventually become tensile. Thus, any failure-induced or pre-existent fracture opens up.

In the case of sylvania, the beds deformed following the subsidence shapes of the stiffer underlying Bois Blanc dolomite, however, the sylvania became separated at or near the top of the bed from the overlying stiffer Detroit River dolomite (Figure 17.3.8). Evidence of that separation has been reported by Dowhan. In a borehole drilled in a rapidly developing bowl of subsidence on the south end of Point Hennepin, he described a 2-m vertical gap in the sylvania about 8 m from the top. A sonar survey measured lateral extent of up to 60 m in at least one direction.

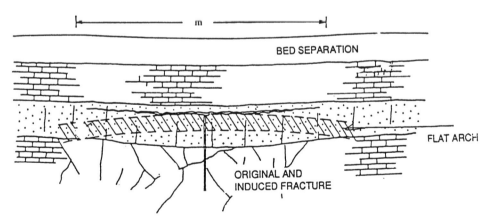

Figure 17.3.8. Formation of bowl subsidence at the location of a long producing well (flat arch effect induces additional compressive stress)

The sylvania then continued to deform under essentially no vertical confinement until, as will be explained later, the upper half failed under horizontal loading conditions. The horizontal stresses are a combination of the in situ stress and the stresses generated by subsidence. A significant portion of the sylvania then became cohesionless sand.

As the underlying Bois Blanc continued to subside and either collapse or crush, cracks began to open up (Figure 17.3.9). These cracks (as well as broken casings) allowed the cohesionless sylvania sand to flow unhindered as a slurry downward toward the deeper solution openings. It should be emphasized that the width of these cracks need not be very large. It has been concluded that when the width of the fracture or fissure is about twice the maximum grain size diameter, an interrupted downward flow of the slurry takes place.[13]

As a result, a significant portion of the sylvania became cohesionless sand, which created a void by the migration of slurry through open cracks into the lower sylvania and the Bois Blanc. Formation of the void resulted in the removal of support from the base of the Detroit River dolomite (Figure 17.3.10).

The unsupported area of void increased gradually, until a large plate of this formation failed progressively upwards to the base of the unconsolidated glacial sediments. This

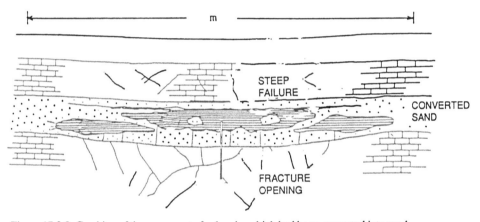

Figure 17.3.9. Crushing of the upper part of sylvania, which had been converted into sand

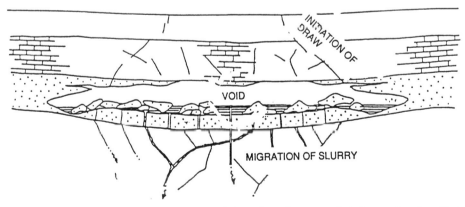

Figure 17.3.10. Void creation by migration of sand slurry through open cracks in lower sylvania and bois blanc

material failed essentially as a plug and slid into a cavern in the sylvania (Figure 17.3.11). Actually the sink hole facilitates a slide of the clay and boulders of the glacial till, as well as the plant waste, to the sides where large volumes of this material can be accommodated.[13]

This mechanism of development of sink holes of unexpected size and depth in the Windsor-Detroit area, as illustrated from the representation above, is to a great deal influenced by the sylvania failing under high lateral stresses, converting into sand, flowing downward through cracks towards deeper solution mining caverns, and creating a shallow void that generates the sink hole subsidence. It could be concluded that sink holes are likely to occur in the areas where the sylvania is close to the ground surface (less than 20 m) and the subsidence bowl being generated has a surface gradient of a few millimeters per meter.[13]

17.3.3 *Palaeo-subsidence development at the Delaware basin*

The palaeo-subsidence located on the northern margin of the Delaware basin in New Mexico was developed hudreds of thousands of years ago. The subsidence at this location was, of course, due to the natural solution of the salt beds, and it has been exposed in the potash mine. The mine workings reveal a transition over a distance of approximately 55 m, from undisturbed sub-horizontal evaporite beds to breccia of the subsidence chimney, as Davis further discussed.[15]

At ground surface, the subsidence chimney appears as a circular hill with a central core of down-dropped, brecciated material (Figure 17.3.12). Including the outward sloping flanks, the diameter of the hill is approximately 365 m and the diameter of the brecciated core is approximately 24 m. In 1975, a potash mine drift intersected the breccia zone at 365 m below ground surface. The drift was driven into the south-east chimney margin. The mine workings offered the possibility to examine and map in detail the transition from undisturbed, sub-horizontal evaporite beds to the breccia of the subsidence chimney. The mine exposure of the subsidence chimney and the adjacent transition zone which is located approximately 110 m below the top of the Salado salt, indicated that the chimney had a deep-seated origin rather than being rooted in the upper salt beds.

The surface subsidence has all the characteristics of solution mining subsidence. In this

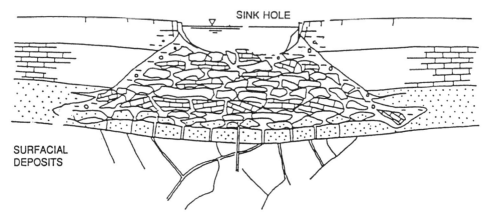

Figure 17.3.11. Sink hole subsidence formation, with collapse of Detroit River beds and surfacial deposits

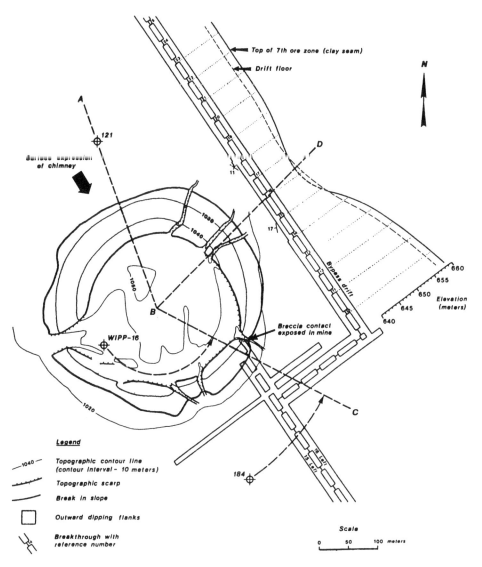

Figure 17.3.12. Ground surface expression of the subsidence chimney and the underlying potash mine drifts

case also there had been a slide of gravel beds of the Gatuna formation (600 000 years BP), cutting across the peripheral collapse contact and resting unconformably on the brecciated central core. This could be observed as gravel filled channels on the north east rim of the subsidence chimney. Apparently, gravel was also deposited in the sink hole depression, which was formed when sub-surface caving intersected the ground surface. Since that time, the sink hole topography has been inverted, as the terrain surrounding the chimney subsidence has been lowered by shallow regional salt solution at the top of the evaporite beds. Tilted columnar soil structures on the flanks of the chimney indicated that this regional subsidence terrain postdates the development of this 500 000 year old caliche.[15]

Sub-surface structural characteristics of the subsidence chimney can be observed in the potash underground mine. A cross-section of the geology of the strata in relation to the subsidence chimney is represented in Figure 17.3.13. Within the transition zone, the bedding of the sedimentary strata dips toward the chimney, reaching a maximum dip of 30° at the breccia contact. The breccia includes clasts of halite, anhydrite and polyhalite. The salt solution induced morphology of halite clast and the breccia matrix composition indicate that ground water flowed through the chimney during and/or following subsidence the sub-surface subsidence structure formed by the salt solution indicates its association with groundwater flow in the Capitan reef aquifer dissolving salt deposits above. The position of the sub-surface subsidence structure suggests that the chimney was formed by salt solution at the base, and within the lowermost portion, of the Salado formation. This salt solution caused incremental subsidence rather than catastrophic collapse of the overlying strata, which is in agreement with the steep-walled chimney containing the downdrop of breccia, evaporite and post-evaporite strata.

The mechanics of the palaeo-subsidence chimney at Delaware may be related to the mechanics of a solution mining subsidence sink hole. K. Terzaghi on the basis of experience from European brine-fields, concluded that sink holes were formed by stoping

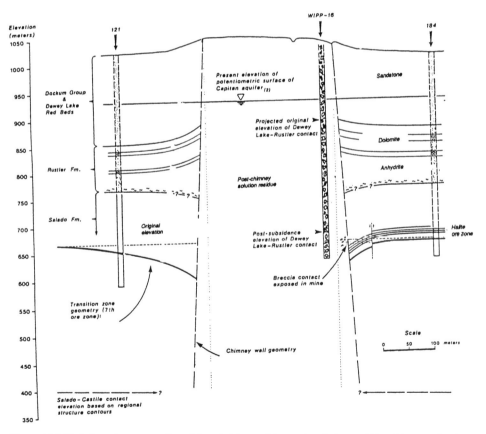

Figure 17.3.13. Sub-surface expression of the subsidence chimney in relation to drifts of the potash mine (modified after Davis)

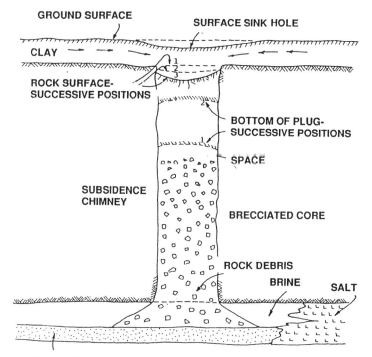

GROUND SURFACE

SURFACE SINK HOLE

CLAY

ROCK SURFACE-
SUCCESSIVE POSITIONS

BOTTOM OF PLUG-
SUCCESSIVE POSITIONS

SPACE

SUBSIDENCE
CHIMNEY

BRECCIATED CORE

ROCK DEBRIS

BRINE   SALT

Figure 17.3.14. Terza-
ghi's model of sink
hole development

INSOLUBLE RESIDUE

which propagated through overburden strata to the ground surface.[16] A cylindrical shape of subsidence occurs in competent rock (deeper) and conical shapes tend to develop in incompetent rock (close to surface). He also assumed that at the Windsor brine-field (Canada) subsidence sink holes most likely occurred in a cylindrical shape similar to the chimney (Figure 17.3.14).[16] Bulking was not evaluated in his discussion, since his main concern was to determine whether ground subsidence was caused by the lateral migration of glacial clay to the depression formed by the collapsed rock, or whether surface subsidence simply reflected the configuration of the subsided rock, which is more likely. Ruth Terzaghi, on the basis of more recent research data, suggested that sink holes were the result of a rapid increase of subsidence within an area of general subsidence.[6] She further postulated a gradual slumping of more or less intact beds to explain the ineffectiveness of bulking in preventing surface collapse, which occurs incrementally rather than catastrophically.

17.4 SURFACE SUBSIDENCE DAMAGES

For a better understanding of surface subsidence damage, it is necessary to briefly comment on the mechanical elements of ground movement due to vertical and horizontal displacement. The effects of vertical and horizontal ground movement on surface structures generally result in damage to combined displacement. In fact, both ground movements are likely to take place simultaneously, but in this chapter the effects are considered separately. In addition, the characterization of subsidence damage and classification for surface structure damage are also briefly described. The discussion in this section is based on the work of Ariogly[17] and Jeremic.[7]

17.4.1 *Simplified elements of ground movements*

The principles of interaction between the sub-surface excavated void and surface subsidence are based on the elements of ground movement and settlement. The basic elements of subsidence are illustrated in Figure 17.4.1 and are briefly discussed here.

1. Vertical subsidence is evaluated as a function of the relationship between the width (w) and the depth (h) of the excavation (w/h) and the height of the excavation (H):

$$\frac{S}{H} = f\left(\frac{w}{h}\right)$$

The maximum subsidence (S_{max}) is at the centre line of the excavation, and can be calculated by the following expression:

$$S_{max} = K \cdot H$$

where:

K = coefficient which depends on the method of mining (room-and-pillar with yielding pillars 0.30–0.60);

H = average height of excavated void.

2. The subsidence profile represents the curvature of vertical displacement at each point of ground settlement. As a result of this a surface building is exposed to a differential vertical movement between the points of the structure. If the strain, due to maximum differential settlement, exceeds a limiting or critical strain in a given material of the structure, then visible cracks will occur. For example, if a building of length L undergoes a curvature of ground surface deformation, and assuming that, with a radius of curvature of R, the deflection of the centre relative to the end is ... then it can be written in terms of the length:[18]

$$R = \frac{L^2}{8}$$

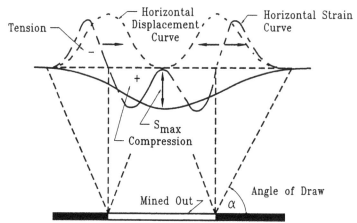

Figure 17.4.1. The elements of surface ground movements

From this equation, the deflection ratio is:

$$\frac{\Delta}{L} = \frac{L}{8R}$$

The radius of curvature can be related to the subsidence factors by the expression:

$$R = A\frac{h^2}{S_{max}}$$

where:

A = coefficient which takes into consideration the local subsidence conditions (in the UK it is in the order of 13.5).

3. Horizontal movements have a differential character, and are primarily responsible for damages to surface structures. Damages result from the transmission of differential ground movement to structures, causing them to strain, and if excessive, to crack.[18] The horizontal displacement is represented by a curve which can be written as:

$$E_{max} = K \cdot S_{max}$$

where:

E_{max} = maximum lateral displacement ($+E_{max}$ is a contraction; $-E_{max}$);
K = function of the dimension of the mine excavation and type of displacement.

4. Surface gound strains are also illustrated in Figure 17.4.2. The subsidence produces regions of ground extension and compression by virtue of its relative displacement towards the centre of the trough. The generalized relationship is represented by the equation above, but in terms of the ground strain it can be written as:[20]

$$\varepsilon = f\left(\frac{v}{h}\right)$$

or

$$\varepsilon = k\frac{v}{h}$$

The maximum strain due to compression is:

$$+\varepsilon = k\frac{v}{h}$$

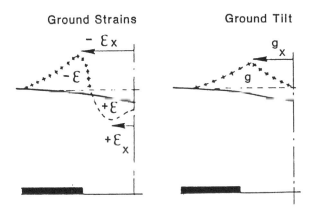

Figure 17.4.2. Subsidence strain development

The maximum strain due to extension is:

$$-\varepsilon = k' \frac{v}{h}$$

where:

k = factor for compressive strain (for longwall mining in the UK it is between 0.5 and 2.5);

k' = factor for tensile strain (for longwall mining in the UK it is between 0.45 and 0.65).

The relationship between tensile strain and compressive strain is represented by ground tilt, reaching its maximum at the transition point between $-\varepsilon$ and $+\varepsilon$. This can be written as:

$$g = k'' \frac{v}{h}$$

where:

k'' = factor for tilt (for longwall mining in the UK it is between 1.9 and 3.6).

The magnitude of the factors k, k' and k'' depends on the position from the centre line, the dimension of the mine excavation and the mine depth.

The understanding of the elements movement is essential for an evaluation of the degree of structure damage, as briefly discussed in the next subsection.

17.4.2 *The effects of ground movements on the structure*

Structural damage results from ground subsidence in relation to the length of the structure, whereas the damage arising from curvature is related to the height of the structure. Both elements share in the damage in a definite expression. Both values under consideration can be combined in the following empirical formula:[19]

$$L \times 10^{-3} \left(\varepsilon \times 10^{-3} + \frac{H}{R} \right) < a \text{ (mm)}$$

where:

L = length of structure;

ε = strain of ground surface due to subsidence;

H = height of structure;

R = radius of subsidence curvature;

a = an empirical factor of permissible degree of damage to the structure (for slight damage it is between 100 and 150 mm).

Making use of Burland and Wroth's approach, the deflection ratio Δ/L'' can be combined to express terms of either limiting direct tensile strain, ε_b, or diagonal tensile strain, ε_d, giving rise to the initiation of visible cracking on a structure. For an isotropic elastic material ($E/G = 2.6$) the limiting value of Δ/L may be expressed in terms of strain:[21]

$$\frac{\Delta}{L} = \left[0.167 \frac{L}{H} + 0.65 \frac{H}{L} \right] \varepsilon_b$$

$$\frac{\Delta}{L} = \left[0.25 \frac{L^2}{H^2} + 1 \right] \varepsilon_d$$

where H is the height of the surface structure.

Substituting the deflection ratio equation and the radius of curvature equation, the strains can be related to subsidence factors and building dimensions as follows:

$$\varepsilon_b = \frac{S_{max}}{A \cdot h \cdot \left[1.33\frac{1}{H} + 5.2\frac{H}{L^2}\right]}$$

$$\varepsilon_d = \frac{L \cdot S_{max}}{A \cdot h^2 \left[2\frac{L^2}{H^2} + 8\right]}$$

For brickwork and blockwork set in cement mortar limiting 'ε' lies between 0.05 and 0.1%, while for reinforced concrete having a wide range of strengths, the values are in the order of 0.03 and 0.05%. For reinforced concrete member an average value of limiting direct tensile strain, b, can be assumed to be 0.035%.

From the foregoing equations it is clear that with increasing maximum subsidence S_{max}, the deformation of the structure becomes more severe (Figure 17.4.3).

17.4.3 *Characterization of zones of damage*

The characterization of zones of damage to surface structures is represented by the case history from a brine-field at Tuzla, former Yugoslavia. Subsidence caused by solution mining resulted in a variety of damages to surface structures which have been observed and recorded.

The surveyed data of horizontal displacement and vertical displacement by Smailbegovic[10] have been used to characterize the zones of damage and to make a delineation of their lateral extension, as illustrated in Figure 17.4.4. The characterized zones of damage in this figure correspond to the geometry of curvature of the surface subsidence given in Figure 17.2.3. The characterization of zones of damage is based on the mechanics of deformation and failure of the surface structure rather than on the classification of structure damage, which is discussed in the next subsection.

On the basis of subsidence curvature, subsided ground strain and the character of damage to surface structures, four concentric zonal areas of structural damage can be inferred.

1. *Collapsing deformation* coincides with the area of maximal subsidence (S_{max}). This deformation can be related to abrupt subsidence when the centre or the core of the subsided surface collapses into a salt gallery below. The depth of the collapsing subsidence is

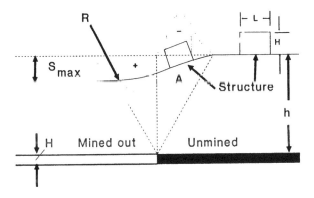

Figure 17.4.3. Surface structure rotation and extension in the zone of tensile stress field

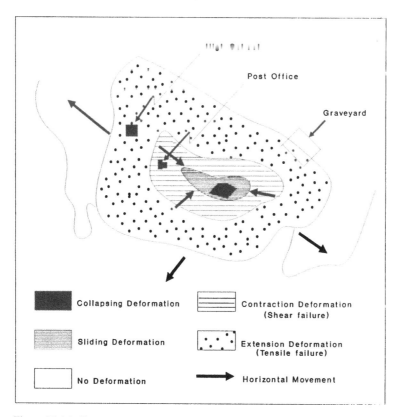

Figure 17.4.4. Characterization and delineation of the zones of deformation

undetermined, because its bottom is composed of the rock debris. This type of subsidence has been observed as crators filled with water and rock debris (Figure 17.4.5). The contours of this subsidence deformation are delineated by vertical strata displacement, which forms like a chimney structure, in which any house above collapses and vanishes.

2. *Sliding deformation* coincides with the area of flat-bottomed troughs. The boundary of this subsidence is controlled by the downward movement of the rock strata prism, which corresponds to the rim of the trough. The displaced downward area of sliding deformation forms more or less the central part of the ground subsidence, and it is very well defined by a sharp boundary and a flood of water (Figure 17.4.6). On the basis of the cracks and fractures in the damages houses, it can be concluded that they are the product of axial compression force. The deformations of the houses are negligible in comparison with the magnitude of the vertical displacement. The zone of collapsing deformation is delineated by a contour of vertical displacement of 70 to 170 cm. It is possible that horizontal displacement was of a lesser magnitude and that surface objects had not been subjected to appreciable horizontal stress.

3. *Shear deformations* are evident in the area around the zone of sliding subsidence. Triaxial compression was the primary cause of the structure deformation. In this area the horizontal displacement and vertical displacement produced the compressive strains. The compressive strains increased towards the zone of collapsing deformation. Under this combination of stress, the surface objects were damaged by shear movements (Figure

Figure 17.4.5. Collapsing subsidence
(crater filled with water and debris), Tuzla
salt basin

Figure 17.4.6. Sliding subsidence (boundary of slided surface marked by water), Tuzla salt basin

17.4.7). The damage evident to the house walls is very similar to the fracture of brittle rock produced by in situ shear stresses. The zone of predominately shear type deformations are delineated by contours of 20 and 70 cm of vertical ground surface displacement. Generally, it is considered that damage in this area is not very severe for the buildings and that significant damage is limited to railway tracks, pipes, and others.[22] This is probably true for subsidence caused by underground mining, but it is not true for subsidence caused by solution mining due to the most intensive displacement.

4. *Extension deformations* caused the most intensive building damage in the Tuzla brine-field. The tensile fractures have appreciable propagation and widening, and some of them are vertically oriented. The tensile stress deformations are more damaging than shear fracturing, because many buildings with tensile fractures are in a semi-collapsing or collapsing state (Figure 17.4.8). This is quite understandable given the ground surface tilting (Figure 17.4.9), which is responsible for severe tensile deformations of buildings as noted in the previous subsection. The intensity of fracturing depends on the magnitude of the horizontal extension displacement. Surveying has established a relationship between the magnitudes of horizontal strain and fracture openings of the buildings, as illustrated in Figure 17.4.10. For a horizontal strain of 12 mm/m, the width of cracks is 20 mm and greater. Under this condition the buildings are in a semi-collapsed state.

Subsidence of a residential area with family houses, which are located in the periphery zone of extension deformation, did not significantly affect their stability. Cracking of the houses was limited, not only because of a lower magnitude of tensile strain but also due to their small size, which was up to 120 m^2 in area and up to 10 m in height.

Figure 17.4.7. Shear failure (post office building), Tuzla salt basin

Figure 17.4.8. Tensile failure (high school building), Tuzla salt basin

Figure 17.4.9. Tilting of the grave at the boundary of extension deformation, Tuzla salt basin

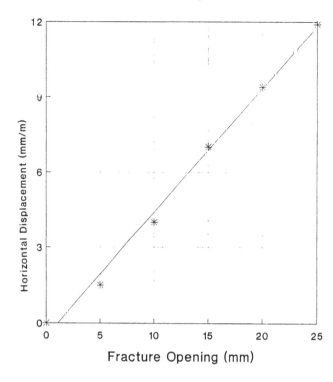

Figure 17.4.10. Diagram of the relationship between horizontal displacment and fracture openings of the buildings

17.4.4 *Classification of structure damage*

At the present time there are several classifications for damage induced by subsidence due to mining operations.[17, 22, 23]

The general classification of structure damage in continental Europe is based on the value of permitted strain of the extension zone of the ground subsidence. The permitted strain (ε_p) is determined by the distance from the point of inflection of the horizontal strain curve to the surface structure (X_p), as illustrated in Figure 17.4.11. The four classes of subsidence damage are:

Class I: The strain is 0.01-1.50 mm/m. Damage in this class consists of cracks which are invisible. The limited strain of this category prohibits the presence of mine shafts, industrial objects (preparation plants, steel plants, power plants and others), highways, railways, dams, bridges, hospitals, rivers and lakes.

Class II: The strain is 1.5-3.0 mm/m. Damage in this class consists of visible cracks, whose openings may be between 2 and 5 mm. The strain of this category prohibits the presence of drill holes, warehouses, pipelines, apartment buildings up to two levels, transformer stations, high voltage electric lines, irrigation projects, agricultural and other industrial structures.

Class III: The strain is 3.0-6.0 mm/m. Damage cracks are open between 10 and 20 mm. Damage to this class prohibits the presence of secondary railways and roadways, bungalows, underground workings and local airports.

Class IV: The strain is 6.0-12 mm/m. Damage cracks have openings greater than 20 mm. This class of surface damage prohibits the temporary buildings such as trailers as well as agricultural land and forest.

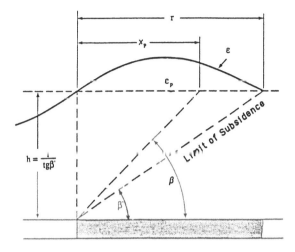

Figure 17.4.11. Diagram of determination of permitted strain (ε_p)

If the strain is greater than 12 mm/m, any structure or cultivated land is prohibited from the damage area. For example, this strain magnitude makes buildings unusable, because they are heavily damaged with collapsed walls.

Ariogly proposed a classification for house damage due to mining subsidence based on two parameters which are listed here.[17]

Crack parameters (density, intensity and width of cracks), with the following ratings:

(I)	Number of total cracks	Rating
	1 - 5	2
	5 - 10	4
	10 - 20	6
	20 - 30	8
	> 30	10
(II)	Intensity of cracks	Rating
	> 20	1
	20 – 40	2
	40 – 60	3
	60 – 80	4
	> 80	5

Note: The intensity of cracks is defined as the ratio of the number of rooms where cracks occurred to the total number of rooms.

(III)	Width of crack (mm)	Rating
	> 0.5	1
	0.5 – 3	2
	3 – 10	3
10 – 20	4	
20 – 40	5	
	40 - 60	6
	60 - 80	8
	> 80	10

Building damages parameters, with the following ratings:

(IV)	Type of damage	Rating
	Architectural (wall, floor)	2
	Structural	4
	Both	6

(V)	Loss of performance in windows and doors	Rataing
	None	0
	Slight	1
	Appreciable	2
	Severe	3

(VI)	Rain-wind penetration through walls	Rating
	None	0
	Slight	1
	Severe	2
	Very severe	3

(VII)	Socio-economic conditions	Rating
	Low	1
	Medium	2
	High	3
	Very high	4

Note: Especially the level of income.

From the total rating by both groups, Ariogly proposed a class of damage for buildings as given in Table 17.4.1.

He gives the following example of the proposed classification for damaged houses:

	Ratings
Number of cracks: 20	8
Intensity of cracks: 100%	5
Width of cracks (average): 10 mm	3
Type of damage: structural	4
Loss of performance in windows and doors: none	0
Rain-wind penetration through walls: none	0
Socio-economic conditions; low	1
Total rating	21

Table 17.4.1. Damages corresponding to total rating

Rating	Class of damages
0 – 5	Very slight
5 – 15	Slight
15 – 25	Appreciable
25 – 30	Severe
> 30	Very severe

Figure 17.4.12. Damage to a house rated as 'severe damage' (Tuzla salt basin)

On the basis of the total rating, this can be classified as 'appreciable damage' (see Table 17.4.1). The houses with appreciable damage displayed diagonal cracks of a width of 30 mm and a tilt of 1/30. From this type of damage, one can conclude that the primary factor contributing to this phenomenon was differential settlement of the structure. The diagonal cracks occurred as a result of diagonal tensile strain. In Figure 17.4.12 is illustrated a house classified as 'severe damage' (Table 17.4.1).

REFERENCES

1. Simic, D. & Z. Klecek 1989. Influence of underground mining on ground surface. In *Principles of rock mechanics*, pp. 161-191. Sarajevo: Posebna Izdanja (in Yugoslav).
2. Jeremic, M.L. 1985. Mining subsidence. In *Strata mechanics in coal mining*, pp. 343-353. Rotterdam/Boston: Balkema.
3. Steel, C. et al. 1985. Subsidence prediction for Saskatchewan potash mines, pp. 163-170. In *Proc. of 26th US Symp. on Rock Mech.*, Rapid City, S.D., 26-28 June, 1985.
4. Dreyer, W.E. 1981. Crude oil storage caverns in a system of salt caverns, pp. 629-660. In *Proc. of 1st Conf. on Mech. Behaviour of Salt*, Penn. State Univ. Clausthal: Trans Techn. Publ.
5. Coates, G.K. et al. 1983. Closure and collapse of man-made cavities in salt. In *Proc. of 6th Int. Symp. on Salt*, The Salt Institute, Virginia, Vol. 2, pp. 139-157.
6. Terzaghi, R.D. 1969. Brine-field subsidence at Windsor, Ontario. In *Proc. of 3rd Symp. on Salt*, Vol. 2, pp. 298-307.
7. Jeremic, M.L. 1975. Subsidence problems caused by solution mining of the rock salt deposits. In *Proc. of 10 Canadian Rock Mech. Symp.*, Kingston, Dept. of Min. Engng., Queen's Univ., Vol. 1, pp. 203-223.
8. Young, C.M. 1926. Subsidence around a salt well. In *Proc. of New York Meeting*, pp. 810-817.
9. Jeremic, M. 1964. Exploitation methods for rock salt deposits in Poland. *Min. & Metall. Bulletin* (Belgrade) 2: 264-274 (in Yugoslav).

10. Smailbegovic, F. 1965. The results of surveying vertical and horizontal displacement at Tuzla town area. Fond for Subsidence Tuzla Town, former Yugoslavia.

11. Otovic, M. et al. 1979. Arching of the overlying strata above leached galleries in rock salt. In *Proc. of 4th Int. Congr. on Rock Mechanics*, Montreux, Switzerland, Vol. 1, pp. 745-753,

12. Jasarevic, J. et al. 1975. Failure processes due to leaching of rock salt deposits, pp. 17-26 (in Yugoslav). In *Proc. of Int. Symp. on Subsidence*, Faculty of Mining and Geology, Tuzla, former Yugoslavia.

13. Nieto, A.S. & A.J. Hendrow 1977. *Study of sink holes related to salt production in the area of Detroit, Michigan*, pp. 1-50. Solution Mining Research Institute.

14. Dorohan, D.J. 1976. *Test drilling to investigate subsidence in bedded salt*, pp. 1-13. Wyandotte, Michigan: BASF Wyandolt Chem. Corp.

15. Davis, P.D. 1983. Structural characteristics of a deep seated dissolution subsidence chimney in bedded salt. In *Proc. of 6th Int. Symp. on Salt*, The Salt Institute, Virginia, Vol. 1, pp. 331-350.

16. Terzaghi, K. 1970. Report on the subsidence of February 19, 1954 in Windsor, Ontario. Unpublished report of October 27, 1970, pp. 1-37.

17. Ariogly, E. 1984. Classification of house damages due to mining subsidence. *Housing Science* 8 (4): 361-372.

18. Gardner, F.P. & G. Hibbert. Subsidence – The transference of ground movement to surface structures. *Min. Engng.*, October 1961: 19-34.

19. Yokel, F.Y. et al. 1982. Housing construction in areas of mine subsidence. *J. Geotechn. Engng. Div.* (Proc. of Am. Soc. Civ. Eng.), 109, GT9, September 1982: 1133-1149.

20. Whittaker, B.N. & M.L. Jeremic 1979. Longwall mining potential of plains region coal deposits in Western Canada. *Coll. Guard Int.*, April 1979: 31-39.

21. Burland, J.B. & C.P. Wroth 1974. Allowable and differential settlements of structures, including damage and soil-structure interaction, pp. 611-654. In *Proc. of Conf. on Settl. and Struc.*, Cambridge, April 1974.

22. Kovalczyk, Z. 1966. Effect of mining exploitation on the ground surface and structures in heavily industrialized and populated areas. *Can. Min. & Metall. Bulletin* (Montreal), October 1966: 1201-1207.

23. Jeremic, M.L. 1985. Protective pillars. In *Strata mechanics in coal mining*, pp. 530-548. Rotterdam/Boston: Balkema.

Subject index

531

Milton Keynes UK
Ingram Content Group UK Ltd.
UKHW051929141024
449569UK00027B/1410